Control Valves for the Chemical Process Industries

Other McGraw-Hill Engineering Books of Interest

BRUNNER • *Hazardous Waste Incineration*

COOK, DUMONT • *Process Drying Practice*

CHOPEY • *Handbook of Chemical Engineering Calculations*

CHOPEY • *Environmental Engineering in the Process Plant*

CROOM • *Filter Dust Collectors*

DEAN • *Lange's Handbook of Chemistry*

DEAN • *Analytical Chemistry Handbook*

DILLON • *Materials Selection for the Chemical Process Industries*

FREEMAN • *Hazardous Waste Minimization*

FREEMAN • *Standard Handbook of Hazardous Waste Treatment and Disposal*

FREEMAN • *Industrial Pollution Prevention Handbook*

KISTER • *Distillation Operation*

KISTER • *Distillation Design*

KOLLURU • *Environmental Strategies Handbook*

LEVIN, GEALT • *Biotreatment of Industrial Hazardous Waste*

MANSFIELD • *Engineering Design for Process Facilities*

MCGEE • *Molecular Engineering*

MILLER • *Flow Measurement Handbook*

PALLUZI • *Pilot Plant Design, Construction and Operation*

PALLUZI • *Pilot Plant and Laboratory Safety*

PERRY, GREEN • *Perry's Chemical Engineers' Handbook*

POWER • *Steam Jet Ejectors for the Process Industries*

REID ET AL. • *Properties of Gases and Liquids*

REIST • *Introduction to Aerosol Science*

SANDLER, LUCKIEWICZ • *Practical Process Engineering*

SATTERFIELD • *Heterogeneous Catalysis in Practice*

SHINSKEY • *Process Control Systems*

SHINSKEY • *Feedback Controllers for the Process Industries*

SHUGAR, BALLINGER • *The Chemical Technicians' Ready Reference Handbook*

SHUGAR, DEAN • *The Chemist's Ready Reference Handbook*

SMITH, VAN LAAN • *Piping and Pipe Support Systems*

TATTERSON • *Fluid Mixing and Gas Dispersion in Agitated Tanks*

TATTERSON • *Scale-up of Industrial Mixing Processes*

WILLIG • *Environmental TQM*

YOKELL • *A Working Guide to Shell-and-Tube Heat Exchangers*

Control Valves for the Chemical Process Industries

Bill Fitzgerald

McGraw-Hill, Inc.

New York San Francisco Washington, D.C. Auckland Bogotá
Caracas Lisbon London Madrid Mexico City Milan
Montreal New Delhi San Juan Singapore
Sydney Tokyo Toronto

Library of Congress Cataloging-in-Publication Data

Fitzgerald, Bill.
Control valves for the chemical process industries / Bill Fitzgerald.
p. cm.
Includes bibliographical references and index.
ISBN 0-07-021176-0
1. Chemical process control—Equipment and supplies. 2. Valves.
I. Title.
TP155.75.F53 1995
660′.2815—dc20 94-23201
CIP

1 2 3 4 5 6 7 8 9 0 DOC/DOC 9 0 9 8 7 6 5 4

ISBN 0-07-021176-0

The sponsoring editor for this book was Gail F. Nalven, the editing supervisor was Nancy Young, and the production supervisor was Pamela A. Pelton. This book was set in Century Schoolbook by McGraw-Hill's Professional Book Group composition unit.

Printed and bound by R. R. Donnelley & Sons Company.

This book is printed on acid-free paper.

This book is dedicated to my parents, who always encouraged me in my academic pursuits.

Contents

Preface

This all started about 4 years ago. I had just completed an article for *Chemical Engineering Magazine* on control valve maintenance that focused on the message that control valves were an important part of the process control system and that too many customers were letting the valve's performance degrade to the point where their plant's efficiency was suffering. The article went over very well, and as a result of the response, I was approached by the editors of McGraw-Hill to see if I would be interested in expanding this theme into a book on control valves. Initially I wasn't too sure that I wanted to proceed with the project, but after talking with a number of industry contacts, it did seem like there was a real need for a book that treated the control valve as an engineered product and not a commodity.

If you get one thing from this book, it should be that the control valve is a vital part of your overall plant's performance. Until you recognize this and take proactive steps to make sure that you select, install, and maintain valves properly, your plant will never operate at peak efficiency, and that will hurt your bottom line and your competitive position in what is becoming a very demanding worldwide market. One of the other objectives that I set for myself in putting this together was to write it in such a way that it would be of maximum benefit for the group that has to live with control valves day in and day out: the end user. I have been active in the valve industry for almost 20 years, but it wasn't until I got involved in field service about 8 years ago that I really began to understand how valves are used and abused in the real world. There are a lot of demands put on the people who have to run the plants, and they don't have time to sift through filler material to get to the information they really need to make a valve work. With this in mind, I stayed away from telling the reader about how a control valve and its related equipment are designed unless it had some bearing on one of the three primary things that they might be dealing with: selecting, installing, or maintaining a valve. I also took some steps to make the book easier to use as a reference by making the table of contents much more detailed than normal, and by highlighting key phrases and figure references. I also made a point of addressing some issues that are very important to the end user of control valves but that are typically ignored by previous works focusing on design. Subjects in this category include a description of hazardous area use, a detailed discussion on packings and gaskets, and an extensive coverage of proper setup and calibration techniques.

I hope that you find that I was successful in my quest to create a reference book that takes a unique approach to helping readers ensure that their valves are working as well as they can. I would welcome any comments regarding the contents and/or style so that I can better determine if I've hit the mark.

Bill Fitzgerald

Acknowledgments

I would like to thank Delores Seifert, Peggy Hazen, and Lisa Brazile for all their help in preparing the manuscript and need to make special mention of the following people who acted as advisors to help ensure that the book would be of maximum benefit to its intended audience:

Jimmy Cerrato, Texas Eastman

Jim Rodda, Dupont

Cullen Langford, Dupont

Stan Weiner, Monsanto

Phil Bruder, Dow

And finally, I'd like to thank my family, who put up with me and the long hours while the book was being written.

Chapter

1

Introduction

1.1 The Chemical Industry Today

Of all the major process industries, the chemical industry is the hardest one to characterize because, unlike power or pulp and paper, there is a very wide range of end products that are being produced and each end product requires a fairly unique process. End products in this group include industrial chemicals, rubber, explosives, agricultural chemicals, fibers, paint, plastics, and soap. Nevertheless, if we take a closer look at the 10,000 or so chemical plants in North America, some common trends begin to develop. The first is the fact that environmental and safety considerations are being increasingly regulated all over the world and are having a serious bottom-line effect on the plants. Second, this is a very global industry where national borders are becoming less important as a consideration of where production takes place. Instead, feedstock and operating costs are becoming the primary drivers, so for a plant to remain viable in this world economy, it needs to make sure that its production facility is truly world class.

This of course means that the plant must produce the end product efficiently, but it also means that its process needs to be flexible in terms of what is produced, how much is produced, and when it is produced. At the same time, economic conditions have resulted in many plants downsizing their staffs; therefore, they are faced with achieving this world-class operating status with fewer people.

The only way out of this dilemma appears to be a new generation of process control equipment that emphasizes flexibility and efficiency through improved communication between the operators and the process.

Unfortunately, no matter how sophisticated these new systems are, the real interaction between them and the process occurs at the control valve. If the valve is not selected, installed, and maintained to provide optimum response, many of the advantages associated with these new control schemes will not be realized. This will be the overriding theme of this book: A process plant that wants to stay competitive must take valve selection and maintenance very seriously. Every loop should be considered critical, and for every loop there is one valve whose operating characteristics and dynamic response best fits the process. The world-class plant will take the time to determine which valve it really needs to optimize this fit and will then set up a cost-effective maintenance plan aimed at ensuring that valve performance is optimized over the life cycle of the valve.

1.2 Typical Chemical Plant

The most insightful comment that can be made about the typical chemical plant is that it doesn't exist. There are so many different processes with so many different end products that it's very difficult to compare one to the next or to talk about similarities. This is further complicated by the fact that many of the processes involved are proprietary, so public knowledge of operations is restricted. About the only common thread that does exist is a tendency toward erosive/corrosive applications where the use of exotic alloys for trim and body parts is more prevalent than in other industries.

1.3 How to Get the Most Out of This Book

If you're like me, you've relied very heavily on books throughout your formative years. Books have been and will continue to be one of the primary tools involved in the transfer of knowledge from one individual to the masses. What is surprising is the way that we go about using these very important tools. When confronted with a book in a learning situation, most of us dive in with the primary goal of "getting through the material." It's a little like starting a race with no idea of the course layout, the ground rules, or where we're going. We just jump in, start the engine, and drive. While we will eventually complete the course with this approach, think of the wasted time and effort. Most of us would agree that a few minutes spent studying the map would be time well spent.

What I'm going to suggest is that you apply this same logic to this book. The following sections are designed to be your road map as you negotiate your way through it. I urge you to take a few minutes to scan this material. I am convinced that "studying the map" will make the book easier to read while improving overall retention.

1.3.1 Author's background

First of all, I need to explain that I am first and foremost a valve person. I do have a smattering of knowledge regarding what goes on in the control room end of the plant, but there are much better sources of information on process control and how it works. The control valve, as I've defined it, includes the valve body assembly, the actuator, and any related accessories normally found on or near the valve. These might include the positioner, the current- (or voltage) to-pressure transducer, volume boosters, limit switches, etc.

I have experience in a number of areas with a major valve manufacturer, including design, research, and marketing. This should permit me to address the typical subjects covered in a control valve reference work such as design considerations, valve types, and the like. However, I believe I can offer a fresh approach because I have spent 8 years in field service and can better relate to the challenges and concerns that the average end user has to deal with when working with control valves.

I should also point out that while I am an employee of a major valve manufacturer, my field experience has exposed me to many different types of valves, and I've solicited input from the other major manufacturers so that the treatment of this subject would be from a balanced perspective.

Finally, I have a healthy disrespect for the conventional approach. As a result, I see no need to adhere to the technical style of writing that is so prevalent in most texts just because mine is a technical subject. My job is to convey information on the care and feeding of control valves for people in the real world. The best way to do that is to attempt to make the material as easy to read as possible. I hope you find this to be true, particularly when compared to earlier works on this subject. I should also point out that there are many issues covered in this book that are not cut and dried. In these cases, I have tried to present a balanced perspective, but I have not shied away from taking a stand where necessary and expressing my opinion. You need to recognize that the opinions expressed as such are strictly mine and no guarantees are given or implied as to the accuracy of the statements.

1.3.2 What is a control valve?

As the title indicates, this book is focused on control valves. As covered herein, a control valve will be defined as a valve that is called upon to throttle, regulate, and/or control the flow of a fluid through a valve. This implies that the valve will normally see a high degree of movement and that the relationship between the input to the valve assembly and the flow through the valve will be highly repeatable with a minimum of error.

This definition leaves out a group of true on-off valves (butterflies and plug valves) that have been "converted" to control valves through the addition of a power actuator. As a matter of fact, I see the increasing use of on-off valves in control applications as part of a disturbing trend in which valve dynamic performance has been discounted in deference to the performance of the electronic portion of the control system. The primary theme of this book, which will be hammered at repeatedly, is that the overall performance of the process control system is ultimately linked to control valve performance. Let valve performance drop off through poor valve selection or ineffective maintenance, and process control will suffer, as well.

1.3.3 Purpose of the book and the target audience

As with any "product," there is a need to decide what benefits customers will derive from it and just who those customers are. In this case, they mostly will be plant personnel, people who are usually referred to as end users. In general, the staff at most of these plants has been downsized, redeployed, reorganized, or whatever other euphemism is used to reflect the fact that they will be trying to accomplish the same goals with fewer resources. They still need to keep the plant operating efficiently, and a key part of this is related to control valve performance and maintenance costs.

This book was written with this audience in mind. It addresses problems that they might face on any given day and gives them enough information to solve the problem without burying them in useless information or duplicating material that they could more easily obtain from one of their local valve suppliers. In other words, this is not a design text. It is more of an application manual that will permit the readers to select, install, and maintain their valves to give peak performance at minimum life-cycle cost.

This raises two other sets of issues that have not been traditionally addressed in previous books on control valves. The first is valve performance. Valves are inherently robust. They can have significant operating problems and still function. This ability to continue to function is generally considered an advantage, but it can pose problems for plants trying to optimize operations. If the valve has not "failed" completely, it may be overlooked as a source of process control problems. I have seen many examples of valves whose dynamic response was very poor, resulting in less than satisfactory process control, even though the control system was state of the art. To address this type of problem, I will take a novel approach here and discuss the control valve as part of the control system, with recommendations on how to optimize its dynamic response so that the control system can do its job. I will also talk at length about the relatively wide range of possible performance, between just getting by and true peak performance. This will include recommendations on what level of performance the user can expect from various types of valves.

The other issue that should be of paramount importance to plant personnel is how to evaluate true return on investment for their installed base of control valves. Most plants are working with control valves that were purchased based on up-front prices. If you compare that expenditure to the operating costs such as maintenance, or the cost of lost production due to poor performance or valve downtime, you can see that the purchase price is really insignificant and should not be the primary consideration during purchase. Unfortunately, this is the easiest element to identify in valve selection, so it still plays a dominant role. I will lay out a framework that should aid in better identifying total life-cycle costs and should, in turn, provide a better rationale for valve selection.

1.3.4 Layout of the book

A few words about the layout of the book. First of all, while the book can be read cover to cover, many of us in the real world don't have the time to read through all our reference books. As a result I have taken several steps to aid the reader who needs to consult the book from time to time. There is a traditional index, but there is also a table of contents that is much more detailed than a typical text so that the occasional user can consult it to quickly zero in on that area of the book that addresses the problem at hand. I've also put the figure references in bold print to make them easier to find; and keywords are in italics. From an organizational standpoint, the book is divided into four separate sections as someone from a plant might be faced with them: valve selection, valve installation and setup, valve maintenance, and general topics and new developments.

1.3.5 Featured sections of the book

In comparing this book to earlier works, there are a number of sections that need to be commented on. First of all, Sec. 5.14 gives a good summary of the different types of hazardous areas that might be encountered in a chemical plant and how to go about selecting control valve equipment based on the classification. This is the first time, to my knowledge, that this subject has been covered in a control valve reference work even though it is something that the end user will routinely face as part of his or her daily activities.

In my experience, the soft parts (gaskets, packing, and seals) are one of the biggest problem areas when it comes to control valve maintenance and performance. In light of this, all of Chap. 7 is devoted to understanding how they work and what measures need to be taken to ensure optimum performance.

Another major concern with control valves is making sure that they are set up properly, including things like travel and benchset. Research has shown that many operational problems can be traced back to incorrect setup procedures. To help address this, all of Part 2 has been devoted to recommendations on how to install, set up, and calibrate a control valve so that it will operate as it should. In particular, the infamous term *benchset* is covered in detail since it seems to be a big point of confusion for most end users.

Chapter 9 is also a departure from previous works in that it attempts to define how an average control valve should perform in a typical application. This reinforces the idea that world-class plants cannot continue to accept anything less than top performance from their control valves if they want to optimize process control performance.

In Chap. 10, several new forms are suggested that will help the end user and vendor identify all the information that needs to be taken into account for proper valve selection. Valve dynamic response, in particular, is a featured part of these new considerations.

In Chap. 13, a new concept called preemptive maintenance is introduced. The idea behind preemptive maintenance is that many performance problems can be avoided altogether if proper care is taken in installing and setting up a control valve.

And finally, Chap. 14 deals with how control valve performance can directly affect the level of process control that can be achieved. The basics of process control are covered and then valve characteristics, such as speed of response and positioning accuracy, are tied back into loop performance.

1.4 Reference

1. Bialkowski, Bill, "The Pneumatic Control Valve: It's Not Just 'Pig Iron'", *The Entech Report,* vol. 5, issue 2, Entech Inc., Toronto, Ontario, Canada, Oct. 1993.

Part

1

Valve Selection

As an end user, valve selection is not normally a big part of your job. However, you need to understand the basis behind the original selection of the valves you're working with so that you can determine if advances in technology or changes in the application have provided you with an opportunity for better performance with a different type of valve or valve trim. You'll also be faced occasionally with the need to replace a control valve, and this section should provide you with the basic considerations you'll need to make an intelligent choice. This is not meant, however, to cover every possible issue involved with valve selection. That could fill a book on its own. You are encouraged, instead, to utilize the services of the valve vendors in your area to cover the details that are beyond the scope of this book.

As a side note, this section of the book will introduce a number of technical terms that are unique to the control valve industry. It will make for easier reading if you first review the Glossary at the end of the book to get a better understanding of the meaning of these terms.

Chapter

2

Sliding-Stem Valves

This chapter will cover the basic design characteristics of the sliding-stem family of control valves including body styles, trim balancing, shutoff, guiding, flow considerations, end connections, and pressure-temperature ratings. It also includes an in-depth discussion of fluid forces inside the valve and how they affect actuator sizing.

Control valves are generally split into two major groups: sliding stem and rotary. The names reflect the primary action of the valve stem or shaft as the valve performs its regulating function. We'll look at sliding-stem valves first since they represent the majority of applications in a typical chemical plant, although there is a gradual trend toward increasing the percentage of rotary valves used, for reasons discussed later.

As the name implies, on this group of valves the valve stem slides in a reciprocating motion through the stem seal as the valve is operated. The stem is connected to the primary flow control element inside the valve body which, in turn, moves with the stem to provide a variable flow restriction. In general, the sliding-stem valve is still considered the standard in the chemical industry because it can handle a very broad range of applications, including high temperature, high pressure drop, and, with proper material selection, a wide range of chemical exposure. However, rotary valves are gaining market share as new designs enable them to cover some of these tough applications. Part of the push for rotary designs is based on the fact that they generally offer a higher flow capacity per dollar and that the rotary stem seal is usually more reliable since debris that collects on the stem is not drawn through the stem seal as the valve operates.

There *is* a downside to the inherent high capacity of the rotary valve that will be discussed in Chap. 6.

The types of sliding-stem valves available will be discussed in terms of their primary design features including body styles, balancing, seating, guiding, flow characteristics, flow direction and action, end connections, and pressure and temperature ratings.

2.1 Body Style

The most common body style for sliding-stem valves is the *globe,* named for the general shape of the internal flow cavity. An example is shown in **Fig. 2.1.** One of the flow paths, either inlet or outlet, is normally perpendicular to the axis of the stem. Flow can be in-line or offset as shown in **Fig. 2.2.**

Another globe body style that is seen frequently in plants is the *split body* (see **Fig. G.4** in the Glossary). The original idea behind this approach was to limit the bolted joints to a single one in the middle that captured the seat ring while also allowing for disassembly of the

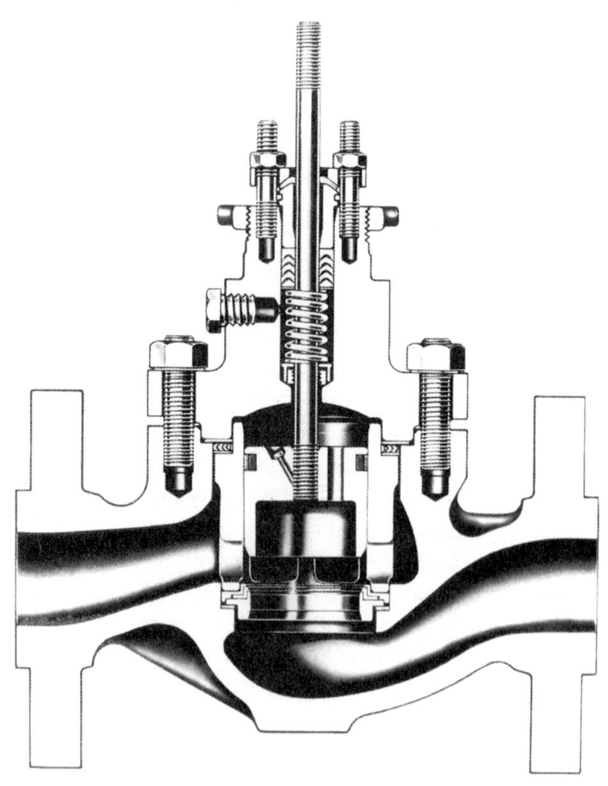

Figure 2.1 Globe-style valve. (*Courtesy of Fisher Controls International, Inc., Marshalltown, Iowa.*)

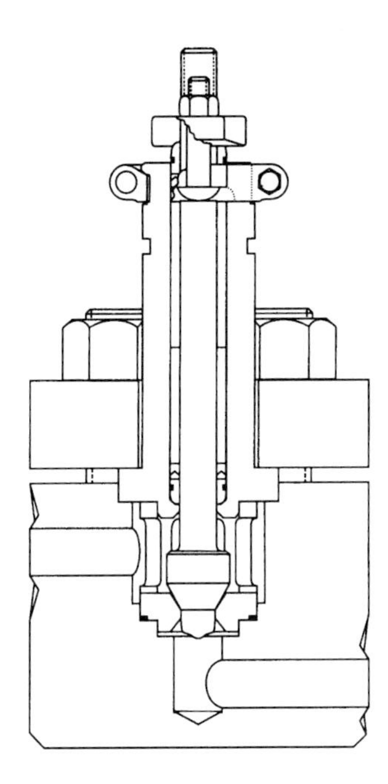

Figure 2.2 Offset-style globe valve. *(Courtesy of Valtek International, Inc., Springville, Utah.)*

valve for maintenance. Other advantages include a streamlined flow path with a minimum of parts and no irregular cavities so that plugging and fouling can be avoided. Disadvantages include leakage problems with the central joint due to thermal cycles or piping loads, the inability to weld the valves in-line since maintenance requires that the valve body be split, and the additional work associated with even minor maintenance since this requires that the end connections be broken. This last problem has become more important recently with the advent of increased concerns over line flange leakage. For all these reasons, fewer of these valves are used now.

The *angle body* is a variation on the globe where the exiting flow path is at 90° to the inlet flow. The obvious advantage is that an elbow can sometimes be eliminated. In general, these valves will also have slightly higher flow capacity than a conventional globe since the fluid makes fewer turns as it passes through the body. Also, if flashing or cavitation is occurring, it tends to manifest itself downstream of the valve where it will be less detrimental to valve performance and maintenance life. An example is shown in **Fig. G.5** in the Glossary.

The *Y-pattern* valve actually has the seat ring and plug tilted at a 45° angle to the flow path. As shown in **Fig. 2.3,** this means that the flow has to make fewer turns as it goes through the body; therefore,

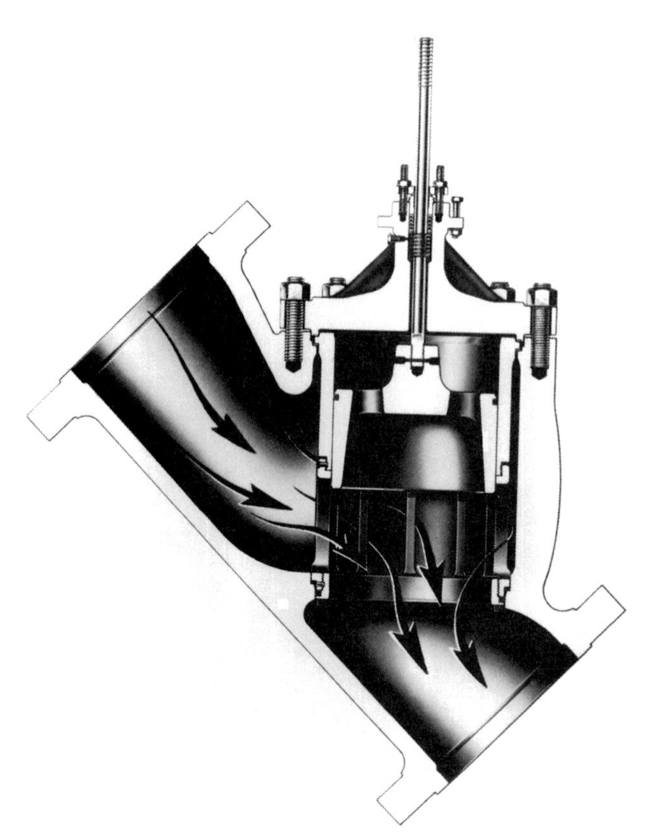

Figure 2.3 Y-pattern valve. (*Courtesy of Fisher Controls International, Inc., Marshalltown, Iowa.*)

these valves, at least in theory, should have higher capacity than a standard globe. In truth, most valves are grossly oversized for reasons discussed later, so capacity is not usually an overriding concern. If you have ever had to work on one of these valves installed in a horizontal run of pipe with the actuator laying over at 45°, you'll understand why they are not very popular with maintenance crews. There's no easy way to lift the parts out of the valve body when the extraction angle is not vertical. These valves also tend to be high-maintenance items since the moving parts have an inherent side load due to gravity.

Three-way valves are double-ported valves that have a total of three flow paths and can be either mixing (two in, one out) or diverging (one in, two out). Examples of each are shown in **Figs. 2.4** and **2.5.** A common application for a diverting valve might be a heat exchanger where one outlet goes to the exchanger and the other to a bypass. A mixing valve might typically be used in a blending opera-

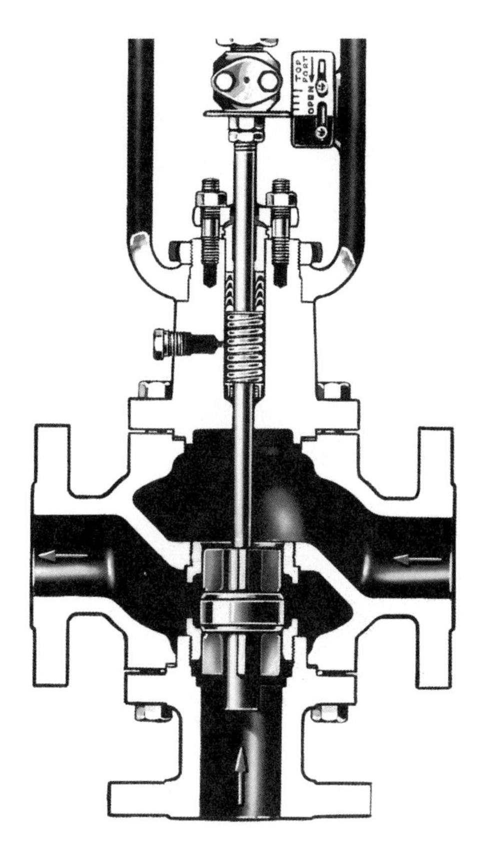

Figure 2.4 Three-way valve: mixing flow. (*Courtesy of Fisher Controls International, Inc., Marshalltown, Iowa.*)

tion to control the incoming ratios to produce the desired end product. Generally, these valves are not pressure balanced and stem forces required to operate them are comparable to those required for single-port valves.

Boot-style, or *pinch,* valves use an elastomeric (or plastic) boot or diaphragm to control the flow. The boot has a hollow cylindrical shape with the flow passing through its center. Flow is controlled by forcing the boot inward, gradually reducing the cross section until the opposite sides of the boot come together, shutting off flow. The boot can be forced closed mechanically or by using pressure as depicted in **Figs. 2.6** and **2.7.** These valves are good for corrosive chemicals and slurries since the boot material can be chosen to be compatible with the chemical being handled. In addition, there is no valve stem and no dynamic seal, so packing problems are eliminated. Temperature limits are fairly low and limited by the elastomer used, and line pressures are limited to about 300 psi for the mechanical pinch and to only about 60 psi for the pneumatic designs.

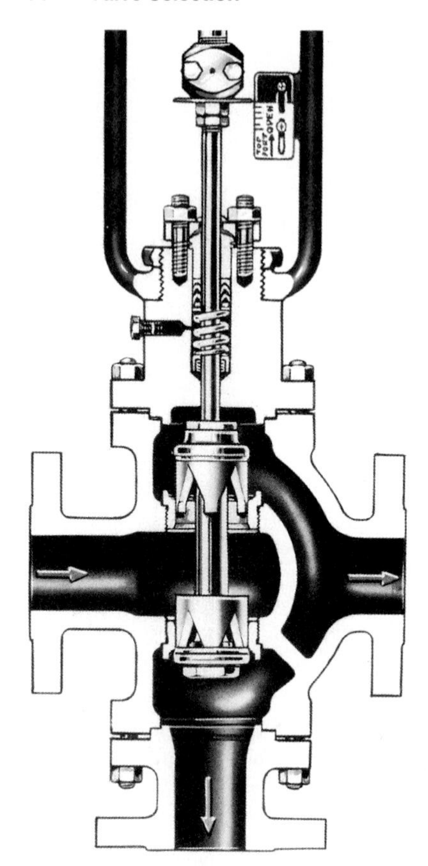

Figure 2.5 Three-way valve: diverging flow. (*Courtesy of Fisher Controls International, Inc., Marshalltown, Iowa.*)

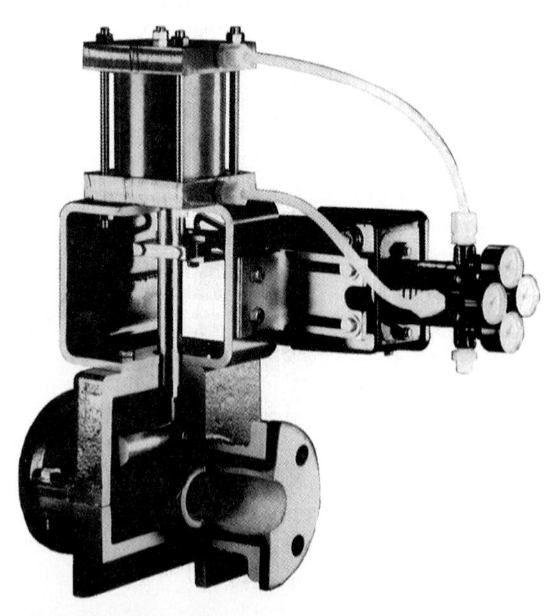

Figure 2.6 Mechanical pinch valve. (*Courtesy of Red Valve, Pittsburgh, Pa.*)

Figure 2.7 Pneumatic pinch valve. (*Courtesy of Red Valve, Pittsburgh, Penn.*)

2.2 Balanced and Unbalanced Valves

Balancing is a general term used to indicate that the static pressure force on the plug from the fluid has been neutralized. Balancing can be accomplished by either reducing the pressure drop across the plug or by reducing the effective area that the pressure drop acts upon. For instance, on a double-seated valve like the one in **Fig. 2.8,** the pressure drop across the plug may be significant but, as we can see, the incoming pressure acts downward on the lower part of the plug and upward on the upper part of the plug, so the forces tend to cancel out.

In a single-seated valve (**Fig. 2.1**), balancing is achieved in a different way. In this case, since we have only one plug member in the flowstream, we have to add balancing holes across the plug so that the pressure drop across the plug is essentially eliminated.

The advantage associated with balancing is that the required force necessary to actuate the valve is reduced, so a smaller, cheaper actuator can be used. Given this advantage, why aren't all valves balanced? First of all, all double-ported valves are, but for reasons covered in Sec. 2.3, they are being used less and less in the chemical industry. Single-seated valves are becoming the industry standard, and the reason that they are not all balanced can be determined by returning to **Fig. 2.1.** Note that once we add the balancing holes, a secondary leak path is created between the plug and the I.D. of the cage, so we have to add a seal of some kind at that point to keep the flow from passing through the balancing holes and continuing downstream.

This second leak path is the problem. Balanced valves, in general, will not shut off as tightly as valves that are unbalanced because the

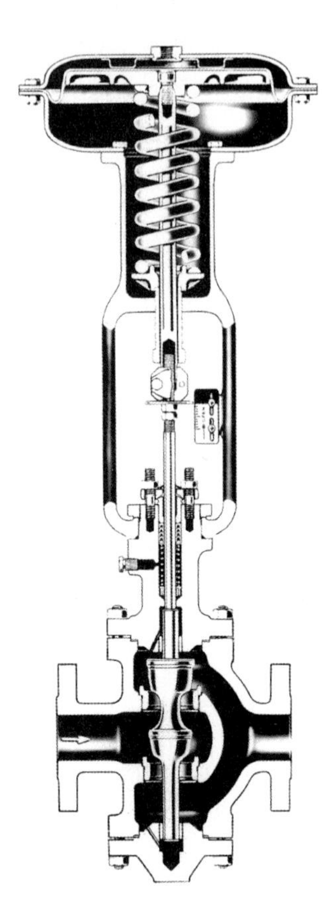

Figure 2.8 Double-ported valve. (*Courtesy of Fisher Controls International, Inc., Marshalltown, Iowa.*)

total leakage through the valve is the sum of the leakage between the seat and plug *and* the leakage past the cage-to-plug seal. Our choice of material for the seal *can* help minimize leakage. A Teflon-based material with an integral metal spring is a common option and provides very good shutoff. However, it is a dynamic seal that is constantly moving with the plug, so maintenance and service life is a concern. In addition, Teflon is limited in temperature range to about 450°F, so if the temperature is higher than this limit, we have to switch to a carbon-graphite piston ring that has much better temperature resistance but does not seal off very well. In general, the tradeoff is between the lower actuator force of a balanced design and the higher leakage that can result, particularly if we have to choose a graphite seal due to temperature. If we need tight shutoff at elevated temperatures, the best choice may be an unbalanced design even though it requires a larger actuator.

There is another point to make before leaving this subject. The end user needs to be aware of the shutoff limitations of balanced designs that were just discussed. There are cases where a plant found the leakage of a valve to be too high for the application and routinely tore it apart to remachine and repair the primary seating surfaces, thinking that this would improve shutoff. In reality, what it had was a high-temperature balanced design whose shutoff properties were determined by the graphite piston ring and not by the primary seat. No amount of maintenance on the seat would have corrected the problem. Finally the trim was changed to a Teflon seal that provided the required shutoff and would still hold up under the actual temperature conditions.

Another important concept that needs to be understood is that many valves that are referred to as balanced are really only partially balanced. There is nearly always some type of unbalance force on any given valve. For instance, the double-ported valve (**Fig. 2.8**) would normally be referred to as a balanced design since the flow acts down on the lower plug and up on the upper plug. However, for assembly purposes the lower plug has to be slightly smaller than the upper plug to make sure that it will fit through the upper seat. As a result, there will be some unbalance due to the difference in effective area between the two plugs. In this case, the net force acts in the upward direction.

In the same way, a single-seated balanced design actually has a small unbalance force associated with it. In the balanced valve shown in **Fig. 2.1,** there is no pressure drop across the valve plug because of the balancing holes. However, if we look at it schematically as in **Fig. 2.9,** it's apparent that the pressure under the plug acts on the sealing diameter between the plug and the seat ring, but the pressure on the top of the plug acts on the sealing diameter between the plug and the cage. The cage diameter has to be larger than the seating diameter to permit the plug to stroke down through the cage and contact the seat. The net result is a small unbalance force equal to the pressure drop across the plug multiplied times the annular area between the two diameters just described.

The direction that the force acts in is just as important as the magnitude. In an unbalanced valve like the one shown in **Fig. 2.10,** the static pressure force on the plug is equal to the pressure drop across the plug times the area defined by the sealing diameter between the plug and the seat. If the flow is up through the seat ring, the flow force acts in the upward direction and has to be overcome with a downward force from the actuator. This is fairly obvious. For a double-ported valve, the situation is also straightforward. As mentioned earlier, because the upper port is larger than the lower to permit assembly, the net force will act in the upward direction.

In the case of the single-seated balanced valve, however, it gets a little more confusing. As described above, there is an unbalance force

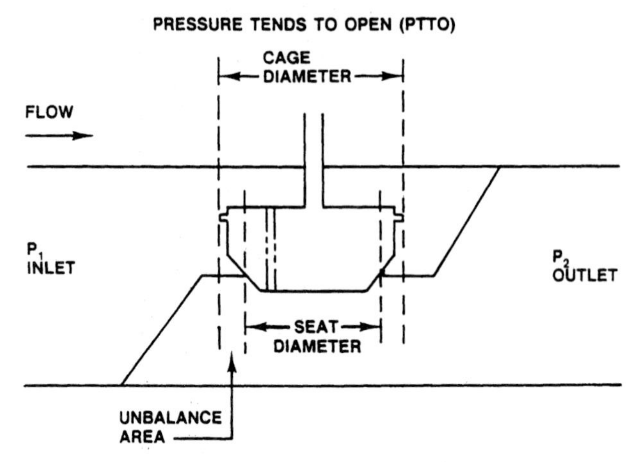

Figure 2.9 Single-seated balanced valve schematic. (*Courtesy of Fisher Controls International, Inc., Marshalltown, Iowa.*)

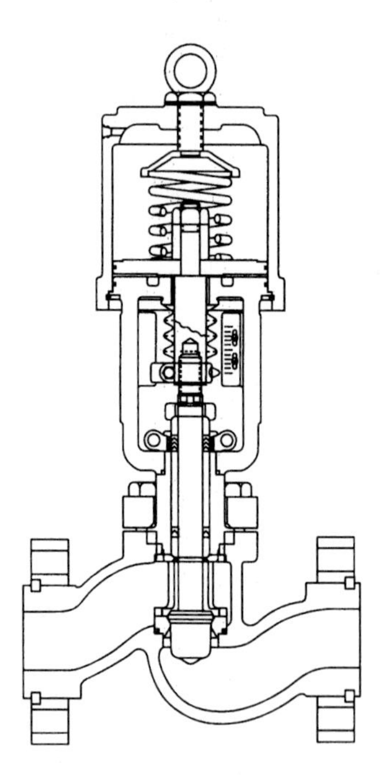

Figure 2.10 Single-seated, unbalanced valve. (*Courtesy of Valtek International, Inc., Springville, Utah.*)

present due to the difference between the seating diameter and the cage I.D. Because the cage diameter is larger, the net force will actually act in the direction opposite to the flow (i.e., if the flow is up through the seat ring, the net force on the plug in the closed position will be down, which seems to be contrary to the conclusion most users would draw). Again, be sensitive to this fact whenever you review shutoff because you may actually have the plug being forced off the seat by pressure even though the valve is flowing down through the seat ring.

2.3 Valve Seating and Shutoff

From a geometry standpoint there are two basic designs used in control valves. The first is the *double-port* approach where there are actually two seats and two plugs working inside the valve body to control flow. Advantages in using a double-port valve include the fact that it is inherently balanced, is easy to change from push down to open to push down to close (reversible), and usually has higher flow capacity than an equivalent single-seated valve. On the downside, shutoff is poor due to difficulties in getting the two plugs to contact the seats at the same time, it is heavy and large, and it has a relatively large number of internal parts that can be high-maintenance items.

Single-port valves are the most common valves used today for a number of reasons. They have fewer internal parts, they are lighter and smaller than their double-ported counterparts, shutoff is relatively good, and maintenance is easy with the top entry, quick-change trim approach.

Another important consideration in seating geometry is how the seat ring is retained in the body. There are three primary methods employed for seat ring retention. The first is what is generally referred to as *drop-in*. In this case, the seat ring is retained in a recessed area of the body web (**Fig. 2.11**). Since it is not mechanically fastened in any way to the body, it is easily removed for maintenance or replacement. It does require some type of seal between it and the body to prevent leakage and some method for loading the seal. The seal is typically a gasket and the standard way of loading the seal is to transmit a portion of the bonnet-to-body bolting load through the cage and the seat ring and, in turn, to the seal. So, while this approach does have significant maintenance advantages, care must be exercised during assembly to ensure that the proper amount of load is transmitted to the seat ring seal to obtain proper shutoff. One common problem with this construction occurs if the dimensions of the body cavity or the internal parts are not correct and, as a result, the load transmitted to the seal is lower than that required to maintain shutoff. This can occur, for instance, if any of the parts are machined or ground-on during maintenance, say, to clean up a gasket surface. Insufficient loading can also occur if the valve experiences high rates of thermal cycling that can unload the seal due to differential thermal expansion between parts. Special constructions may have to be used in this case.

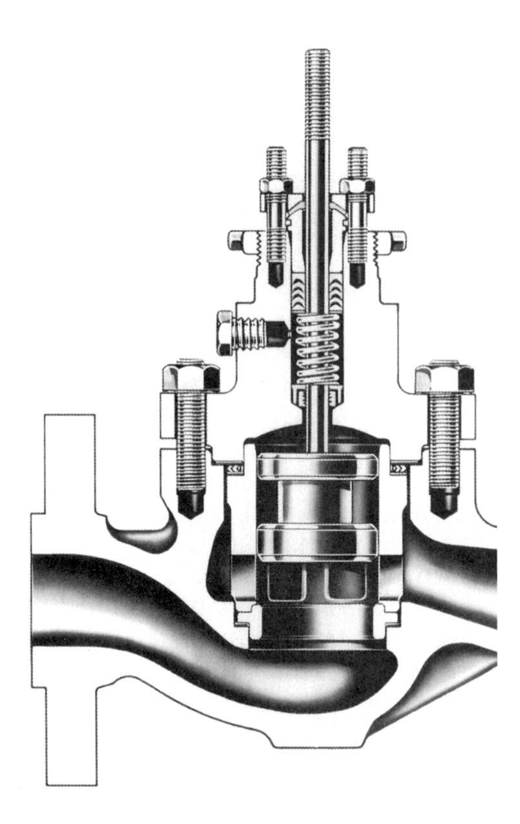

Figure 2.11 "Drop-in" seat ring. (*Courtesy of Fisher Controls International, Inc., Marshalltown, Iowa.*)

A second method for retaining the seat ring is to provide threads on the body web and ring so that the ring can be screwed into the body. This gets away from the loading problems just mentioned for the drop-in construction but carries a heavy penalty in the maintenance arena because the seat ring can be nearly impossible to remove after being in service for even a short period of time. Heat, corrosion, and contact force tend to "weld" the seat ring to the body, requiring excessive torque levels to break it loose. A primary concern in this case is for employee safety since the application of these high torque levels can result in an injury-causing energy release if something suddenly breaks loose.

The third method involves inserting the seat ring into a valve cavity much like the drop-in approach but instead of adding a seal and loading it up, the seat ring is welded to the body to eliminate the leak path around the seat ring. Once again, the design is simpler, but maintenance is difficult, requiring grinding or in-line machining to extract the ring.

Aside from geometry, the other primary consideration regarding seating is that of *shutoff* or, put another way, the leakage through the valve from the upstream to the downstream side. The seat configuration is the primary factor in determining how well a valve will shut off.

There is an industry standard that is used to define how well a valve needs to shut off to meet the requirements of a particular application. The standard is ANSI/FCI 70-2-1976(R1982) and is summarized in **Table 2.1.** For a given valve, the leakage class needs to be selected as a function of how tightly the valve should shut off in service. The classes run from I to VI, with class I applying to a valve that really doesn't have to shut off at all up to a class VI, which is very close to zero leakage. The table shows that the leakage class is defined as a percentage of total flow through the valve for classes II through IV, and the test is run with air as the test fluid at 50 psid pressure drop or lower. Class V is unique in that it is run at actual service pressure drop, uses water as the test medium, and is dependent on the pressure drop and the size of the port, not the rated capacity of the valve. The test approach for class VI returns to the 50 psid or lower guideline with air as the test medium, but the accept-

TABLE 2.1 Leakage Specifications

ANSI B16.104.1976*	Maximum leakage*	Test medium	Pressure & temperature
Class II	0.5% valve capacity at full travel	Air	Service ΔP or 50 psid (3.4-bar differential), whichever is lower at 50 to 125°F
Class III	0.1% valve capacity at full travel	Air	Service ΔP or 50 psid (3.4-bar differential), whichever is lower at 50 to 125°F
Class IV	0.01% valve capacity at full travel	Air	Service ΔP or 50 psid (3.4-bar differential), whichever is lower at 50 to 125°F
Class V	5×10^{-4} mL/min/psid /in. port diameter (5×10^{-12} m³/s/bar differential/mm port diameter)	Water	Service ΔP at 50 to 125°F

	Nominal port diameter		Bubbies per	mi per	Test	Pressure &
Class VI	in	mm	minute	minute	medium	temperature
	1	25	1	0.15	Air	Service ΔP or 50 psid (3.4-bar differential), whichever is lower, at 50 to 125°F
	1.50	38	2	0.30		
	2	51	3	0.45		
	2.50	64	4	0.60		
	3	76	6	0.90		
	4	102	11	1.70		
	6	152	27	4.00		
	8	203	45	6.75		

*Copyright 1976 Fluid Controls Institute, Inc. Reprinted with permission.

able leakage is defined by the nominal port diameter not the valve capacity.

This whole question of shutoff can be confusing for someone who is only occasionally exposed to it. From a practical standpoint, it needs to be pointed out that class II leakage is normal for a balanced, single-port graphite piston ring design or for a standard double-ported valve. In other words, class II would normally be specified for a valve that only occasionally closes and for which shutoff is not a big concern. Class III has half as much leakage and requires that the valve be unbalanced or, if it's balanced, that the seating surfaces be specially prepared if a graphite piston ring design is used or that a Teflon-based piston ring be employed. As a side note, the special preparation of the seating surfaces is commonly called lapping and is done to ensure metal-to-metal contact around the circumference of the seat. The lapping procedure is covered in more detail in Sec. 13.7 on valve maintenance.

Class IV leakage is one-tenth of that of class III and requires a lapped seat and either a Teflon-based piston ring or multiple graphite piston rings. Class IV leakage is usually specified for those control valves that have to shut off on a regular basis. Probably 90 percent of control valve applications are covered in the range from class II to class IV.

Classes V and VI are reserved for those control valves for which shutoff is one of the primary considerations. To obtain shutoff to these levels some very special constructions are required. General guidelines regarding valve construction and shutoff are summarized in **Table 2.2.**

One of the factors that is identified in this table and plays an important role in shutoff is seat load. Every valve manufacturer has

TABLE 2.2

Shutoff class	Typical constructions required
Class II	Balanced, single port, single-graphite piston ring, metal seat, low-seat load Balanced, double port, metal seats, high-seat load
Class III	Balanced, double port, soft seats, low-seat load Balanced, single port, single-graphite piston ring, lapped-metal seats, medium seat load
Class IV	Balanced, single port, Teflon piston ring, lapped-metal seats, medium seat load Unbalanced, single port, lapped-metal seats, medium-seat load Balanced, single port, multiple-graphite piston rings, lapped- metal seats
Class V	Unbalanced, single port, lapped-metal seats, high-seat load Balanced, single port, Teflon piston ring, soft seat, low-seat load Unbalanced, single port, soft-metal seat, high-seat load
Class VI	Unbalanced, single port, soft seat, low-seat load.

its own guidelines regarding seat load, but in general the higher the leak class required, the higher the seat load required. What this means in practice is that if the end user underestimates the pressure unbalance on the plug, either because of not knowing the actual pressure drop across the valve (a very common occurrence) or the unbalance area, the actual seat load can be considerably less than the assumed load, and the shutoff capability of the valve will not meet the specifications. In many cases, a simple comparison of the actual unbalance against the unbalance that was assumed for the valve will reveal a discrepancy that can be addressed by increasing the seat load to a value that will meet the requirements of the application in terms of shutoff.

There is another interesting aspect of shutoff that needs to be covered here if you are to understand the potential impact of the leak class specification cited earlier. Note that most of the leak classes specify a relatively low test pressure of 50 psi. As long as the actual application pressure is equal to or less than this pressure, there isn't a problem applying the results of these tests to what you might find in actual service. However, if the actual service pressure is higher than the test pressure called for in the test standard, a potential problem can surface, particularly for valves that have a significant unbalance area. Let's say we have a valve with a pressure drop of 500 psid and an unbalance area of 4 in^2. This means that the unbalance force will be equal to 2000 lb under actual service conditions. This unbalance, if it acts to open the valve, will reduce the seat load by 2000 lb. If we perform the seat load test per the ANSI/FCI standard at a pressure drop of 50 psig, the unbalance will be only 200 lb, and the valve may very well pass the seat leak test and still leak a considerable amount in service due to the higher unbalance. The message here is that actuator sizing should take the actual pressure drop into account regardless of what may be required to meet the leak class standard under its specified test pressure drop. It's also important to note that even though a valve may meet the test limits specified for a given leak class, this doesn't guarantee that it will meet the leak class requirements under actual service conditions unless the guidelines mentioned in the preceding sentence are followed.

2.4 Guiding

Guiding for a control valve is defined by the method that is used to keep the primary flow control element (the valve plug) in the proper position as it moves to control flow. Guiding is extremely important to the long-term life and reliability of a control valve. As the valve plug moves in and out of the flow, the fluid is constantly exerting forces on it in all directions. The guiding is provided by bearing surfaces that resist the fluid forces to keep the plug properly positioned and operating in a stable manner. If the guiding is not good enough or fails, high levels of vibration can result, giving unstable control and even complete failure of the internal parts.

If the choice of the bearing materials is incorrect, another possible consequence is that the guiding surfaces will break down and galling will result. Galling is characterized by the tearing of the metal in one or both of the bearing surfaces and generally results in very high levels of friction that can interfere with proper valve operation or even cause the valve to seize altogether. Standard rules of thumb used to avoid galling call for the use of bearing materials with different hardness levels and staying away from materials with high levels of nickel. Stainless steels in general are not good bearing materials unless they have been hardened in some way. All things being equal, higher-hardness materials are usually better than lower-hardness materials for bearings.

There are four types of guiding generally used in control valves. The first and most common is *cage guiding*. In this case, the O.D. of the plug rides inside the I.D. of the cage. This has become the standard approach for most valves because it provides a relatively large bearing area that acts directly on the valve plug and in the same general area where the fluid forces act. It also provides for self-alignment among the bonnet, cage, seat ring, and plug during assembly so that the pieces fit together and no inherent side loads exist. An example is shown in **Fig. 2.12.** It is not generally recommended when the fluid is

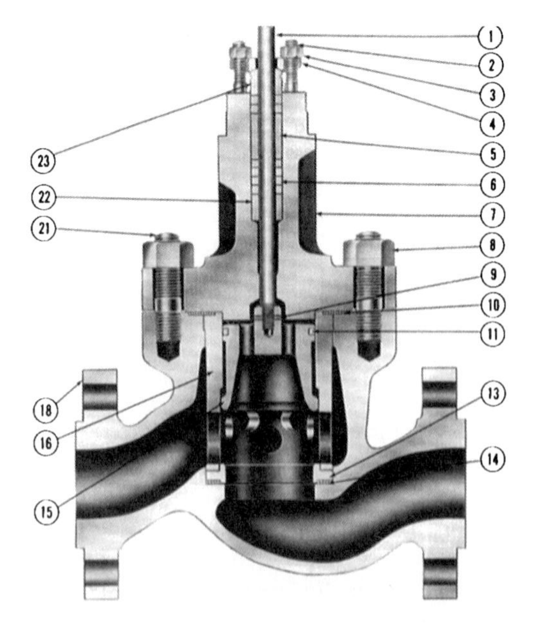

Figure 2.12 Cage-guided valve. (*Courtesy of Masoneilan-Dresser, Inc., Houston, Tex.*)

sticky, gummy, highly viscous, or contains entrained solids since the fluid can tend to build up between the plug and cage and cause operational problems. This problem is commonly called *fouling*.

The type of guiding most often used, if there is a risk of fouling, is called *post guiding*. The "post," as it is referred to, is a recessed area of the plug whose diameter is smaller than the full port of the valve but larger than the stem diameter. It can be above the plug or both above and below. These two types are called top guided or top and bottom guided, respectively, and are shown in **Fig. 2.13.** Post guiding tends to keep the bearing surfaces out of the flow path so that there's less chance of fluid buildup. Certain designs also get the bearing surfaces close to where the fluid forces are acting. The bearing diameters are smaller than for its cage-guided equivalent, but it's a good compromise if the fluid meets the conditions described above.

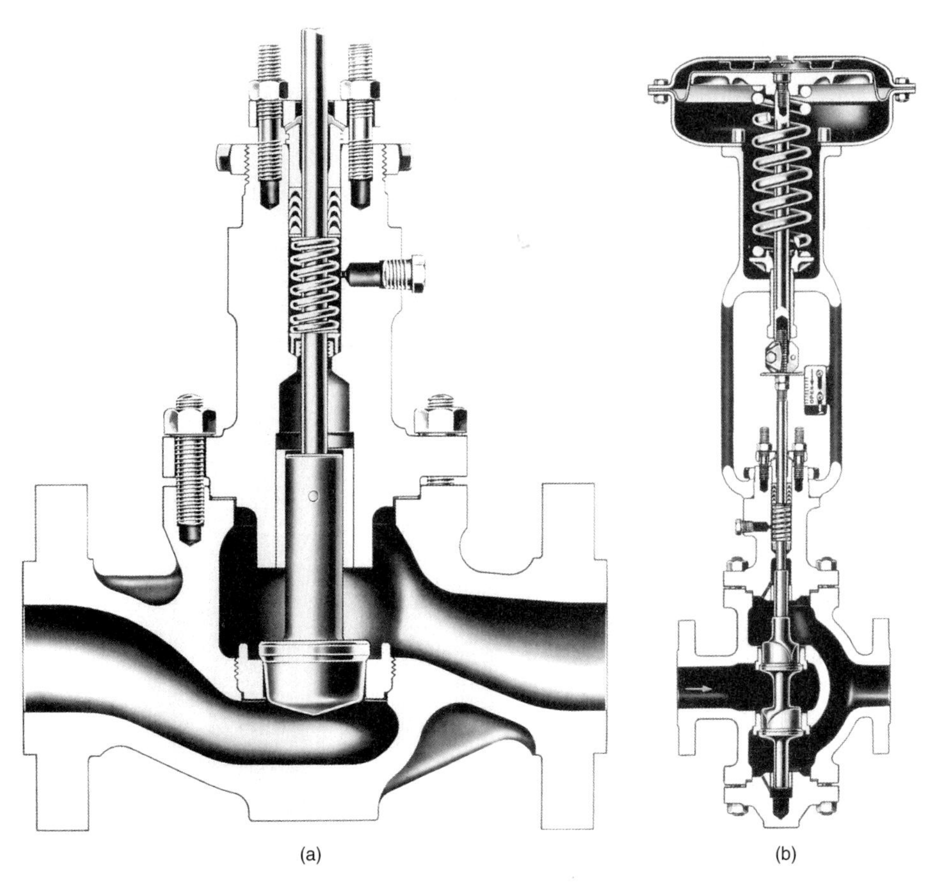

Figure 2.13 (*a*) Post guiding—top; (*b*) post guiding—top and bottom. (*Courtesy of Fisher Controls International, Inc., Marshalltown, Iowa.*)

Stem guiding is an approach that is very similar to post guiding except that the bearing area is slightly farther away from the point where the fluid side loads act, and the stem itself acts as the bearing. Guiding performance is relatively poor for these reasons, but this design is cheaper to manufacture, and bearing replacement is also easier than for other alternatives, so for certain applications it may be the best choice. An example is shown in **Fig. 2.14.**

The fourth type of guiding that one might come across is called *port* or *skirt guiding.* In this case, the port is used to guide the plug as the valve strokes. This design has all the disadvantages of the cage-guided approach with regard to fouling and relies on a relatively small bearing surface provided by the seat ring. As a result, it is not normally used in modern designs, particularly as drop-in cage guiding has become more common. An example is shown in **Fig. 2.15.**

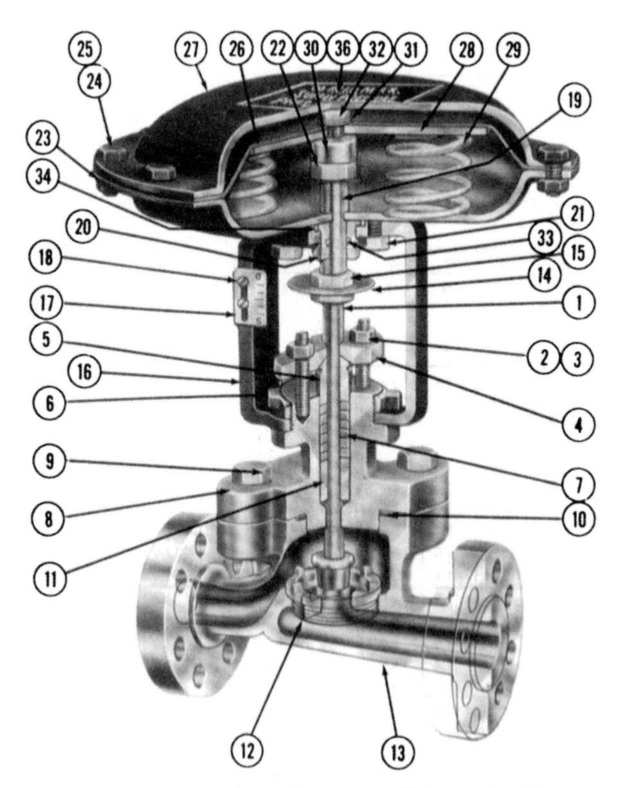

Figure 2.14 Stem guiding. (*Courtesy of Masoneilan-Dresser, Inc., Houston, Tex.*)

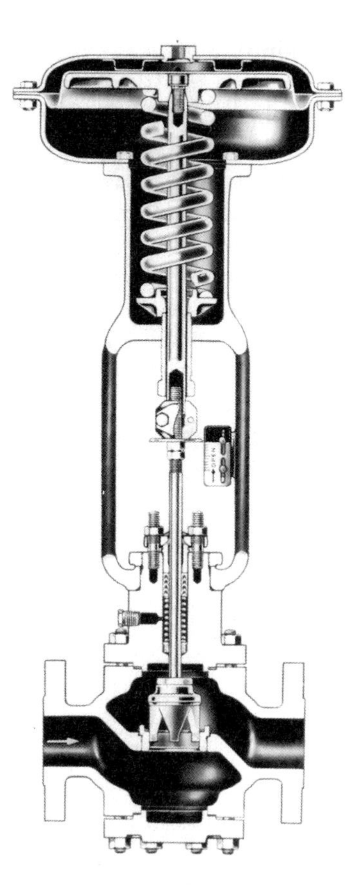

Figure 2.15 Port or skirt guiding. (*Courtesy of Fisher Controls International, Inc., Marshalltown, Iowa.*)

2.5 Flow Characteristics

There are two types of flow characteristics that we normally deal with in working with control valves: *inherent* and *installed.* The inherent characteristic refers to the relationship between the valve position (also referred to as stroke or travel) and the flow through the valve given a constant pressure drop. It is usually defined as valve C_V (see Glossary) versus travel and is determined experimentally. Most valve vendors publish tables of inherent characteristics for their valves that can be used to determine the capacity for a valve for a given set of service conditions and a given travel. The tables are approximate and actual flow may vary up to 10 percent due to such things as trim part tolerances and piping configuration. The shape of the curve that defines the characteristic can change from one valve to the next and can be customized to fit the application, but there are three generally recognized shapes of curves: *equal percentage, linear,* and *quick opening.*

Equal percentage is the most commonly used for control and is illustrated in **Fig. 2.16.** As you can see, it starts out slowly with relatively small changes in flow versus travel near the seat, gradually increasing as the valve reaches full open position. The name comes from the fact that the change in flow rate versus travel at any point in the travel is a constant or "equal percentage" of the flow at that particular travel. Expressed another way, the slope of the flow versus travel curve is a linear function of the flow at any point in travel. What this type of curve provides in practice is a valve that throttles well because it provides relatively small changes in flow versus travel through the first 50 percent of the stroke but still gives good capacity since the valve opens very quickly after that. Equal percentage trim also has an advantage in that it provides for high rangeability in a control valve. Rangeability is defined as the ratio of maximum-to-minimum controllable flow for a valve and is a relative measure of the range of flows that can be controlled by a given valve. In general, high rangeability is desirable since it increases the potential applications for which a valve can be used. The equal percentage characteristic also improves the control resolution and repeatability in the first 50 percent of travel because the capacity change versus position change is small over this range. And lastly, because it opens rather slowly, it tends to keep the valve plug away from the seat, reducing the potential for erosive damage at low flows.

A *linear* characteristic is illustrated by a straight line for the flow versus travel curve. It is shown in **Fig. 2.16.** While throttling capability and rangeability are not as good as for the equal percentage trim,

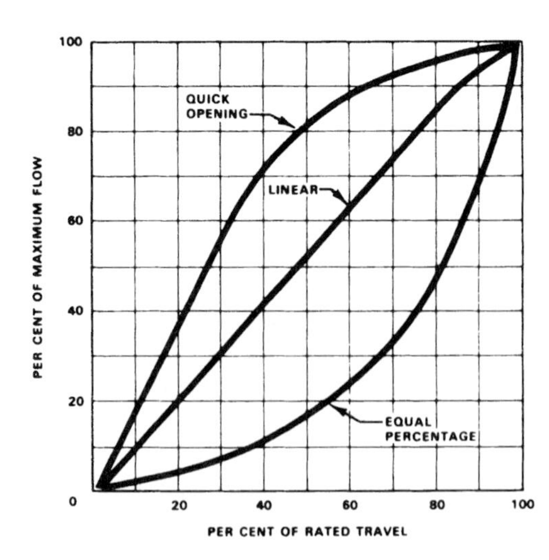

Figure 2.16 Equal percentage, linear, and quick-opening characteristics. (*Courtesy of Fisher Controls International, Inc., Marshalltown, Iowa.*)

it does provide higher capacity. If maximum flow capacity is a concern, this may be a better choice.

Quick-opening trim, as the name implies, is aimed at providing the maximum amount of flow as quickly as possible in the valve travel. It is normally used for on-off service since rangeability and throttling resolution are not very good. It does provide high capacity for a given trim size, which can save some money if maximizing capacity is the primary concern. Its inherent characteristic is linear up to about 70 percent of the maximum flow rate, which occurs at that point when travel equals about 25 percent of the port diameter. It is illustrated in **Fig. 2.16.**

Trim can be characterized in many different ways. With cage-guided valves the openings in the cage walls can be shaped in such a way that the flow area versus travel provides the desired flow characteristic (**Fig. 2.17**). In valve designs where there is no cage or where the cage serves only as a cage retainer, the plug is usually employed as the means for varying flow area versus travel. Examples of plug and port characterization are shown in **Figs. 2.18** and **2.19,** respectively. The important thing to remember here is that the characteristic can be changed relatively easily with a change in the plug or cage, as applicable.

The *installed characteristic* differs from the inherent in that it takes into account the potential for the pressure drop to change as the valve strokes. In many applications, as the valve opens, the resistance to flow drops off, and as a result, the pressure drop across the valve also decreases. In this case, a valve with an inherent equal percentage characteristic will exhibit a more linear flow curve when installed because the reduction in pressure drop as the valve opens counteracts the increasing flow area, making the curve flatter. Linear installed characteristics are desirable because they provide constant gain regardless of travel, making the loop easier to tune and improving control performance. We will come back to this in Sec. 9.3.

If we start with a linear inherent curve under these circumstances, it will resemble a quick-opening characteristic when installed. A quick-opening inherent curve will also change shape in this case, but the shape of the curve for a quick-opening valve is of little consequence since it is not normally used for throttling service. The difference between inherent and installed characteristics for all three types of trim is shown in **Fig. 2.20.**

In deciding what kind of trim characteristic is needed for a given application, do not lose track of the fact that not all applications have decreasing pressure drops with travel. If the pressure drop is constant, the inherent and installed characteristics are essentially the same, so a linear inherent curve may be the right choice in this case.

Figure 2.17 Characterized cage. (*a*) Quick opening; (*b*) linear; (*c*) equal percentage. (*Courtesy of Fisher Controls International, Inc., Marshalltown, Iowa.*)

Flow capacity must also be a consideration. Linear and quick-opening trims may not be as good for certain throttling applications, but they do provide higher capacity, so they might be the logical choice if capacity is the overriding concern for a particular service.

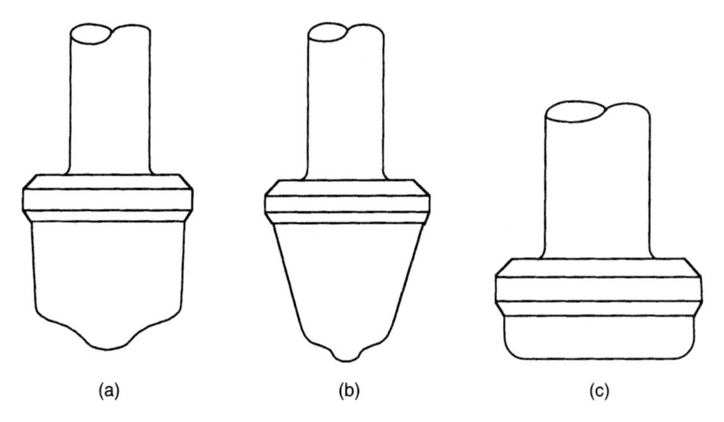

Figure 2.18 Plug characterization. (*a*) Equal percentage; (*b*) linear; (*c*) quick opening.

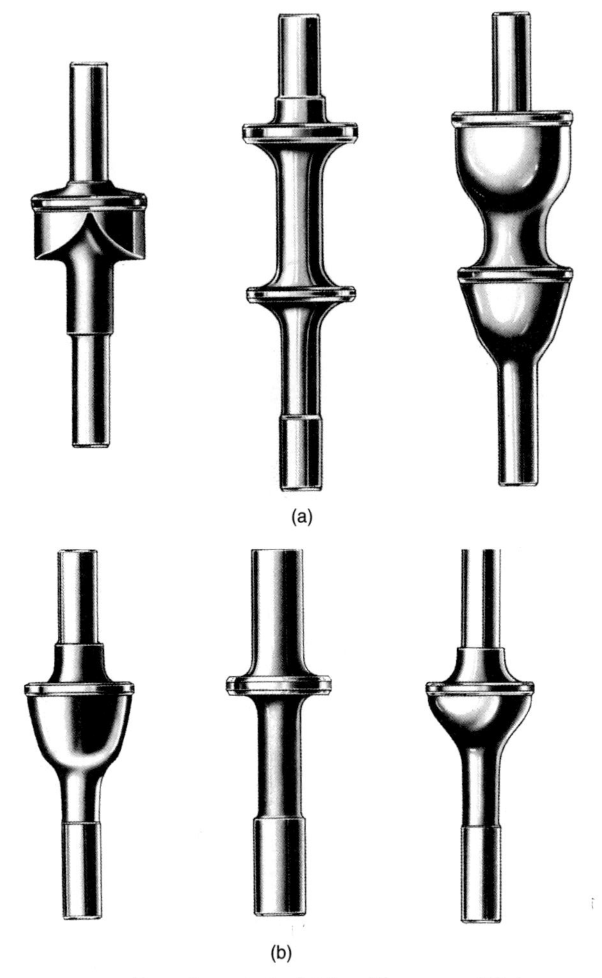

Figure 2.19 Port-characterized trim. (*Courtesy of Fisher Controls International, Inc., Marshalltown, Iowa.*)

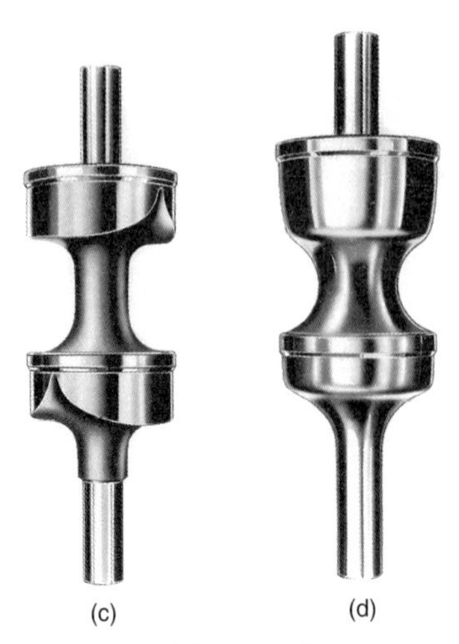

Figure 2.19 *(Continued)*

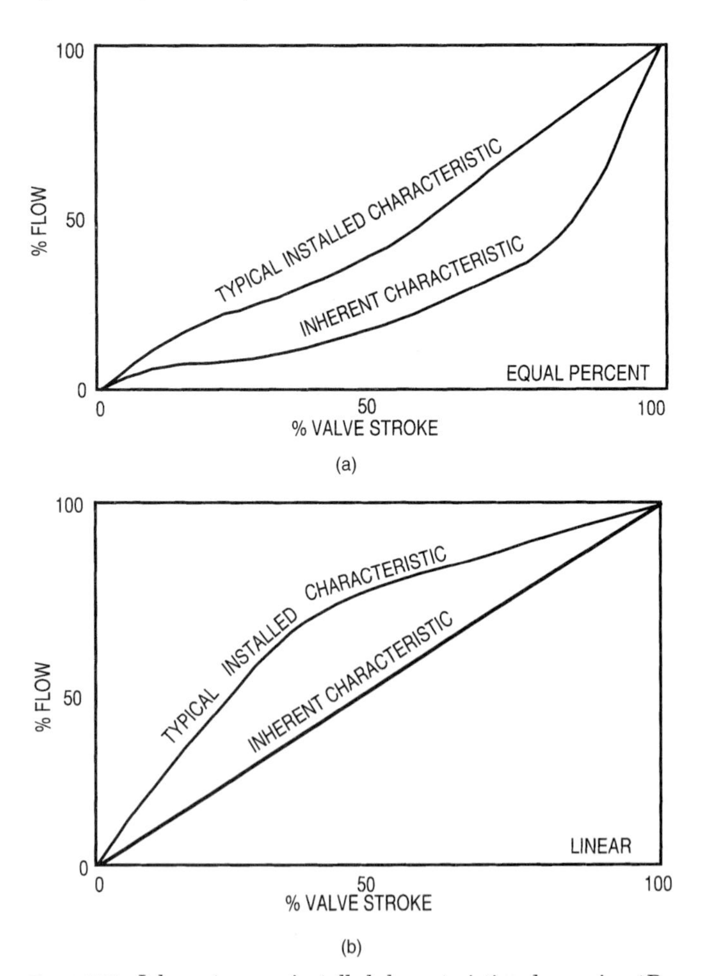

Figure 2.20 Inherent versus installed characteristics; decreasing ΔP.

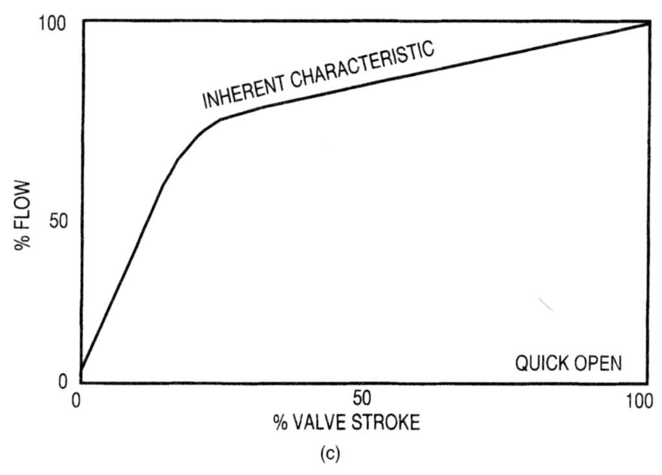

Figure 2.20 *(Continued)*

2.6 Flow Direction and Valve Action

Two additional considerations regarding valve selection involve the direction that the flow takes as it passes through the valve port and whether the valve is push down to open or push down to close. The flow direction is important because, as discussed earlier, the fluid exerts a force on the valve plug and the direction in which it acts is dependent on the flow direction. The magnitude and direction of the force enter into the actuator sizing considerations. If we select a valve where flow tends to open, the required force to close the valve will be comparatively larger than if we selected a flow-tends-to-close valve. (The force required to close a valve and keep it closed is usually larger than the force required to stroke the valve at any other point in valve travel.) Don't forget that for a balanced valve the net flow force is not completely eliminated and that it acts in the direction opposite to flow. (See Sec. 2.2 for a more complete discussion.) Larger actuating forces usually mean larger actuators, and that means more expense.

On the other hand, if a flow-tends-to-close valve is selected, the required actuator force will be reduced, potentially allowing a smaller, cheaper actuator to be utilized. However, some manufacturers recommend against this because of the potential for the plug to be "sucked" down into the seat by the fluid forces present, causing seat damage, control problems, and in some cases, water hammer. The effect is not unlike a sink stopper being held just above the drain while the water runs out. If the stopper gets too close, it will be pulled down into the opening before the person holding it can react.

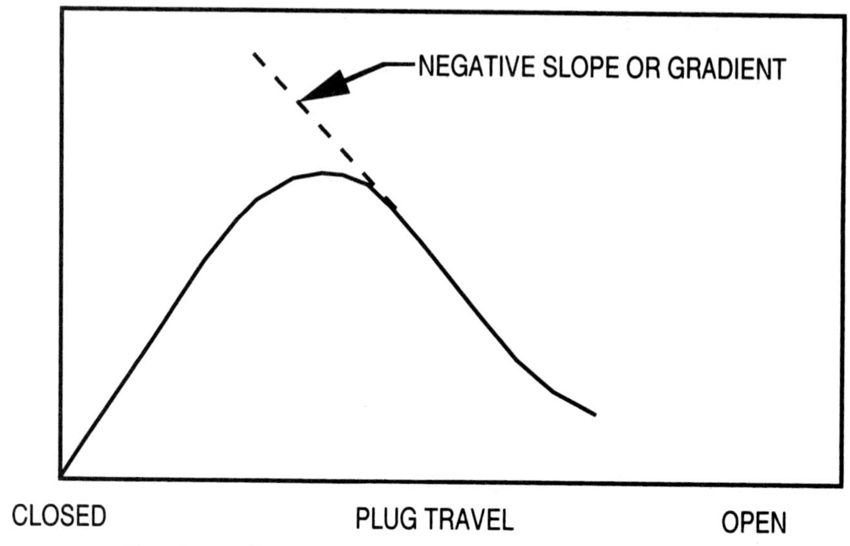

Figure 2.21 Negative gradient.

This phenomenon is commonly referred to as a *negative gradient.* The name comes from the fact that if we plot the stem force required to counteract the fluid force on the plug versus the plug travel for a valve that exhibits this behavior, the curve will show a downward slope moving from left to right (**Fig. 2.21**). It is formally defined as the case in which plug movement in either direction results in a change in force which tends to move the valve plug farther in the same direction. It is most bothersome when it occurs near the seat as described above and is most common in flow-tends-to-close valves, but it can occur at any point in the travel and is sometimes present with flow-tends-to-open configurations, although this is rare.

To counteract this phenomenon, the valve flow direction can be changed to flow tends to open or the stiffness of the actuator or actuator/positioner combination can be chosen so that the inherent spring rate is higher than the gradient expressed in the same units, usually pounds per inch. There is no way to calculate a gradient for a given valve. It must be determined experimentally and then expressed as a function of pressure drop. Proper actuator sizing should include an estimate of the gradient, and a check to make sure that the stiffness of the actuator (or actuator/positioner) can resist it. In some cases, the concerns associated with negative gradients can outweigh any actuator sizing benefits from the flow-tends-to-close approach. Note that the addition of a positioner greatly increases the inherent stiffness of the assembly, so this change alone can sometimes cure the problem.

As far as valve action is concerned, there are really only two types: *direct* and *reverse acting.* For a direct-acting valve, the plug moves

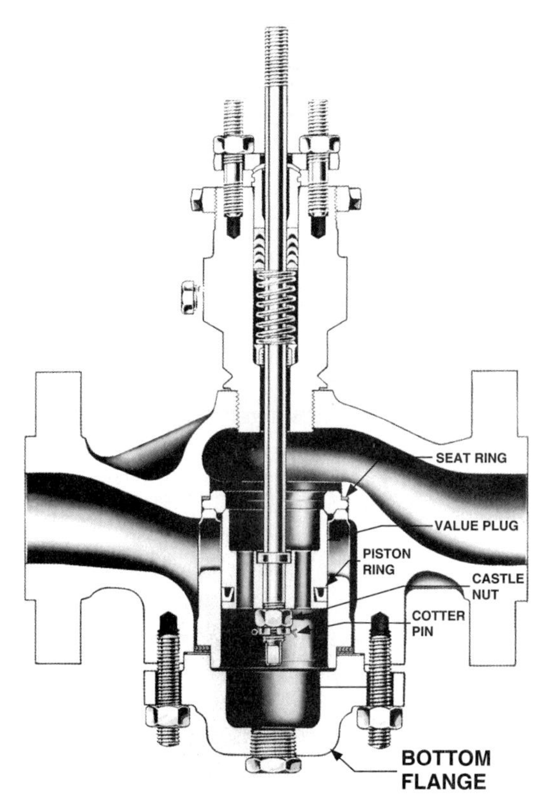

Figure 2.22 Reverse-acting valve. (*Courtesy of Fisher Controls International, Inc., Marshalltown, Iowa.*)

down to close the port. This is referred to as push down to close. For a reverse-acting valve (**Fig. 2.22**), the plug moves out of the seat as the stem moves down into the valve body. This is called push down to open. With the advent of the cage-style valve, reverse-acting valves are becoming less common since the change from direct to reverse action for these valves requires a major construction change. The older valve designs with top and bottom openings were many times touted as being easily "reversible," although there appears to be little advantage associated with this feature, particularly since the action of the assembly can be more easily changed in the actuator or in one of the accessories. A reverse-acting valve does permit inspection and/or removal of the trim parts without removing the actuator, which may be advantageous in some cases.

2.7 End Connections

There are four basic types of end connections used in the chemical industry: *clamped, screwed, bolted,* and *welded*. With the recent enactment of the EPA regulations limiting fugitive emissions, the selection of end connections has become even more important than in the past. Any valve handling a hazardous substance as defined by the EPA is now subject to emissions monitoring that includes measuring leakage from the end connections. As a result, if proper attention is not paid to selecting and adjusting the end connections to minimize leakage, a chemical plant may have difficulties in meeting the EPA regulations, which can have serious monetary consequences.

In a *clamped* design, the valve has no connections per se and is simply clamped in between two pipe flanges with appropriate gasket materials between the valve body and the ends of the pipe. This technique is relatively inexpensive and is usually reserved for bar stock or wafer-style rotary valve bodies. The bolting is longer than with other approaches and is prone to relax with changes in temperature, resulting in leakage. It can also be difficult to properly position the body between the pipe flanges to ensure a tight seal. An example is shown in **Fig. 2.23.**

Screwed end connections are, again, relatively inexpensive but are not used very often in the chemical industry because they can be prone to leakage. The seal is essentially established between the male and female threads on the pipe and the end of the valve body, and even though NPT threads are used, leaks are still common. Maintenance is also a headache because if the valve body has to come out of the line, you either have to cut the pipe or employ some type of pipe union that permits the threads to be disengaged (**Fig. 2.24**).

The *flanged* design is the most common approach used in the chemical industry because it combines reasonable cost with good sealing capability and is easy to maintain. Standards exist that ensure interchange-

Figure 2.23 Clamped-in valve body. (*Courtesy of Fisher Controls International, Inc., Marshalltown, Iowa.*)

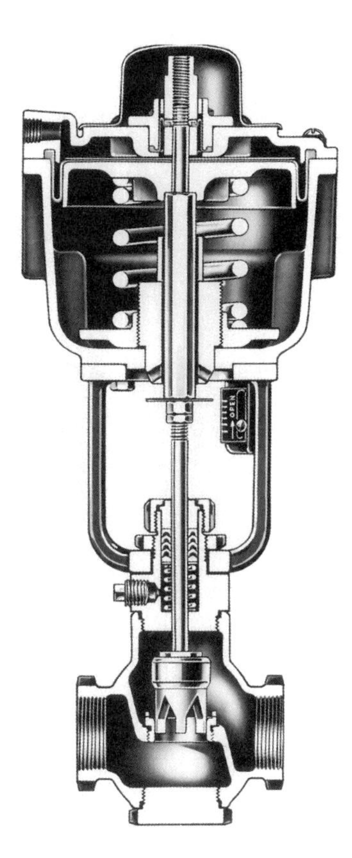

Figure 2.24 Screwed-end connections. (*Courtesy of Fisher Controls International, Inc., Marshalltown, Iowa.*)

ability and design integrity, making the application of flanged end connections relatively painless. In the United States, the most common standard is ANSI B16.5. There are other standards in use such as DIN, and it's important to note that they are *not* interchangeable.

Flanged connections come in three unique styles, raised-face, flat-face, and ring-type joint. The raised-face style has a small stepped or raised area on the face of both flanges inside the bolt circle, where the gasket sets. The raised face reduces the area over which the gasket has to seal when compared to the flat-face design, so the force required to seal the gasket is also lower. It subjects the neck area of the flange to high bending stresses; therefore it is not usually used for materials such as cast iron, brass, and bronze. The sections of the flanges not in contact do act somewhat as springs, helping to counteract the tendency of the gasket to relax with time or due to temperature or pressure cycling (**Fig. 2.25**).

Flat-face flange connections (**Fig. 2.25**), in contrast, have continuous contact between the two flange faces all the way out to the outside diameter of the flanges. It puts less stress on the flange neck but

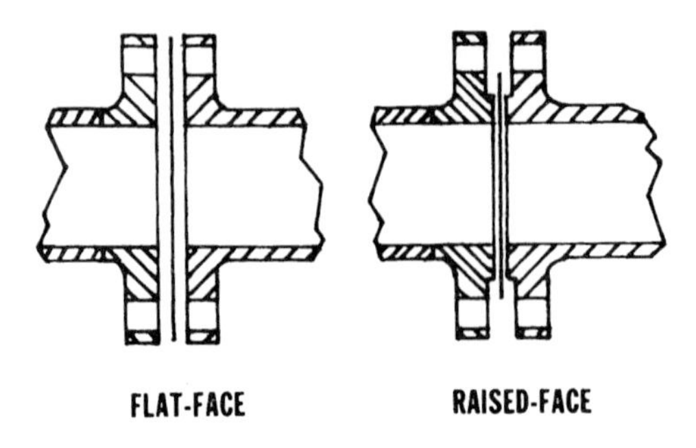

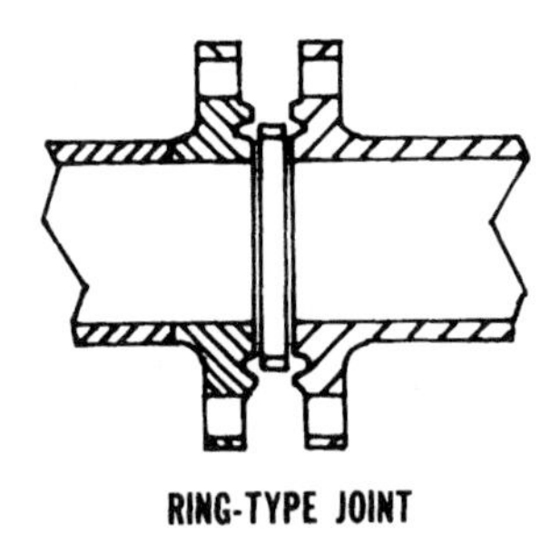

Figure 2.25 Flanged connections. (*Courtesy of Fisher Controls International, Inc., Marshalltown, Iowa.*)

requires higher bolting loads because the gasket sealing surface is larger. As a result, the raised-face construction is more common.

The ring-type joint design (**Fig. 2.25**) is like the raised-face connection except that the gasket is replaced with a metal ring that has an oval or octagonal cross section. The ring makes line contact with the mating flanges to assure a seal and is normally made from a soft metal like Monel to facilitate sealing. It has very good performance at extremely high pressures (up to 15,000 psi), where a gasket might be prone to blow out.

One final note on flanged end connections. Because many of the applications in a chemical plant are erosive or corrosive, they require the use of expensive alloy materials for the wetted parts. Money can be saved by making the flanges as separate pieces from cheaper stronger materials. This is illustrated in **Fig. 2.26.** The sealing approach is essentially the same as for an integral flange, but the design has to be slightly different to account for the lack of a neck in using separable flanges. Because there is no flange neck, the flange itself can bend more easily than its ANSI-based counterpart. As a result, the flange needs to be slightly thicker, and this can result in the studs between the flanges not being long enough. This is something to watch out for when changing from integral to separable flanges. Also be careful with

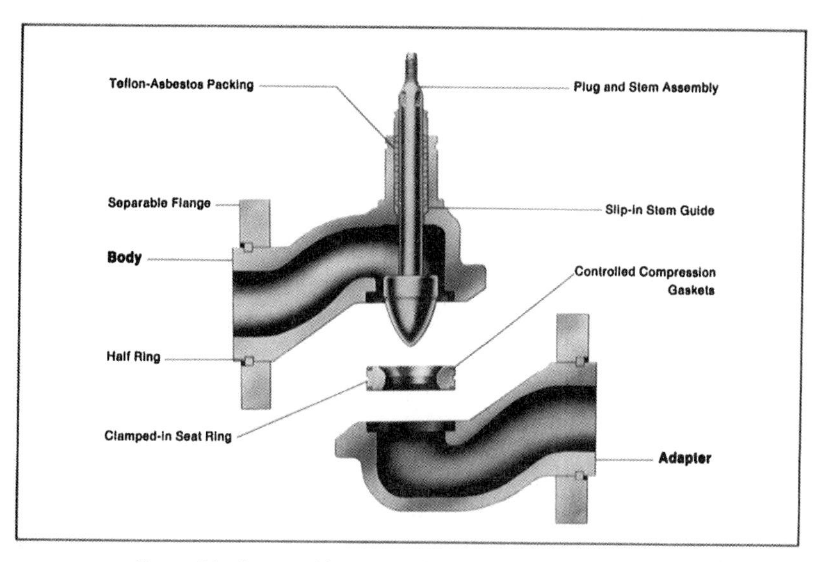

Figure 2.26 Separable flanges. (*Courtesy of Masoneilan-Dresser, Inc., Houston, Tex.*)

separable flanges during installation of the valve to ensure that the retaining rings are properly positioned and can't drop out since this can result in leakage or, in the worst case, injury to personnel. Special attention needs to be paid to supporting the body during installation and removal because once the flange bolting has been loosened, the body can rotate inside the flanges and could fall, resulting in damage to the equipment or in personal injury.

The last type of end connection mentioned is the *welded* connection. This type of integral connection provides a very good seal because there is no leak path if the weld is done properly, and you don't have to worry about line stresses or thermal and pressure cycling loosening the joint as you do with the other types already covered. On the down side, it is not very maintenance friendly since the body has to be cut out of the line if it is to be removed for service. It's also much more expensive to use in the field due to high labor costs associated with welding. It has not been used all that frequently in the chemical industry except in powerhouse applications involving steam, but its use could become more common with the increasing need for zero-leakage pipeline joints given the new EPA restrictions on fugitive emissions.

Welded ends come in two common configurations: *socket welding* and *butt welding*. In a *socket weld,* the body has a recessed area into which the pipeline slides. The body and pipe are then joined together

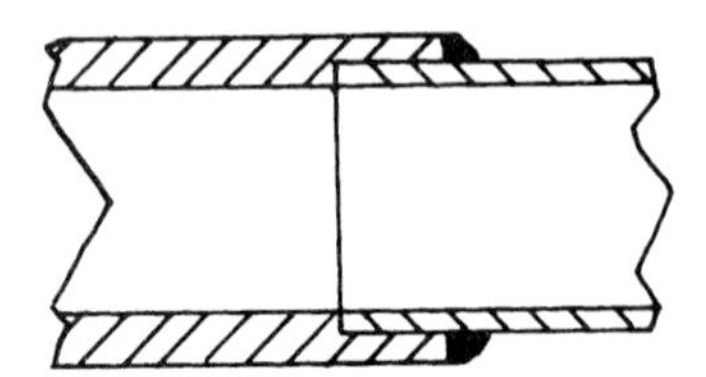

Figure 2.27 Socket-weld end connections. (*Courtesy of Fisher Controls International, Inc., Marshalltown, Iowa.*)

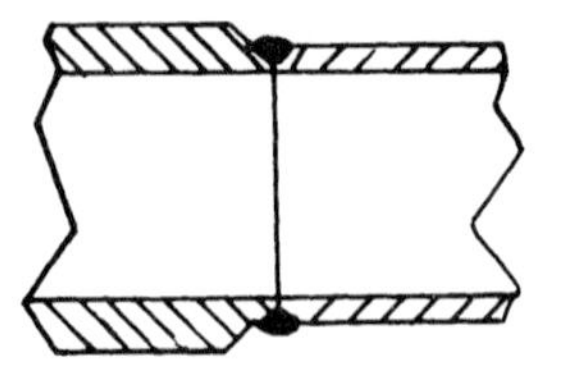

Figure 2.28 Butt-weld end connections. (*Courtesy of Fisher Controls International, Inc., Marshalltown, Iowa.*)

using a fillet weld between the end of the body and the O.D. of the pipe (**Fig. 2.27**). The socket helps position the valve before welding but is not a full-penetration weld because of the crack starter as the base of the weld, so it's not as strong as the butt-weld approach.

The *butt-weld* approach does provide a full-penetration joint, so it's preferable from a strength standpoint. However, the end preparations are more difficult and expensive to make, and positioning the valve between the ends of the pipe can be time consuming. The valve and pipe ends are beveled to match, according to an ANSI standard where the size and the configuration of the ends depend on the pressure and temperature of the internal fluid. See **Fig. 2.28** for an example.

2.8 Pressure-Temperature Ratings

In any given application, two of the primary service conditions to be considered are the pressure and temperature of the internal fluid. In general, the higher the pressure and temperature, the larger the body wall thickness has to be to ensure that the valve will not rupture in service. ANSI B16.34 is the standard that was put together to reflect this relationship and shows graphs for different groups of materials and different pressure ratings that illustrate how the acceptable pressure varies with the service temperature. An example is shown in **Fig. 2.29.** Essentially, if we know the service pressure and temperature, along with the body material, these curves will tell us where we fall as far as the ratings are concerned. For instance, if the pressure is 1200 psi at 200°, a class 600-lb rating or higher will meet the ANSI requirements for the service. Meeting the requirements of the 600-lb rating is then spelled out in a separate section of ANSI B16.34, which covers the design of the pressure-retaining parts and typically centers on minimum wall thickness and end-connection design.

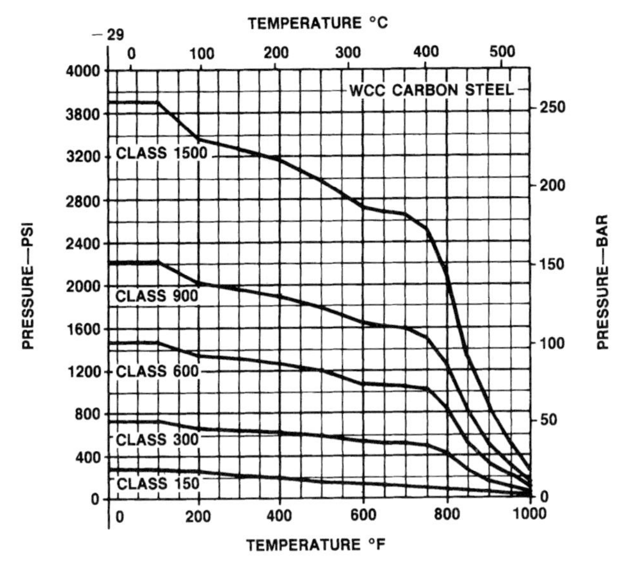

Figure 2.29 Pressure-temperature ratings for normal carbon steels. (*Courtesy of Fisher Controls International, Inc., Marshalltown, Iowa.*)

2.9 Small Flow Control Valves

The chemical industry is unique in that it uses a fairly large number of control valves with very low flow capacity. It's somewhat arbitrary, but any valve with a maximum C_V of 1 or less could be said to fall into this category. They are primarily used in pilot-plant operations where processes are tested on a small scale before going into full production. They may also be seen in regular production plants where they can be used to inject small amounts of chemicals into larger lines.

They may be used in high or low pressure drop applications but are usually not asked to operate at elevated temperatures since even small changes in relative length due to differential thermal expansion can result in large changes in valve stroke given the reduced size of the internals. Conventional control valve constructions work well down to about a C_V of 1, but below this level, it becomes very difficult to control the size of the flow path and the valve travel to give consistent flows. There are a number of alternative designs that can be used to help solve this problem. Keep in mind that most vendors list flow characteristics for these valves, but, in truth, the flow capacities are so low and can change with only minor changes in machining and the like that it is difficult to determine the actual characteristics. In addition, the flow itself can change from turbulent to laminar or a mixture of the two, and this can greatly affect the real capacity. In general, treat published C_Vs and characteristics as ballpark indications only and try to simulate actual service conditions to see what the flow capacity really looks like.

The first example is shown in **Fig. 2.30** where a contoured needle valve plug is used to control flow through a very small orifice. The plug, in turn, is connected to an actuator that is specially designed to provide for small changes in travel. This is accomplished through a lever mechanism that converts the full travel of the rolling diaphragm into a much smaller travel at the valve plug. The lever also provides a mechanical advantage that aids in developing seat load. This particular model features an adjustment knob (key 13, **Fig. 2.30**) that allows the maximum C_V of the valve to be changed, in place, by modifying the effective lever arm in the actuator. (**Figure 2.31** shows a cutaway detail of the actuator.) Note that the valve also features very precise guiding due to the potential for high pressure drops and the importance of positioning the plug in the center of the port to get repeatable results from a flow standpoint. Due to the potential for high pressure drops and high-velocity erosive flows, hardened trims are usually available for these valves.

Another approach is shown in **Fig. 2.32.** This is an anticorrosive model with a Teflon insert that channels the flow past a Hastelloy plug. The plug has a very small slot milled in its side through which the flow passes. This valve can control C_Vs of 1 or less with a conventional actuator but is limited to fairly low pressure drops and nonerosive service due to the use of Teflon. Although this design features the use of Teflon internals, the same type of milled slot or milled flat approach works with a more conventional all-metal trim, but the flow is slightly harder to control due to the inherent leakage between the metal plug and metal port. This clearance flow does not exist with the Teflon port since it can seal tightly against the plug even when it is in the open position.

A third type of valve uses a floating sapphire ball that lifts off the port to throttle flow. The actuator has an adjustable pneumatic feedback mechanism that permits the user to change the stroke obtained for a standard input range like 3 to 15 psi. This model is good for high-pressure and erosive applications. It is illustrated in **Fig. 2.33.**

The fourth and final type of low-flow valve to be covered here is one where rotary actuator motion rotates the valve stem against a fine-pitched thread that transforms the rotary motion into a linear motion that can be used to position the plug with respect to the seat. In a typical design, 60° of rotary motion results in only 0.0052 in of stem travel, so very fine control can be maintained for very low flows. A cross section is shown in **Fig. 2.34.** A common thread of all these designs is the fact that the fluid has to be very clean, or debris will plug the small passages and affect flow capacity.

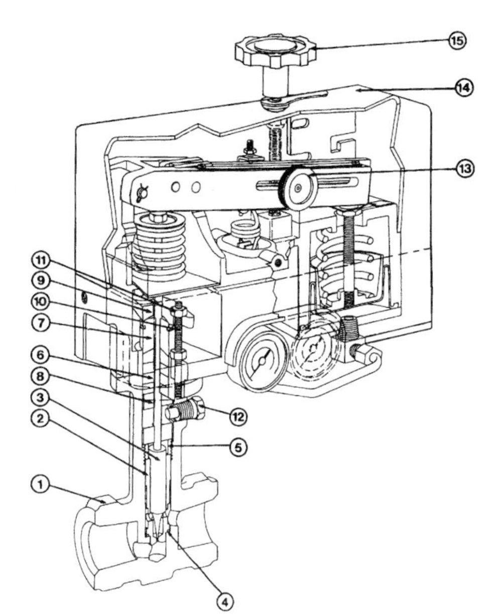

Figure 2.30 Cross section of low-flow valve, actuator, and positioner. (*Courtesy of Masoneilan-Dresser, Inc., Houston, Tex.*)

Figure 2.31 Cutaway of low-flow positioner and actuator. (*Courtesy of Masoneilan-Dresser, Inc., Houston, Tex.*)

2.10 Plastic Control Valves

Another special group of control valves that sees a lot of use in the chemical industry is the plastic control valve. As mentioned earlier, the handling of corrosive fluids is fairly common, and if we stick with metal valves, the special alloys needed to stand up to certain chemicals can become very costly. Modern plastics have been developed that

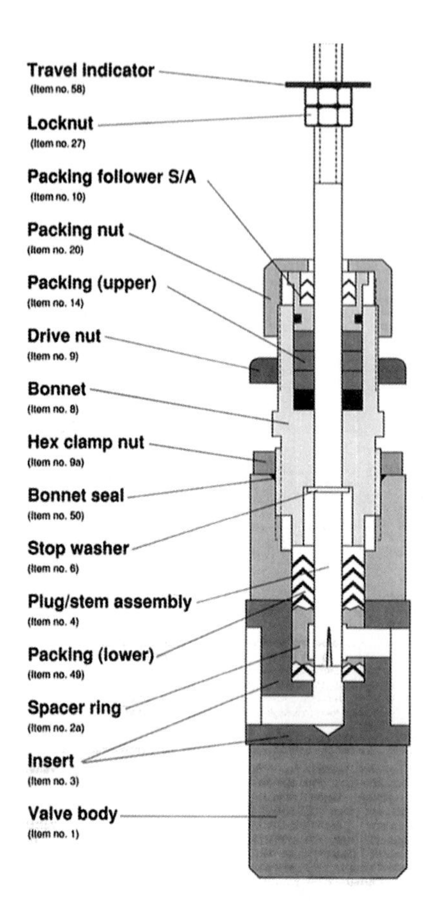

Figure 2.32 Low-flow, corrosive service valve. (*Courtesy of Valtek International, Inc., Springville, Utah.*)

stand up very well to these chemicals, and control valve product lines now exist that are built upon the principle of using plastic not only for the wetted parts but also for the actuator so that the valve assembly can resist corrosive attack from the outside environment as well.

Figure 2.35 shows that plastic valves come in all the standard valve configurations. Typical service would include sulfuric acid, hydrochloric acid, hydrofluoric acid, and caustics. Valve bodies and trim are available in polypropylene, PVDF, EXTFE, and PVC. Actuators and yokes are manufactured from glass-filled UV-inhibited polypropylene for strength and creep resistance and to guard against damage from exposure to the sun. The positioners are usually integrally mounted to eliminate the need for external linkages that can corrode, and the accessories can be hot-wax dipped to protect them from the harsh environments sometimes found in the chemical industry. A typical cross section is shown in **Fig. 2.36.**

The following list of questions and issues has been put together to

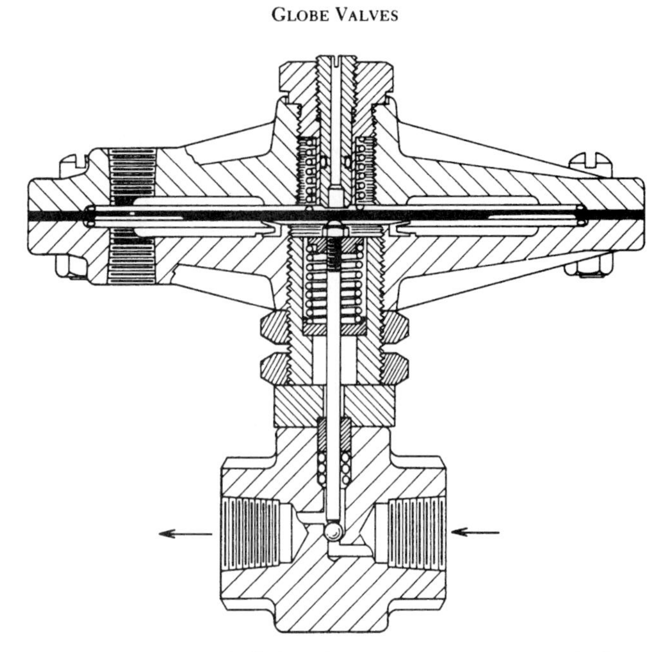

Figure 2.33 Floating ball, low-flow valve. (*Courtesy of IMI Cash Valve, Inc., Decatur, Ill.*)

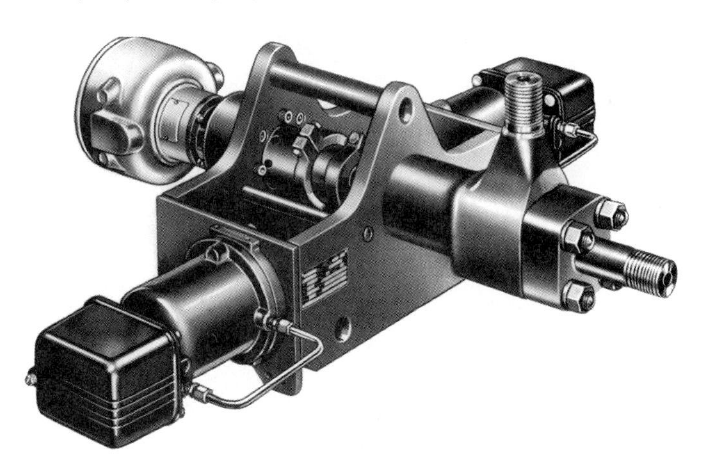

Figure 2.34 Rotary conversion, low-flow valve. (*Courtesy of Fisher Controls International, Inc., Marshalltown, Iowa.*)

help determine whether a plastic valve might be appropriate for a given application:

1. Is the application highly corrosive like one of the services listed above?

2. What are the pressure requirements? Plastic valves are limited to fairly low pressures and temperatures: 300 psi at ambient and 150 psi at 225°F are typical.
3. Is an erosive fluid being handled? The relatively soft parts used in this construction will not stand up to erosion as well as their metal counterparts.
4. Pipe stresses are a concern. Proper alignment before installation is critical to guard against failure due to residual loads.
5. How much flow is required? Valve size is limited, and maximum C_Vs are in the neighborhood of 40 or less.

In other respects, such as stroking speed, plastic valves are very similar to valves of metal construction, and if the answers to the above questions don't turn up any problems, they could be the most cost-effective answer for corrosive applications.

Figure 2.35 Plastic control valve configurations. (*Courtesy of Collins Instrument Co., Angleton, Tex.*)

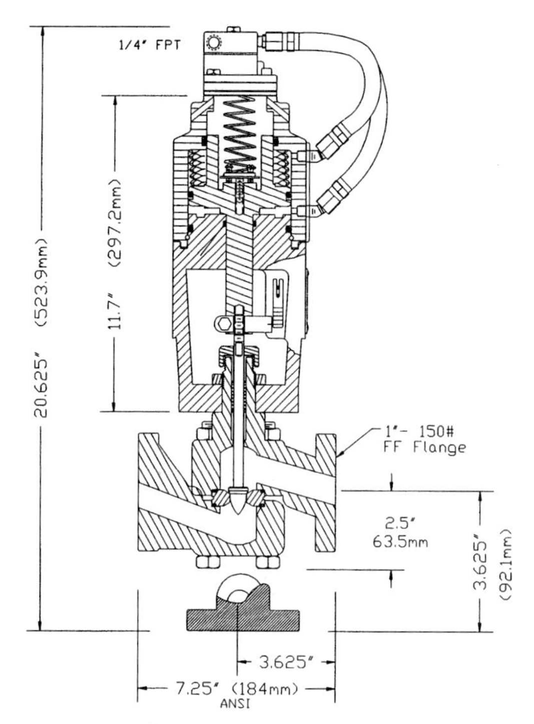

Figure 2.36 Plastic valve cross section. (*Courtesy of Collins Instrument Co., Angleton, Tex.*)

References

1. *Control Valve Handbook,* 1st ed., Fisher Controls, Marshalltown, Iowa, 1965.
2. *Control Valve Handbook,* 2d ed., Fisher Controls, Marshalltown, Iowa, 1977.
3. Liptak, Bela, (ed.), *Instrument Engineer's Handbook,* rev. ed., Process Control, Chilton Book Co., Radnor, Pa., 1985.
4. Hutchinson, J. W., *ISA Handbook of Control Valves,* ISA, Research Triangle Park, N.C., 1971.
5. Smith, William T., "Plastic Control Valves in the Chemical Processing Industry," *Chemical Equipment,* Feb. 1993.
6. Schafbuch, Paul, *Fundamentals of Flow Characterization,* TM 29, Fisher Controls, Marshalltown, Iowa, 1985.
7. Anderson, Gerald D., *ST/FT–4 Valve Characteristics,* Fisher Controls, Marshalltown, Iowa, 1975.
8. Luthe, Fred J., *Proper Sizing of Diaphragm Actuators for Control Valve Service,* TM-25, Fisher Controls, Marshalltown, Iowa, July 1972.
9. Schuder, Charles B., *Understanding Fluid Forces in Control Valves,* ISA, Research Triangle Park, N.C., 1971.
10. ANSI/FCI 70-2-1976 (R1982), American National Standards Institute, 1982.
11. *Red Valve-Control Pinch Valves,* Red Valve Co., Carnegie, Pa., 1990.
12. *Catalog 10–Fisher Controls,* Fisher Controls, Marshalltown, Iowa, 1990.

Chapter

3

Rotary Valves

This chapter covers the rotary family of control valves. It includes discussions on typical configurations, guiding, shutoff flow characteristics, end connections, and pressure-temperature ratings.

Rotary valves by definition are valves where the valve stem or shaft rotates within a sealing area to manipulate the final control element inside the valve body to regulate flow. While the configuration of these valves differs from sliding stem, many of the points covered in Chap. 2 apply to rotary valves as well. In instances where this occurs, the material will not be repeated. You can, instead, refer to the appropriate paragraph in Chap. 2.

In general, rotary valves provide more flow per dollar than their sliding-stem counterparts because the flowstream is more efficient since the flow turns less as it goes through the valve. It is for this reason that rotary valve use has become more widespread in the last 10 to 15 years. However, there are still many applications, particularly severe service, where the sliding-stem valve is still the best choice because of its rugged design. There are also cases where the high-capacity characteristics of a rotary valve may actually be a disadvantage. Because they have highly efficient flow paths, they exhibit a trait called high recovery. In other words, the downstream pressure will be a large percentage of the upstream pressure due to the low pressure drop across the valve. In some instances this can lead to problems with cavitation that can be solved by switching to a sliding-stem design with lower recovery characteristics. This will be covered in more detail in Sec. 6.2.

3.1 Rotary Valve Configuration

Rotary valves come in many shapes and sizes. They are usually characterized by the shape and action of the flow control element. One of

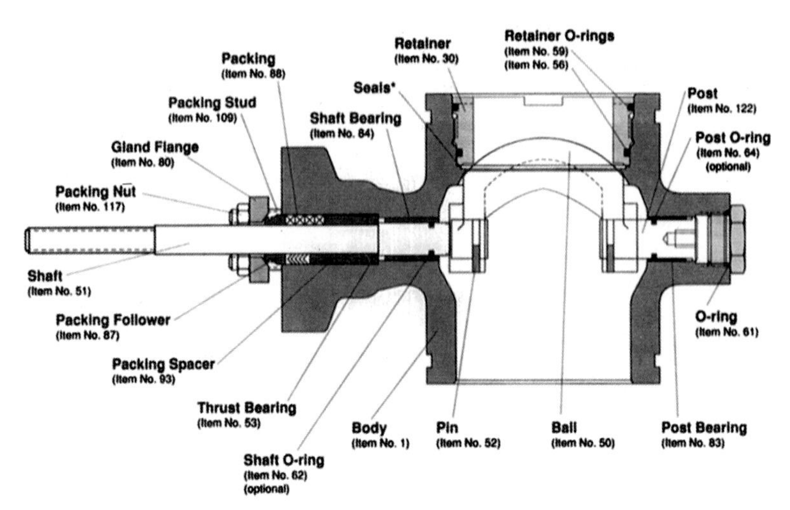

Figure 3.1 Ball-segment valve. (*Courtesy of Valtek International, Inc., Springville, Utah.*)

the most common types is referred to as the *ball valve* or *ball-segment valve* (**Fig. 3.1**). Its name comes from the fact that the valve plug is actually a segment of a ball or sphere that rotates on an axis perpendicular to the flowstream. Its leading edge moves in and out of the flow to open and close an opening between the ball and a valve seat, and the movement is normally a rotation of 90°. It can be supplied with many different seating arrangements that are primarily selected based on the type of fluid handled and/or the shutoff required. Variations in seal design are shown in **Table 3.1,** which includes notes on design characteristics and where each one might be used.

Ball valves can be flowed in either direction but flow into the face of the ball is more common. The opening between the ball and the seal can also be changed to provide for different flow characteristics. **Figure 3.2** shows one example called a V notch that provides a more gradual opening that improves rangeability and throttling capability. This shape also has an inherent shearing action that facilitates closing against fibrous or viscous fluids. At least one vendor offers a micronotch that opens even more slowly. While notching the ball improves the inherent characteristic, the ball valve still opens up much more quickly than its sliding-stem counterpart.

Other important design features relating to ball valve selection include the bearing design and the ball-to-actuator connection. Bearings come in metal, plastic, or plastic-lined metal designs and need to be carefully sized to withstand the full flow force on the ball face and still permit the shaft to be rotated by the available actuator force. Too large a diameter provides long bearing life but also increas-

TABLE 3.1 Ball valve seals

Illustration	Name	Comments
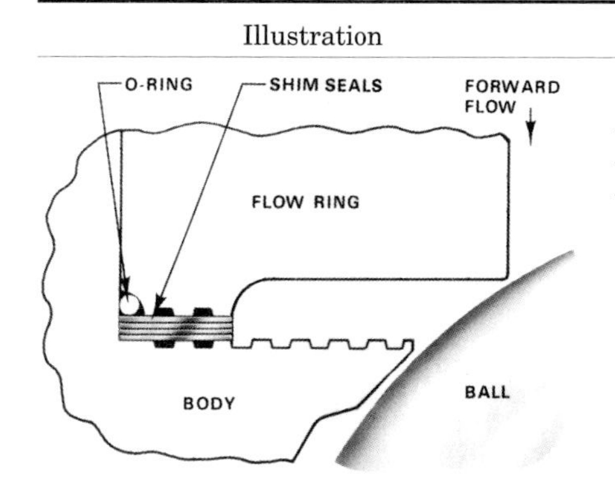	Flow Ring. (*Courtesy Fisher Controls International, Inc., Marshalltown, Iowa*)	A clearance—flow design that never completely suts off due to the seal. Advantages include no seal friction or wear. No better than Class I shut-off. (2% of valve capacity is typical)
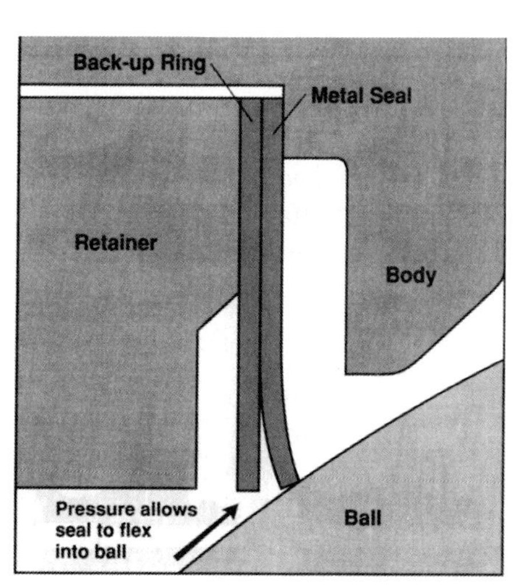	Metal Seal (*Courtesy Valtek International, Inc., Springville, Utah*)	Usually some type of thin metal disc that can flex, allowing constant contact with the ball. Good compromise between durability and shut-off (can maintain Class IV). Also tends to be better than soft seals if fluid is sticky because seal tends to scrape the ball clean. Relatively high friction.
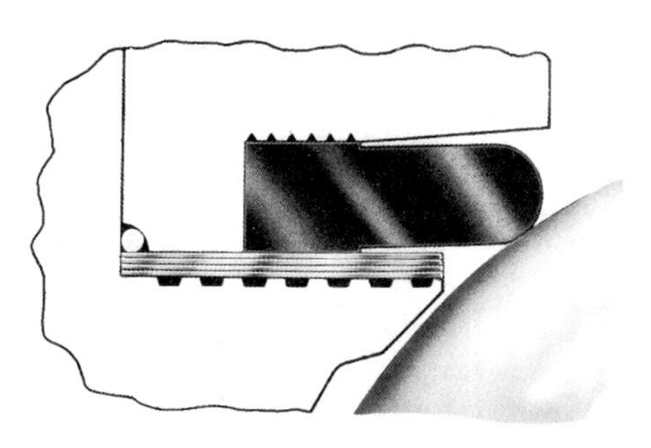	Soft Seat (*Courtesy Fisher Controls International, Inc., Marshalltown, Iowa*)	A flexible disc usually made from teflon-based material. It may include a metal back-up ring for longer life. Provides tighter shut-off (Class V) and lower friction than metal seal but wears out more quickly, particularly with erosive fluids.

Figure 3.2 End view showing V notch. (*Courtesy of Fisher Controls International, Inc., Marshalltown, Iowa.*)

es the torque necessary to rotate the shaft. If the bearing's diameter is too small, it can wear out too quickly. It's the traditional engineering compromise, and it's something that the vendors have to struggle with, depending on the application. In fact, bearing limitations can be the deciding factor in switching to a sliding-stem design for severe service applications, as mentioned earlier.

The connections running from the ball to the actuator are also critical in determining how good the control of the valve will be. Sloppy or weak linkages will contribute to a positioning error, which means poor control. In fact, one of the disturbing trends in the industry today for rotary valves is the attempt to use on-off designs in throttling applications in an attempt to save money. These valves will not control well and any up-front savings are quickly eliminated in the course of operation.

The *full-ball valve* is another type that is encountered in the chemical industry. As shown in **Fig. 3.3,** the ball in this case is a full sphere with a cylindrical hole cut through the middle where the fluid passes. There are many different versions available, but a common one has the ball "floating" or held between two seals that also act as bearings. While this simplifies the design, it can also contribute to friction levels that are higher than conventional bearing designs and hurt control performance. There is also an inherent problem in making the seals act as bearings since bearing wear will hurt shutoff performance. Seal

Figure 3.3 Full-ball valve. (*Courtesy of Fisher Controls International, Inc., Marshalltown, Iowa.*)

options, shutoff capabilities, and actuator connections are similar to those offered for the ball segment valves. Rotation is normally 90°, and the inherent characteristic is equal percentage. Flow capacity at 90° open is very high because of the line-of-sight flow path.

Standard, or *conventional, butterfly valves* are shaped like dampers, with a disc that rotates in the flow path to throttle flow. An example is shown in **Fig. 3.4.** From the side view, you can see that the valve body is relatively narrow and takes little space in the pipeline. The shaft is centered on the axis of the pipeline and is in line with the seal. Once again, there are many seal options, ranging from a swing-through design with very poor shutoff up to adjustable or inflatable elastomeric seals that provide bubble-tight (class VI) shutoff. The disc pulls away from the seal upon opening, minimizing seal wear (except near the shaft) and reducing friction. The "break-out" torque required to pull the valve out of the seat can be high, causing the valve to jump on opening and making control near the seat difficult to achieve. The bearing designs look much like those already described for the ball segment valves and the actuator connection issues are also the same. The flow characteristic tends toward equal percentage, but the rotation is limited to about 60° for a stan-

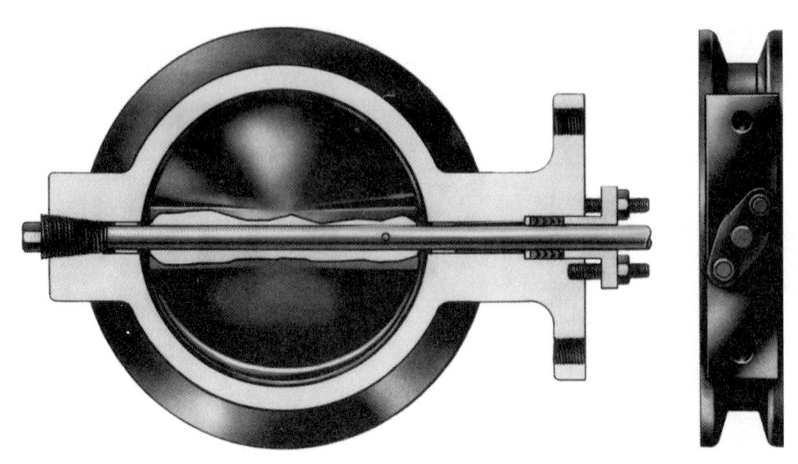

Figure 3.4 Conventional butterfly valve. (*Courtesy of Fisher Controls International, Inc., Marshalltown, Iowa.*)

dard symmetrical disc shape since the leading edges of the disc are hidden within the flow cross section of the shaft for rotations greater than that. This limitation can be addressed through the use of special shapes for the disc, such as the *fishtail* shown in **Fig. 3.5,** which permits effective throttling out to a full 90°. Flow forces on the disc vary greatly with rotation and pressure drop and are relatively high, limiting the range of applications for this type of valve. Many conventional

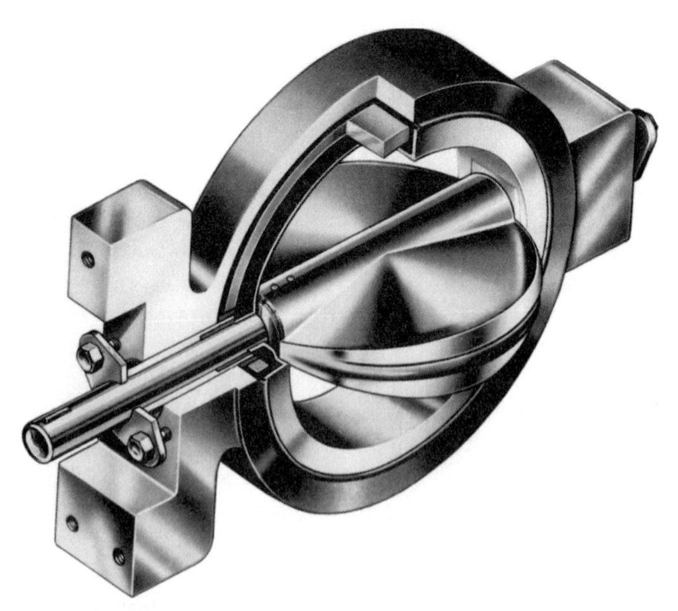

Figure 3.5 Fishtail butterfly disc. (*Courtesy of Fisher Controls International, Inc., Marshalltown, Iowa.*)

butterfly valves in the chemical industry are used with TFE or rubber liners to provide better corrosion resistance and to improve shutoff.

The *high-performance butterfly valve* (HPBV), or the eccentric disc as it is sometimes referred to, is a relatively new variation where the rotation axis for the disc is offset from both the centerline of flow and the plane of the seal (**Fig. 3.6*a* and *b***). This approach yields a number of advantages, including better seal performance, lower dynamic torque, and higher allowable pressure drops. The seal performance is improved because the disc cams in and out of the seat, only contacting it at closure, so wear is reduced. Because the disc can approach the seal from one side, the pressure drop across the valve can be used to provide a pressure-assisted seal, further improving performance. The special shape and contour of the disc helps reduce dynamic torque and drag, permitting higher pressure drops, and the fact that the disc is never hidden in the shaft cross section allows for good throttling control through a full 90° of rotation with a linear characteristic. Because of this increased capability and the relatively high capacity-to-cost ratio, the high-performance butterfly valve is making serious inroads in applications previously handled by sliding-stem valves.

The *eccentric plug valve* is similar to the eccentric disc design except that the flow control element is a rounded plug rather than a disc and that the offset from the seal plane to the center of rotation is greater. **Figure 3.7** shows a typical construction. Note that the seal is more massive than in the HPBV and actually serves as the travel limit for the plug, much like in a sliding-stem valve. This is in contrast to the other rotary valves already mentioned where the rotation is limited by the actuator, not by the internal trim parts. These valves are relatively inexpensive and provide good throttling performance and good shutoff (better than class IV), particularly with erosive fluids. The tight shutoff is obtained by either letting the seat ring center itself on the plug as it seats or, in another approach, the plug actually flexes to conform to the seat. Setting the actuator up to provide proper seating without exceeding load limitations on the trim parts can sometimes be tricky, though. These valves can be flowed in either direction. The flow capacity *is* lower for a given size than the other rotary valves already discussed.

The last type of rotary valve that we're going to cover is called the *eccentric ball valve,* which is actually a hybrid of the eccentric plug and ball valve designs. It is shown schematically in **Fig. 3.8,** which shows that it looks like a traditional ball valve except that there is an offset between the shaft centerline and the flow centerline, enabling the ball to cam into the seat at closure. It has all the performance features of the ball segment valve with the added advantage of extended seal life due to the camming action. The seal designs parallel those of the eccentric plug design and provide the same type of performance.

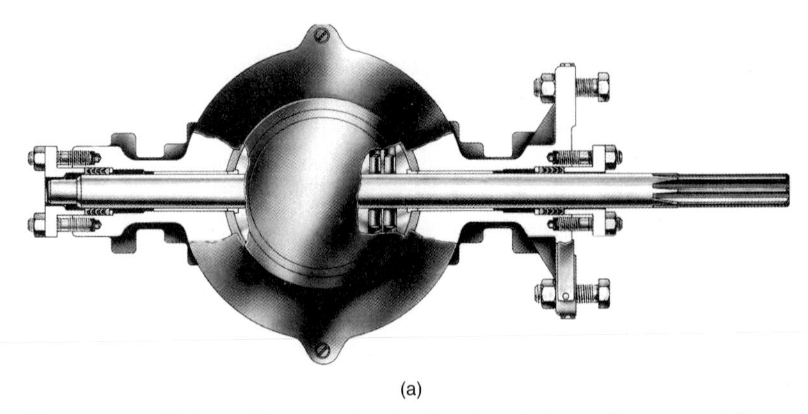

Figure 3.6a. High-performance butterfly—front view. (*Courtesy of Fisher Controls International, Inc., Marshalltown, Iowa.*)

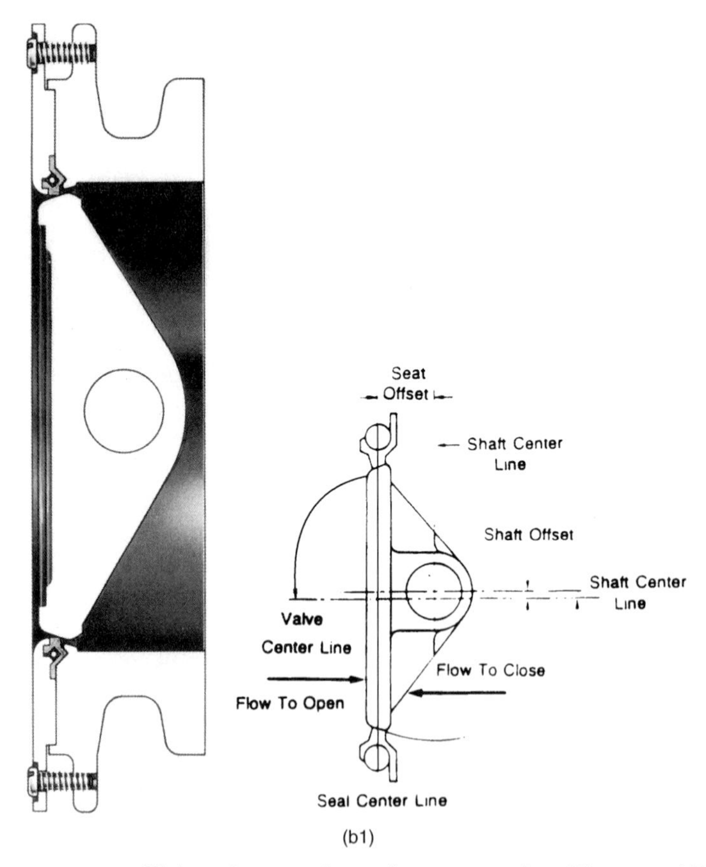

Figure 3.6b. High-performance butterfly—cross section. (*Courtesy of Fisher Controls International, Inc., Marshalltown, Iowa.*)

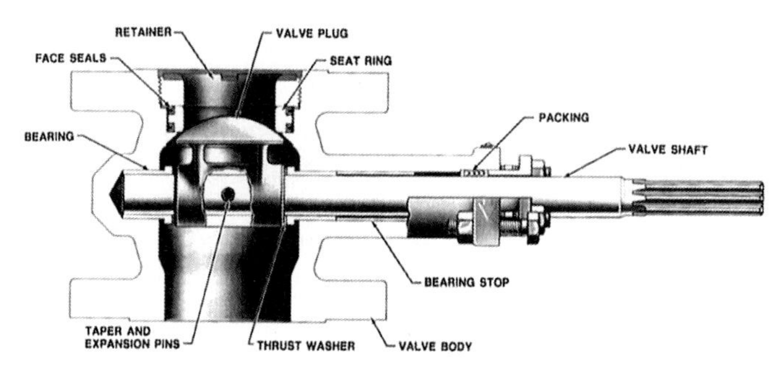

Figure 3.7 Eccentric-plug valve. (*Courtesy of Fisher Controls International, Inc., Marshalltown, Iowa.*)

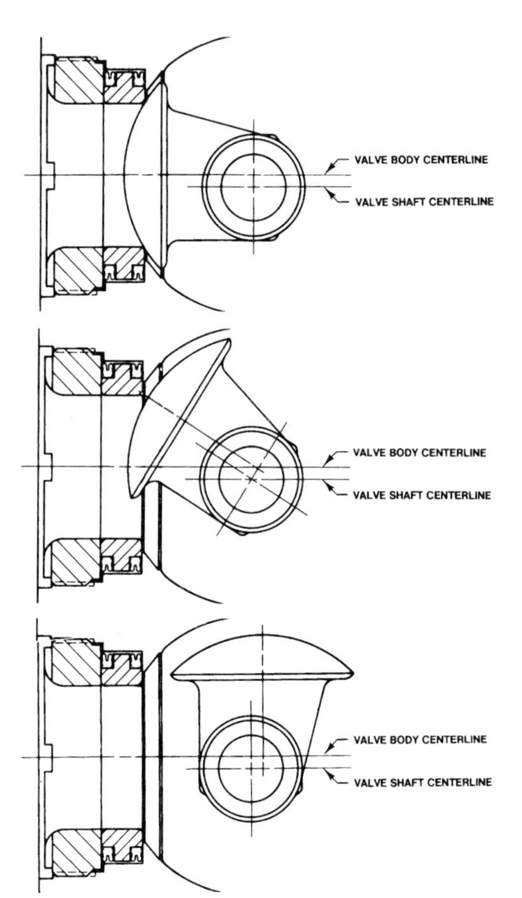

Figure 3.8 Eccentric-ball valve. (*Courtesy of Fisher Controls International, Inc., Marshalltown, Iowa.*)

3.2 Balancing, Guiding, Shutoff, and Flow Characteristics

Unlike sliding-stem valves, there is no way to balance rotary designs, so the static fluid forces on the flow control element are always relatively large. The good news is that the actuator doesn't have to directly overcome this force for a rotary valve. It only has to be able to rotate the shaft in the bearings, overcoming the frictional torque caused by the fluid force on the ball, disc, or plug. There *are* dynamic torques acting on the flow element that also have to be resisted by the actuator. These are a function of flow element position and pressure drop, and these must be experimentally determined by the valve manufacturer. The magnitude of these forces are taken into account in actuator sizing software and won't be covered here. It is important to note the direction that these forces act in to be sure of the fail action of a particular valve in case the actuator loses power. These are summarized in **Table 3.2.**

As far as *guiding* is concerned, either the shaft bearings or the seats perform this function for rotary valves. In either case, the available bearing area is smaller when compared to a sliding-stem valve, and the flow force has to be resisted directly by the bearings. On a sliding-stem valve only the side load on the plug is transmitted to the

TABLE 3.2 Dynamic torque action

	Flow direction		Flow tends to:	
MODEL	Through seat, past shaft	Past shaft, through seat	Close	Open
Ball segment	X		X	
		X	X	
Full ball*	N.A.	N.A.	X	
Butterfly*	N.A.	N.A.	X	
HPBV	X			X
		X	X	
Eccentric plug	X			X
		X	X	
Eccentric ball	X			X
		X	X	

*This design is symmetrical with respect to flow direction.

bearings. These factors, along with the fact that rotary valves cannot be balanced, are the reasons why sliding-stem designs still dominate in severe service. *Shutoff* and *flow characteristic* considerations parallel those already covered in the sliding-stem section.

3.3 End Connections and Pressure-Temperature Ratings

Once again, the *end connection* and *pressure-temperature rating* discussion from the sliding-stem section applies here. There *are* more rotary valves used with flangeless connections where they are simply clamped between the line flanges. This is especially true for butterfly valves.

References

1. Egnew, John, *Rotary Valve Design Options,* Fisher Controls, Marshalltown, Iowa, Feb. 1990.
2. *Neles-Jamesbury Bulletin A110-1 Quadra-Power™ Actuators,* Neles-Jamesbury, Worcester, Mass., 1990.
3. *Neles-Jamesbury—The Right Control Valve,* CVC-2010593, Neles-Jamesbury, Worcester, Mass., 1993.
4. *Camflex II, 35002 Series; Spec. Data CF 5000,* Masoneilan- Dresser, Houston, Tex., Sept. 1992.
5. *Control Valve Handbook,* 2d ed., Fisher Controls, Marshalltown, Iowa, 1977.
6. *Shearstream Control Valves—MI27,* Valtek No. 53938, Valtek, Inc., Springville, Utah, Aug. 1991.

Chapter

4

Actuators

Actuators are the primary energy source used to move and position the flow control element within the valve body. They come in many different forms and can use many different sources of energy including *pneumatic, electric, hydraulic, electrohydraulic,* and *manual operation.* This chapter provides an overview of each type with particular emphasis on pneumatics since they are still the most popular choice for throttling applications. In the ensuing discussion, pneumatics will be used as the standard of reference in discussing the performance and cost of the other types of actuators.

4.1 Electric Actuators

There are two primary methods that are employed to position control valves using electricity as the power source. The first can be referred to as *electromechanical* and involves using a motor through a gearbox to provide either reciprocating or rotary action depending on the type of valve that is being used. An example is shown in **Fig. 4.1.** This particular example shows an actuator set up for reciprocating action where the rotary motion is transformed to linear motion through a power screw. For rotary action, the power screw is not required.

While this type of actuator is capable of generating very high torques or stem thrusts, it is generally slow when compared to more traditional approaches, and its throttling capability is limited, so it sees little service in control applications. It is also more expensive than pneumatics and requires a high-voltage electrical source, which may be difficult to provide at the control valve and can also have safety implications in hazardous environments. To summarize, if high thrusts or torques and/or a high degree of actuator stiffness are

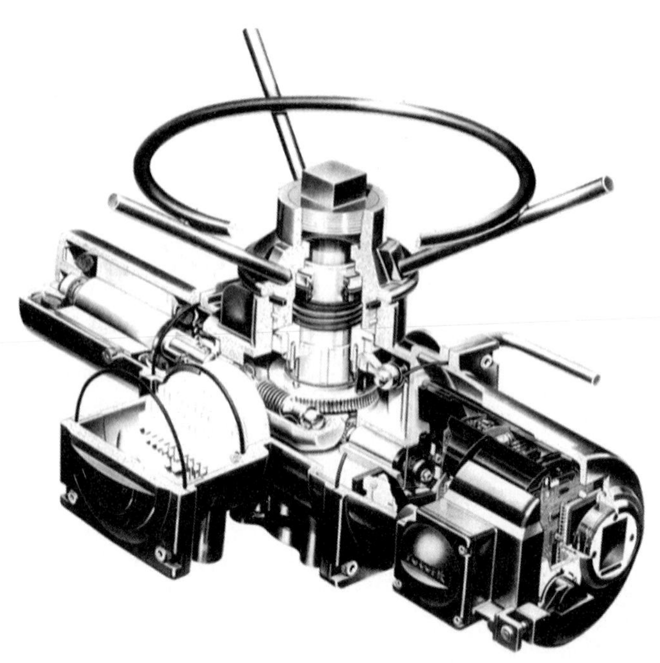

Figure 4.1 Electromechanical actuator. (*Courtesy of Rotork Controls, Inc., Rochester, N.Y.*)

required and the throttling performance is not that critical, this type of actuator could be a good choice.

The second type of electric actuator in use today is based on the *solenoid* principle whereby an electric current sets up an electric field that acts upon a metal plunger. The position of the plunger is determined by the strength of the electrical field and it, in turn, is connected to the valve plug, which regulates flow. Solenoid valves have been in use for some time now for on-off service, but recent developments have made them more practical for throttling service. **Figure 4.2** shows a typical construction. Advantages include the elimination of the stem seal (packing), so the threat of package leakage is eliminated, and the fact that they can interface directly with the electrically based control system (i.e., no conversion from electrical to pneumatic signal is required). They *are* limited in thrust, so the pressure range over which they will work is limited. They are generally more expensive than pneumatics but less expensive than electromechanical actuators. Throttling performance, if you respect the pressure limitations, is comparable to pneumatics.

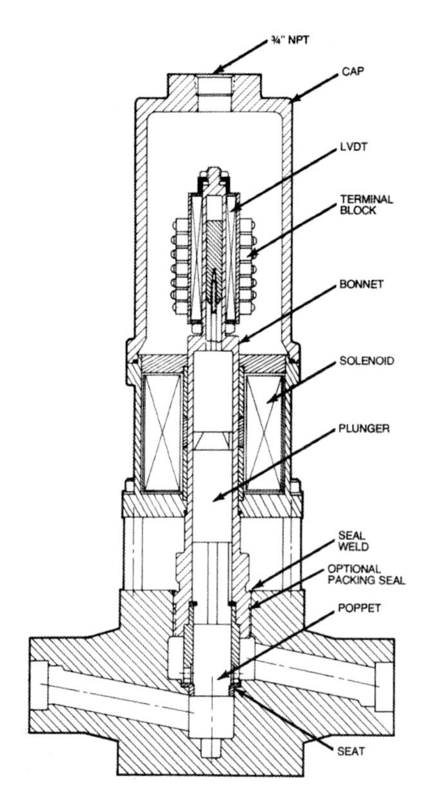

Figure 4.2 Throttling solenoid valve. (*Courtesy of Valcor Engineering Corp., Springfield, N.J.*)

4.2 Hydraulic Actuators

This type of actuator uses an incompressible fluid in a piston configuration to provide the force necessary to position and modulate a control valve (**Fig. 3.3**). It can stroke very quickly and provide very high loads but needs some type of external hydraulic supply that is not always readily available. It can also be made to be very stiff, so if there is an instability problem with a valve, switching to hydraulics may be the solution. In other respects, it is much like a pneumatic actuator since it can be equipped with volume boosters to increase stroking speed and with positioners to improve control performance. It does not normally come with a spring, so fail-safe operation can be a more complicated issue, involving complex trip systems.

4.3 Electrohydraulic Actuators

In this case the actuator is still an hydraulic piston, but the hydraulic power supply is integral to the assembly; therefore no external source is required. An integral pump is run off an electrical line run out to the valve. The pump itself can be one of several different styles and normally draws off some type of sump or reservoir that makes the actuator large and heavy. These actuators can be adapted to linear or rotary valves and are capable of high thrusts and/or torques and have a high degree of dynamic stiffness. The electrical source required to run the system can cause safety complications in a hazardous environment. See **Fig. 4.3** for an example.

4.4 Manual Actuators

As the name implies, this type of actuator is manually operated, so for automatic control applications it is used in conjunction with one of the other actuators described in this chapter as a manual override. It is mentioned here only because it can interfere with the proper operation of the valve if it is not installed and maintained correctly. The details concerning installation and maintenance will be covered in later chapters.

Figure 4.3 An electrohydraulic actuator. *(Courtesy of Fisher Controls International, Inc., Marshalltown, Iowa.)*

4.5 Pneumatic Actuators

Pneumatic actuators have been around for over 100 years, so the technology is not new, but this is a case where new technology has not been able to displace the old. They are still considered the standard for automatic control for a number of reasons. They are inexpensive to manufacture and reliable. They provide good control, particularly when teamed up with some of the new accessories that have been developed. They have enough power and speed to fit most applications. They are easy to maintain and can be made to be relatively light. Fail-safe action is easy to obtain. Their principal power source, compressed air, is easy to generate and transport and is inherently safe from such things as sparking hazards. This type of actuator comes in four principle forms, the *piston,* the *domotor* (which is a special type of piston actuator), the *spring* and *diaphragm,* and the *rotary vane.* The first three can be adapted for use on either rotary or sliding-stem valves. The latter is used only on rotary valves. Each major type will be addressed in the following sections.

4.5.1 Spring and diaphragm actuators

Nearly every major valve manufacturer offers a spring and diaphragm actuator. Typical construction of spring-to-open and spring-to-close actuators and the names of major parts are shown in **Figs. 4.4** and **4.5.** One of the distinguishing characteristics of this

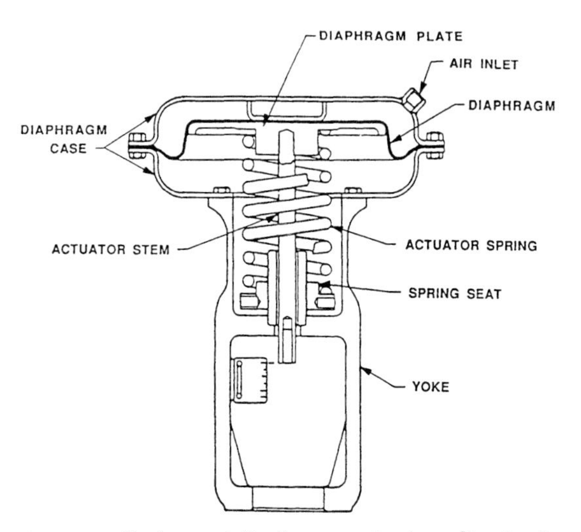

Figure 4.4 Spring and diaphragm actuator—direct acting.

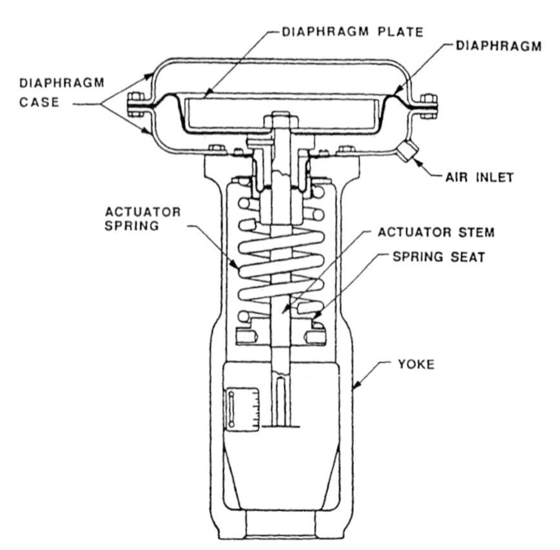

Figure 4.5 Spring and diaphragm actuator—reverse acting.

type of actuator is the spring and the fact that it provides a positive fail-safe action in case of loss of supply air. As shown in the cross-sectional views presented, the fail mode can either move the stem up or down depending on the construction. For a direct-acting sliding-stem valve, this would be fail open or fail closed, respectively. Of course, the failure mode selected depends on the application, but fail closed is more common. The downside to adding a spring is that a portion of the available actuator force must be used to compress the spring and is not available to position the valve.

The other key element in this design is the diaphragm which, in combination with the actuator, forms a pressure chamber in which the compressed air pressure can be regulated to adjust the force output of the actuator. Note that most diaphragms are rated at 60 to 70 psi, so a reducing regulator may be required between the actuator and shop or instrument air. To determine the force output, the air pressure must be multiplied by the effective area of the diaphragm. The diaphragm itself is a flexible element trapped between the two casings and is normally made of a fabric-reinforced elastomer to make it airtight and strong. Unlike a piston, which has a constant area, the effective area of the diaphragm changes as a function of the actuator travel away from the fail-safe travel limit. For standard designs, travel is usually 4 in or less. It's easy to see why the area changes, if we consider the sketch in **Fig. 4.6.** Note that the diaphragm is not flat

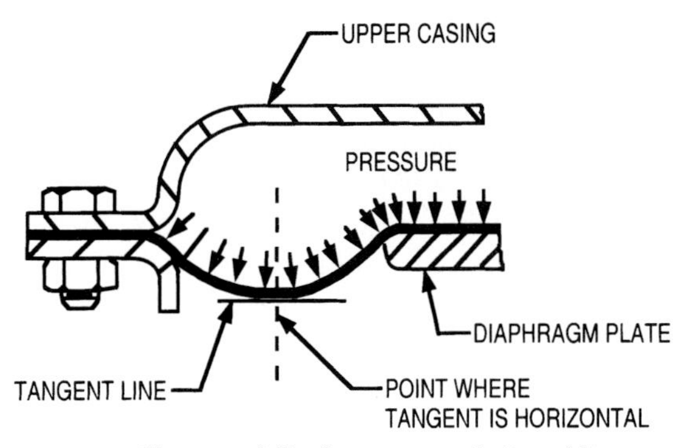

Figure 4.6 Closeup of diaphragm convolution. (*Courtesy of Fisher Controls International, Inc., Marshalltown, Iowa.*)

(flat-sheet diaphragms are not used in control applications because they are severely limited in travel, and their change in effective area versus travel can be greater than 30 percent). It has a molded, convoluted shape that permits it to follow the valve motion without stretching. As the actuator travels, the diaphragm "rolls" to follow it. As the diaphragm rolls, the midpoint of the convolution shifts inward and outward slightly, and it is the position of this midpoint that determines the effective area since the surface area of the diaphragm outside of the midpoint cannot transmit any load to the diaphragm plate under the diaphragm. Practically speaking, the change in effective area is less than 10 percent for most actuators, so the assumption of constant area is a reasonable approximation that significantly reduces the complexity of the actuator sizing calculations.

Note that, unlike a piston, the diaphragm provides a static airtight seal that does not wear out like the O-ring in a piston. However, on the fail-closed model, there is an O-ring bushing required under the diaphragm plate to ensure a proper seal against the actuator stem. This can be a high-maintenance item since the bushing is subject to wear during normal cycling, and if the bushing wears out, the O-ring seal will be lost and the valve will not stroke properly. For this reason, the fail-open model is preferable.

Spring and diaphragm actuators come in many different sizes that reflect the effective area of the diaphragm. Because the effective area is relatively large, the available actuator force can be significant, even with low air pressures. What this means in practice is that a spring and diaphragm actuator can be selected such that the stroking range for the actuator mounted on the valve corresponds to a standard air signal output from one of several pneumatic devices such as a controller or an I/P transducer. These standard outputs are either 3 to 15 or 6 to 30 psi, and if the actuator is selected to work in this range, no intermediate device such as a positioner is required. This approach can save money, and it's also important if the local air supply pressure is limited. A special type of spring and diaphragm actuator is shown in **Fig. 4.7.** In this case, a linkage system is used to increase available thrust near the seat where it is needed most. While in theory this sounds good, in practice the linkage adds a lot of deadband and hurts control performance.

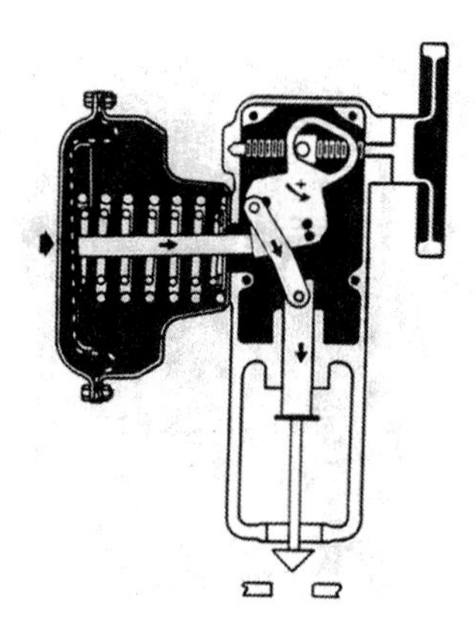

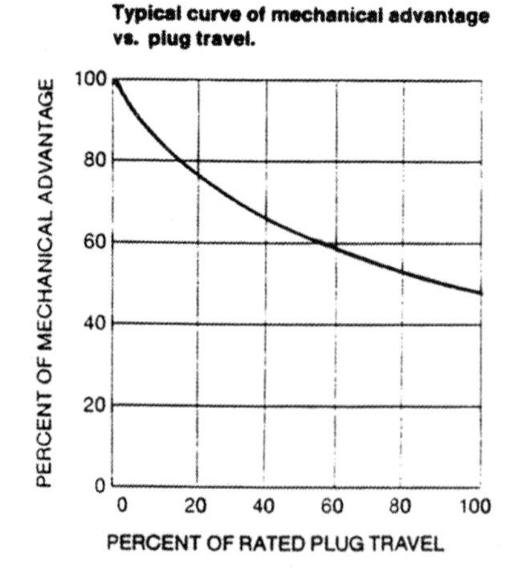

Figure 4.7 Special spring and diaphragm actuator with mechanical advantage linkage. (*Courtesy of Masoneilan-Dresser, Inc., Houston, Tex.*)

4.5.2 Piston actuators

The second type of pneumatic actuator in use today is the *piston*. In this design an O-ring is used to seal between a piston and the cylinder it slides inside (**Fig. 4.8**). Air pressure is then regulated on one or both sides of the piston to provide an actuator force used to manipulate the valve. This requires that the piston and cylinder be precision machined and guided to guard against wear or damage that could affect the ability of the O-ring to seal. Also, improper lubrication of the O-ring can result in it becoming a significant source of friction that can affect the ability of the valve to provide good process control.

If the piston is *single sided,* one side of the piston is open to the atmosphere. In this case, some type of restoring force is necessary to allow the valve to stroke in both directions. The standard approach for the restoring force is to use a spring, in which case the actuator operates in much the same way as the spring and diaphragm actuator already described.

The more common approach for piston actuators is to make them *double sided,* whereby the air pressure is regulated on both sides of the piston. In this case, the spring is not required. This does require a more complicated double-acting positioner and eliminates the possi-

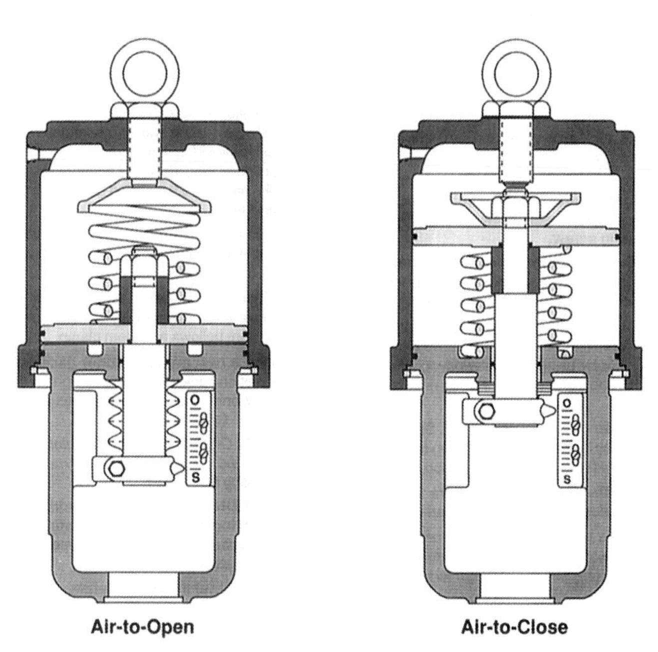

Figure 4.8 Piston to actuator—air to open and air to close. (*Courtesy of Valtek International, Inc., Springville, Utah.*)

bility of working directly with a controller and/or an I/P. Most double-sided pistons contain what are called bias springs that provide a "bias" toward one end of the stroke on loss of air. What this means is that the springs are not always strong enough to provide all the force necessary to stroke the valve to one end of the stroke and hold it there, but they will tend to push it in that direction.

Pistons, in general, are rated for higher pressures than their spring and diaphragm counterparts. They also use a smaller portion of the total force to overcome the bias spring, if one is present. For these reasons, they can be smaller and lighter for a given amount of thrust. This assumes, of course, that the higher supply pressures are actually available. The use of higher pressures also provides the potential for a very stiff system. The higher the stiffness of the control system, the easier it is to overcome control problems due to variations in flow force. The effective air spring rate of the actuator varies directly with the cylinder pressure. The stiffness (spring rate) is equal to the expression:

$$K = \frac{kPA^2}{v}$$

where K = spring rate, (in lb/in)
k = ratio of specific heat (1.4 for air)
P = supply pressure, in psia
A^2 = piston area, in in^2
v = cylinder volume under piston, in in^3

For a 25-in^2 cylinder actuator (typical for a 2-in valve) with a supply air pressure of 100 psi and a ¾-in stroke, the spring rate would be 9333 lb/in at midstroke. Note that as the volume under the piston becomes smaller, the stiffness factor becomes larger in a cylinder actuator. The equivalent diaphragm actuator (46 in^2) on the same valve with a 3 to 15 psi signal has a spring rate of only 920 lb/in at midstroke.

Unfortunately, in practice, the actual cylinder pressure required to position the valve at intermediate intervals depends upon an adjustment called *crossover pressure*. If this is not done correctly, the cylinder pressures will not be optimized and the stiffness will be reduced. In reality, crossover pressure is very rarely set properly. Other advantages associated with pistons include the fact that their travel is virtually unlimited because the seal is dynamic. Be aware, however, that the longer the travel, the lower the actuator stiffness due to the increased actuator volume. Also, because of their high pressure rating, they can be used directly with shop air without a reducing regulator.

TABLE 4.1 Actuator Trade-offs

Piston actuators*	Spring and diaphragm
1. Fail-safe operation can require some type of trip system.	1. Spring supplies simple method of guaranteeing fail-safe position.
2. Generally requires higher pressures because of lower effective areas.	2. Generally requires lower pressure for a given thrust due to large effective area. Can work with lower supply pressures.
3. Smaller and lighter for a given thrust because of higher pressures and lower effective areas.	3. Large and heavy due to high-area–low-pressure characteristics.
4. Normally requires a positioner.	4. Because of lower pressure requirements, can commonly be used directly with I/P or controller without positioner.
5. Normally rated above shop air levels, so no regulator required.	5. Spring partially counteracts diaphragm thrust, so available thrust to valve is lower.
6. Can be supplied for very long valve travels—up to 30 in or more. Watch stiffness on long strokes.	6. Diaphragm is static seal—less chance of leakage and no O-ring friction or wear.
7. Field reversible.	7. Usually limited to 4 in of travel or less.
8. High pressures can increase dead time at either end of travel.	8. Simple, robust construction.
9. Stiffer if crossover pressure properly adjusted.	9. Spring selection can "stiffen" system for good, stable operation.
10. Guiding must be precise to avoid piston or cylinder damage.	10. Normally rated to 60 to 70 psi, so regulator is required.
	11. Spring action is not easily reversed.

*Assumes a piston operator with no spring or a spring that does not completely provide the restoring force.

As you can see, the choice between a piston and a spring and diaphragm actuator can be relatively complicated. To help you make an informed decision, the previous discussion is summarized in **Table 4.1.**

4.5.3 Domotor actuators

The *Domotor* is a special type of pneumatic piston actuator marketed by Masoneilan that can still be found in many chemical processing facilities. Typical constructions are shown in **Fig. 4.9.** It is unique in that a constant loading pressure is applied to the top side of the piston (sometimes called the dome pressure), and a single-acting positioner is used to supply the restoring force on the other side of the piston. It is rated to 100 psi and, as a result, provides a very stiff actuat-

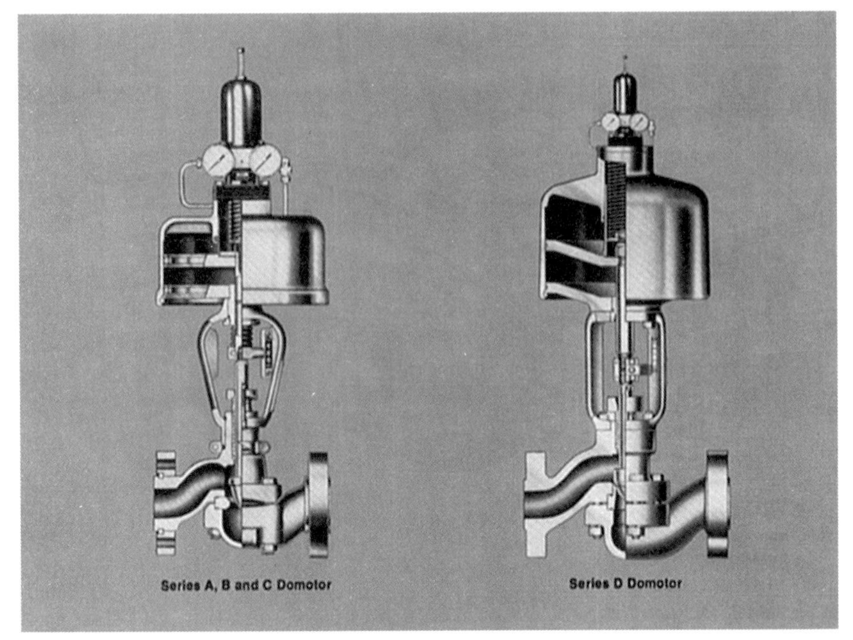

Figure 4.9 Domotor construction. (*Courtesy of Masoneilan-Dresser, Inc., Houston, Tex.*)

ing package that is highly resistant to dynamic problems due to load changes.

The schematics shown in **Fig. 4.10** help to illustrate the operating principles. A reducing regulator and positioner are integrally mounted to the top of the "dome," which helps keep these two devices protected from any harsh environments that might be present around the valve. The regulator is used to adjust the loading pressure to the point where it can properly close the valve and where the restoring force on the bottom of the piston can still provide enough force to get the valve fully open. The regulator does have a relief feature, so the dome pressure should stay constant as the piston strokes if everything is working properly. It also has a built-in check valve on the air-to-open model that locks the dome pressure in, on loss of air, so that the valve will move to the closed position.

On air-to-open models, the positioner relay has a larger diaphragm on the bottom, which means an increasing signal will move the relay down, allowing the pilot valve to open, increasing the lower piston pressure. As the piston moves up, the range spring relaxes, reducing the spring force and permitting the relay to move back to a new equilibrium position corresponding to a new piston position. A "helper"

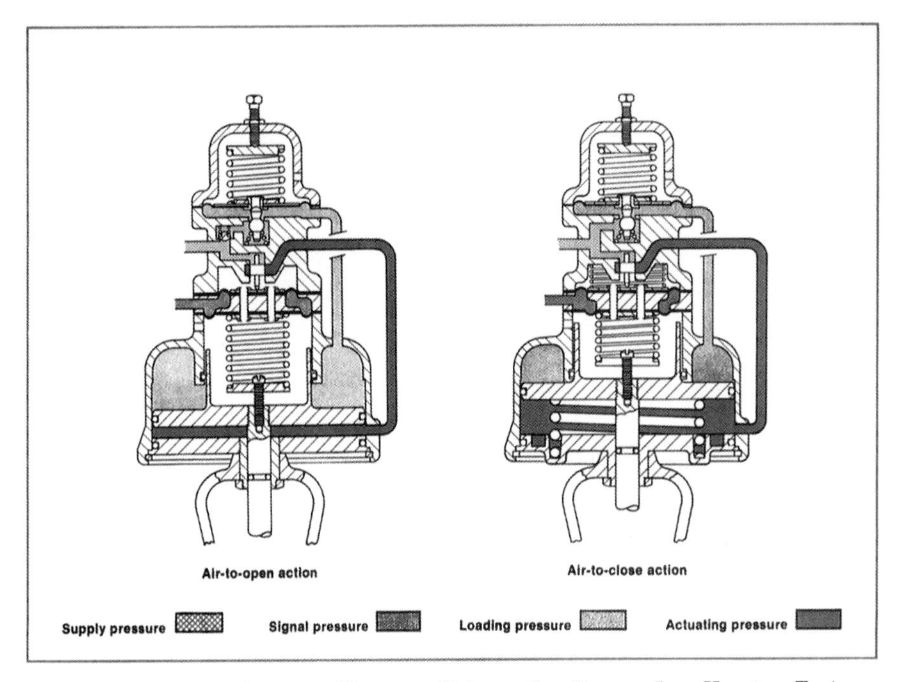

Figure 4.10 Domotor schematic. (*Courtesy of Masoneilan-Dresser, Inc., Houston, Tex.*)

spring may be used in conjunction with the dome air spring in cases where the supply pressure is low.

On air-to-close models, the relay is reversed and an increasing air signal moves it up, closing the pilot valve and reducing the lower piston pressure. The piston moves down until the increased force of the range spring pulls the relay and pilot valve back to a new equilibrium position. On air failure, both piston pressures drop to zero, and the fluid pressure on the plug is counted on to open the valve. A "helper" spring can also be installed on this model to help raise the piston and plug.

While these actuators are theoretically simpler than the more conventional piston operators with two-way positioners, they do have more seals that can wear out and leak, and the dome pressure adjustment can be tricky for those not accustomed to it. If it's too high, the valve won't open. If it's too low, the valve won't close. To be sure that it's right, you have to do an actuator sizing check, utilizing the effective area of the piston, which can sometimes be difficult to determine.

The other major problem with this construction is the adjustment of the zero and span for the positioner. It should be clear from the schematics that the range spring cannot be adjusted from the exterior. The positioner has to be disassembled, the adjustment guessed at, and then everything has to be put back together to see if the adjustment is close to being correct. If not, the process starts all over again and is repeated until the setting is within specification. As you can imagine, this is not a very productive way of calibrating an instrument. There is a retrofit kit now available that helps to simplify this procedure, and it is strongly recommended for those who routinely face the task of recalibrating this type of actuator. The retrofit kit consists of a modified relay assembly with an access hole cut in the side that permits the adjustment of the zero without disassembling the relay. Even with this kit, there is still no way to change the range without changing the range spring.

4.5.4 Rotary vane actuators

With the exception of the electromechanical actuator, all of the actuators discussed so far in this chapter provide a linear motion that must be converted through a lever arrangement to provide rotary action. Section 4.6 covers some of the considerations associated with the conversion of the linear motion to rotary.

The *rotary vane actuator* differs from these other actuators in that it directly converts air pressure into a torque that can be used to position a rotary valve. As **Fig. 4.11** shows, this conversion is accomplished through the buildup of pressure on one or both sides of a vane housed within a pressure chamber. As in a piston actuator, an O-ring or similar device provides the seal between the vane and the chamber. The pressure difference acts on the effective area of the vane to create a force that is then converted to a torque through the action of the bearings holding the actuator shaft in place. The effective

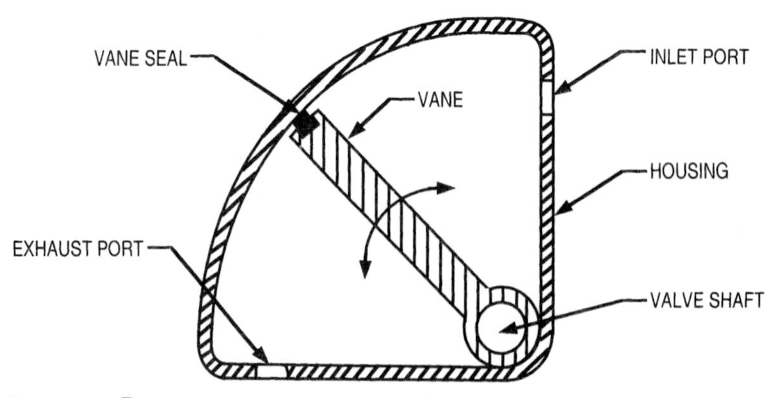

Figure 4.11 Rotary vane actuator—schematic.

moment arm is near the midpoint of the vane, so the torque can be calculated using the vane effective area times the pressure drop across the vane, multiplied by half the radius of the vane.

The big advantage, of course, is the fact that no lever arrangement is required to convert linear motion to rotary. Early design difficulties included the ability to obtain a finish on the inside of the chamber that would provide a good seal for a long cycle life. The seal can also be a source of friction, hurting control performance. The latest designs have addressed many of these problems, and, as a result, this type of actuator is now a viable choice for control applications. See **Fig. 4.12** for current design features. Most of the other advantages and disadvantages discussed for piston-type actuators apply here, with the addition that this type can be difficult to equip with a spring fail-safe mode that will generate sufficient restoring torque. All other typical accessories such as positioners, limit switches, etc., are available for these actuators.

4.6 Rotary Considerations

Most of the actuators already discussed provide a linear output but can be used on rotary valves if the proper adaptations are made. There are many possible choices for converting linear motion into rotary motion. Each one will be covered in the following sections with special emphasis on two characteristics. The first is how much *positioning "error"* is added in the linear-to-rotary linkage.

One of the recurring themes of this book is that we must consider control valves as an integral part of the control system and that to gain maximum performance from a control system, the error added

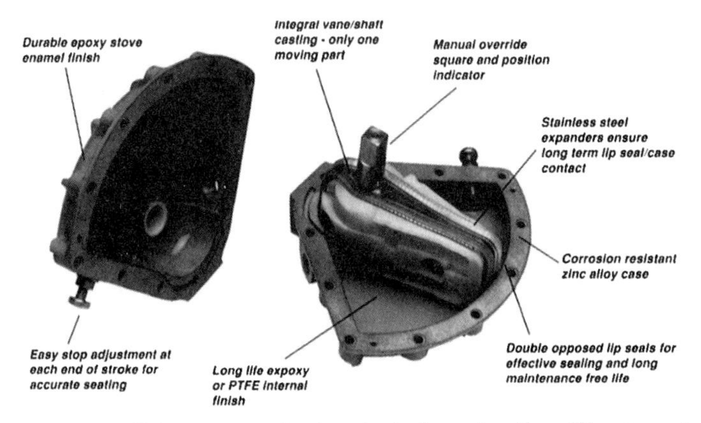

Figure 4.12 Rotary vane actuator—typical construction. (*Courtesy of Kinetrol, Ltd., Surrey, England.*)

by the valve must be minimized. This is particularly true for rotary valves with inherently linear actuators because the link between them and the valve is complex and can add hysteresis and deadband to the control equation if the choice is not made carefully. The potential for added error will be examined in detail for each design.

Secondly, the *force-to-torque conversions* for each approach will be reviewed so that you will better understand how to relate the actuator forces discussed earlier with the amount of torque that is actually available to turn the valve shaft and position the valve.

Note that if a fail-safe spring return is required on a rotary valve with a diaphragm actuator, the choice should always be for a design where the spring pushes up under the diaphragm since this approach eliminates the need for an O-ring seal on the valve stem, as noted earlier. This design can then be made fail open or fail closed by simply flipping the actuator around to the other side of the valve shaft. There is another configuration where the spring is contained in a cartridge on the opposite side of the valve shaft, but the operational considerations really don't change.

One final design feature needs to be discussed regarding actuators used on rotary valves. Because many rotary valves do not have positive stops inside the valve body, you must depend on the actuator to permit proper adjustment of the *travel stops* to position the ball, disc, or plug with respect to the seat. Improper adjustment can result in leakage or even seal damage, so actuator selection should include a look at how easily the travel stops can be adjusted and what locking mechanisms are employed to keep them from coming out of adjustment.

4.6.1 Direct connection

With a direct connection, the actuator is directly connected to the lever on the valve shaft (**Fig. 4.13*a***). This approach is the best from an error standpoint since it reduces the amount of play between the actuator stem and the valve shaft to a minimum. It is relatively difficult to design, however, since it requires that the piston or diaphragm pivot as the valve strokes because the relative position of the connecting point between the stem and lever changes as the shaft rotates. The amount of pivot varies directly with the length of the lever and is inversely proportional to the length of the stem. This pivoting design is particularly complex for a piston actuator since the O-ring seal on the actuator stem has to move laterally with the stem.

From a force-to-torque standpoint, it's easy to see that the torque will be maximized at midstroke where the stem and lever are perpendicular, and the effective length of the lever is the largest. This is illustrated in **Fig. 4.13*b*** where a qualitative graph of torque and

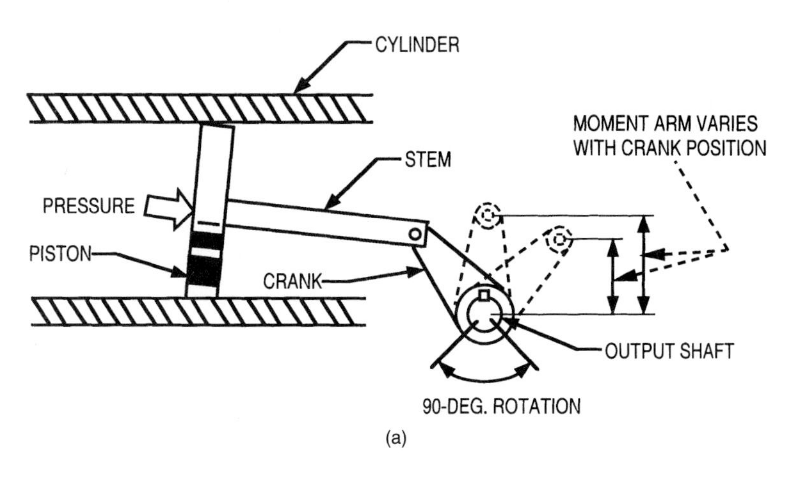

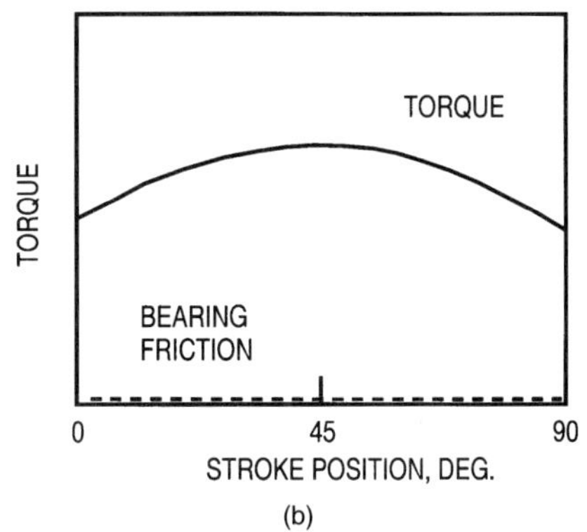

Figure 4.13 Rotary actuator. (*a*) Direct connection; (*b*) direct connection, torque versus travel.

actuator friction versus travel is shown. Note that the available torque is actually the smallest with the valve in the closed position. Many rotary valves have the highest required torque at this point due to the breakout torque associated with pulling the ball, disc, or plug out of the seat. You need to be cognizant of this fact when evaluating actuator sizing.

4.6.2 Linkarm connection

Linkarm connection is similar to the direct connection but a *secondary linkarm* is inserted between the lever and the stem. This arm adds error (hysteresis and deadband) because it adds another joint along with stem bearing friction, but it does eliminate the need for

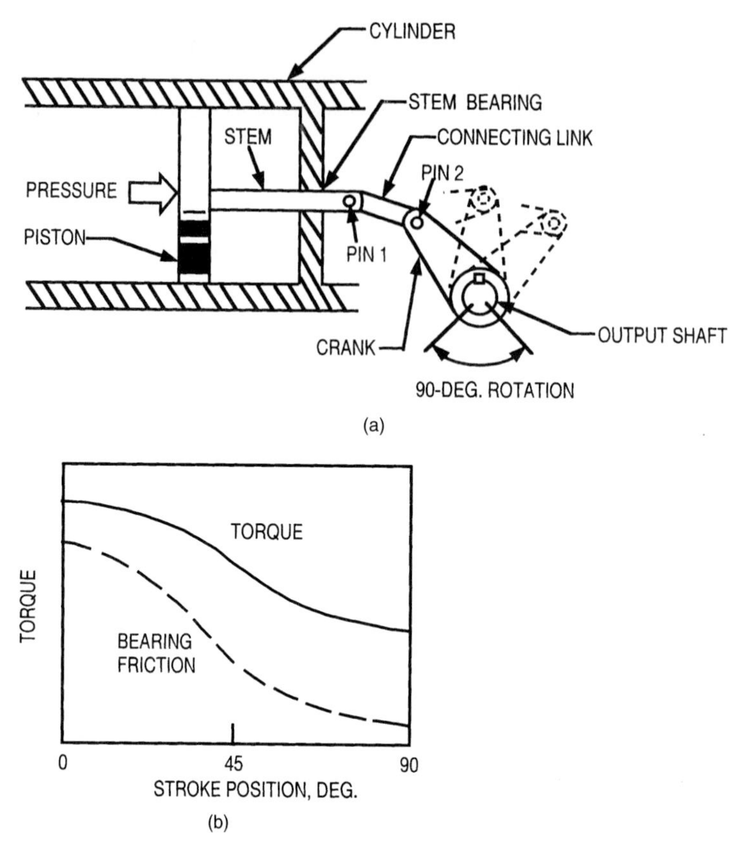

Figure 4.14 Rotary actuator. (*a*) Linkarm connection; (*b*) linkarm connection, torque versus travel.

the pivot in the actuator, simplifying the design. Basically, the trade-off here is cost versus performance. The schematic and torque versus travel for this design are shown in **Fig. 4.14.** Note that this actuator can be installed so that the highest available force is with the valve in the closed position where it is needed most.

4.6.3 Rack and pinion

With a rack and pinion, the actuator stem has a rack on the end whose teeth engage a gear that is fastened to the valve shaft. The effective moment arm doesn't change with travel, so that the torque versus travel curve is flat. This design is not very good for control since the fit between the gear teeth is never perfect and, hence, adds

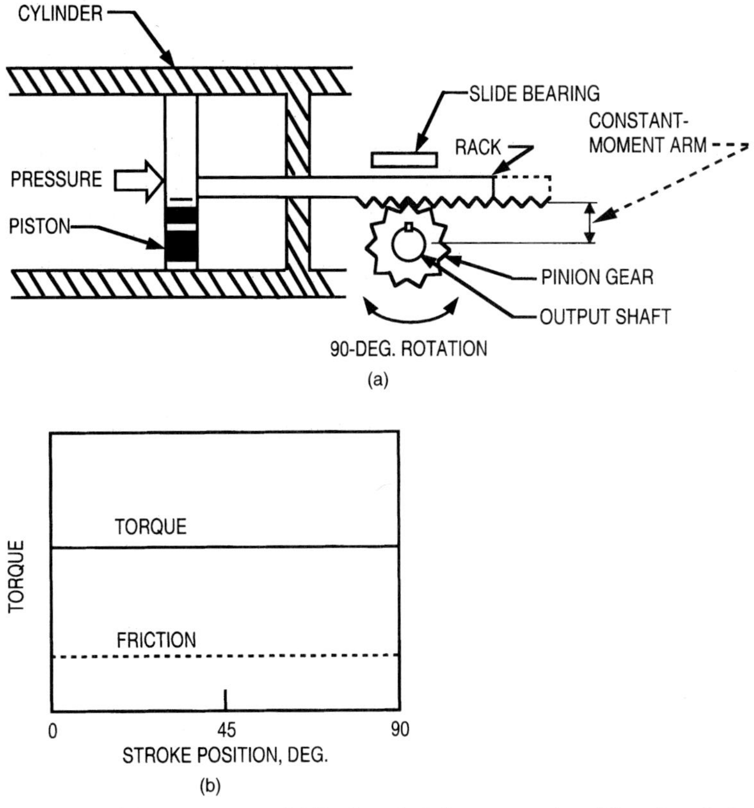

Figure 4.15 Rotary actuator. (*a*) Rack and pinion connection; (*b*) rack and pinion, torque versus travel.

play. In addition, the rack must be held in contact with the gear through the use of a bearing that adds friction. See **Fig. 4.15** for details.

4.6.4 Scotch yoke

As **Fig. 4.16*a*** shows, the Scotch yoke design allows the connection on the end of the lever to slide in and out in a slot so that the actuator doesn't have to pivot. Unfortunately, the slot adds a lot of deadband (play), so this design is not well suited to control applications, particularly as the slot begins to wear. The torque versus travel curve is interesting in that it provides maximum torque at the endpoints where it's usually needed (**Fig. 4.16*b***).

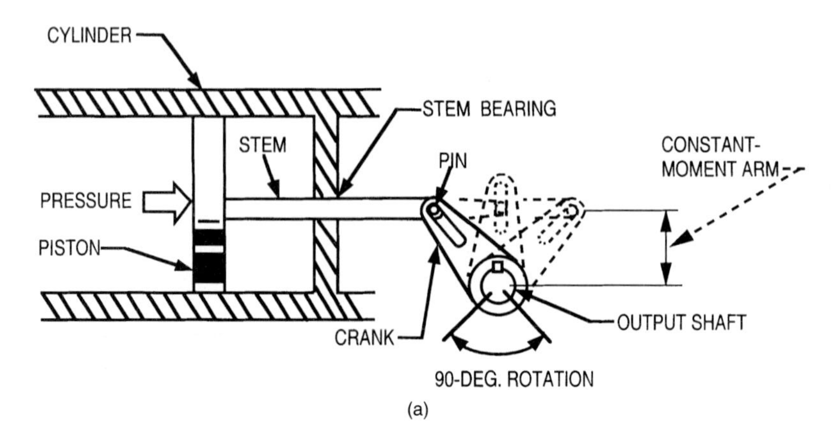

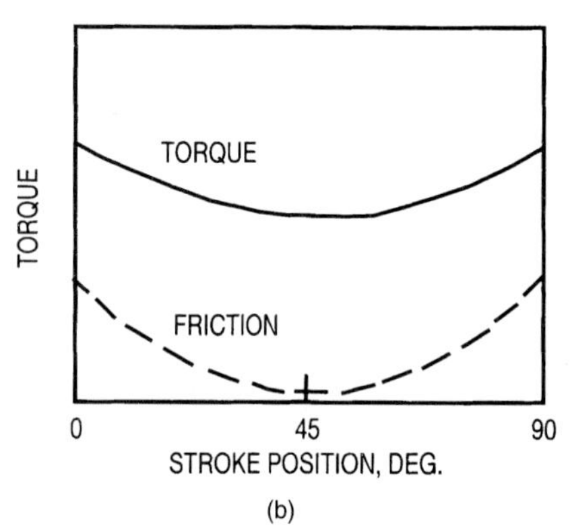

Figure 4.16 Rotary actuator. (*a*) Scotch yoke connection; (*b*) scotch yoke connection, torque versus travel.

References

1. Fitzgerald, William V., *The Basic Fundamentals of Spring and Diaphragm Actuators,* Fisher Controls, Marshalltown, Iowa, 1992.
2. Anderson, Vaughn R., "Rotary Actuators for Quarter-Turn Valves," *Chemical Engineering,* Aug. 15, 1988.
3. Valcor-Technical Data Series V526-6500, Valcor Engineering Corp., Springfield, N.J., Oct. 1984.
4. *Kinetrol Rotary Actuators—KF84,* Kinetrol LTD., Sussex, England, Jan. 1993.
5. Jury, Floyd D., *Fundamentals of Gas Pressure Regulation,* Fisher Controls, Marshalltown, Iowa, 1972.
6. *Valve Instruction Manual #1*; *Mark One & Two Control Valves* (no. 49011), Valtek, Inc., Springville, Utah, 1992.
7. "Group 26 Split Body Control Valve Building Blocks," *Bulletin BK5000E,* Masoneilan-Dresser, Houston, Tex., 1975.

Chapter

5

Valve Accessories

This chapter reviews the basic function of each of the major types of accessories used with pneumatic control valves. Special emphasis is placed on pneumatics since it is still the industry standard. For the accessories covered here that could be used on something other than a pneumatic valve, the comments would apply equally as well to nonpneumatic applications. Once again, the discussion will not go into any great depth about how each one works unless it has a bearing on how it might be used. The accessories covered include positioners, I/Ps, E/Ps, boosters, solenoid valves, quick exhausts, limit switches, handwheels, snubbers, regulators, position feedback devices, and pneumatic controllers. Tubing and connections between accessories along with hazardous environment considerations will also be covered.

It should be pointed out that some types of valve assemblies employ integrally mounted accessories that eliminate the need for tubing and connections. Two examples are shown in **Fig. 5.1.** Having the positioner and/or I/P built into the actuator structure is particularly advantageous in chemical plants where the corrosive atmospheres surrounding the valves can tend to build up on or corrode the linkage and tubing in a conventional assembly. The downside is that maintenance and adjustments to the accessories can be more difficult to accomplish with this approach.

5.1 Positioners

5.1.1 Positioner basics

In simple terms, a *positioner* can be described as a device that adjusts its output to the valve actuator in an attempt to place the valve stem

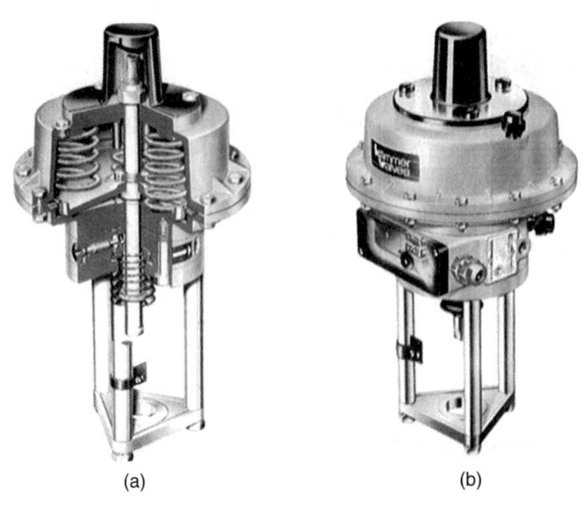

Figure 5.1 Actuator assemblies illustrating integral accessory approach. (*a*) Actuator with integral positioner; (*b*) actuator with integral positioner and I/P. (*Courtesy of Kammer Valve, Pittsburgh, Penn.*)

or shaft in a position that corresponds in some functional fashion to the input signal received from another device. There are several important aspects of this definition that deserve further study. Note that it indicates that the positioner "attempts" to put the valve in a desired position. What this means is that the positioner output range can only vary between zero and full supply pressure. If the desired position cannot be reached given this output range, something else must be changed to permit proper operation. The positioner is doing the best that it can at that point.

Also note that the output is regulated to put the valve in a desired position and that the positioner doesn't care what the output is as long as the requirement regarding position is met. For this reason, it is difficult to correlate output pressure to the input signal. In fact, the output may "wander" somewhat even though the valve stays in the asked-for position. This is especially true if the valve has high friction that permits the output pressure to change appreciably with no change in position.

The other point that is understood in this definition is that the positioner operates on measured valve position, and the performance of the positioner depends upon the accuracy of the *position feedback linkage* employed. The more accurate and reliable feedback is, the better the positioner will perform its duties. Ideally, the position feedback should reflect the actual position of the control element in the flow. This is usually not practical, but you need to be sensitive to the fact that the closer you get to the "real" position, the better the performance will be. On critical control situations, make sure that the linkage is accurate and robust. Loose linkages result in zero feedback over a portion of the stroke and can result in bothersome control problems such as limit cycling or poor repeatability.

And finally, note that the definition states that the valve position corresponds in some preset fashion to the input signal, normally 3 to 15 or 6 to 30 psi. Usually, it is a linear function of the input signal, but it can be changed to other functional relationships, such as equal percentage or quick opening, through the feedback linkage system (**Fig. 2.16**). On positioners with cams, for instance, the shape of the cam is modified to change the feedback. This is usually done in an attempt to characterize the valve assembly to meet the control requirements of the application (see Sec. 2.5). While this can be done this way, it's usually better to characterize a valve by changing the trim characteristic.

5.1.2 Positioner use

When should a positioner be used? This issue has been debated many times over the years, and it would be easy to fill a room with "experts" who could defend both sides of the issue. In general, the positioner provides improved performance in most cases when compared with operating directly off a controller or an I/P transducer. With double-acting actuators, there is no choice since the double-acting positioner is the only way to feed both sides of the actuator. On valves with single-acting positioners, the situation is less straightforward, but the following discussion should help to clarify the choice in most cases.

Positioners generally amplify the pneumatic signal, so stroking speed is improved when they are used, permitting a more rapid response to setpoint or load changes. They also tend to help overcome friction in the valve since they operate on valve position, not output pressure. In high-friction valves without positioners, over 75 percent of the input range can be used up just reversing the valve's stroking direction. (Watch for this in valves with graphite packing.)

Positioners tend to help valves seat and shut off and to reach full travel because their output pressure is not limited to a linear function of the input as it would be with devices such as I/Ps and controllers. If properly calibrated, they will vary the pressure from full supply to zero at the endpoints to ensure that the actuator will do everything it can to seat the valve at one end and provide full travel at the other.

Positioners also operate over a larger possible output range. I/Ps run at 3 to 15 psi or 6 to 30 psi. Positioners can routinely work up to 100 psi and higher, with their output many times limited only by the supply pressure available. Higher pressures mean more available force and a stiffer system that is less sensitive to load changes. Be aware that higher pressures can also mean longer lags at the endpoints since the higher the pressure, the more air that has to be exhausted before the stroke can be reversed. There have been a number of cases where valve stroking time was adversely affected by setting the supply pressure too high.

On the other hand, too much gain or speed in the positioner can result in overshoot and can set up a limit cycle where the process variable jumps back and forth across the setpoint. This seems to be particularly true for fast processes. The magnitude of the jump is a function of positioner gain, friction in the valve assembly, and the effective actuator spring rate. The frequency of the jump also depends on these variables and on the integral setting on the controller. A typical limit cycle is shown in **Fig. 5.2.** Note that the process variable is a square wave with an oscillation superimposed over it. The oscillation may be real, or it may be simply noise. In any case, it's obvious that the square wave represents a crossover of setpoint, and integrating the difference between setpoint and the process variable over time shows the total integrated error. The controller output is also interesting in that it shows a triangular shape typical of a controller with integral action. The integrating function sees the error due to the crossover and gradually builds up the output to the point where the

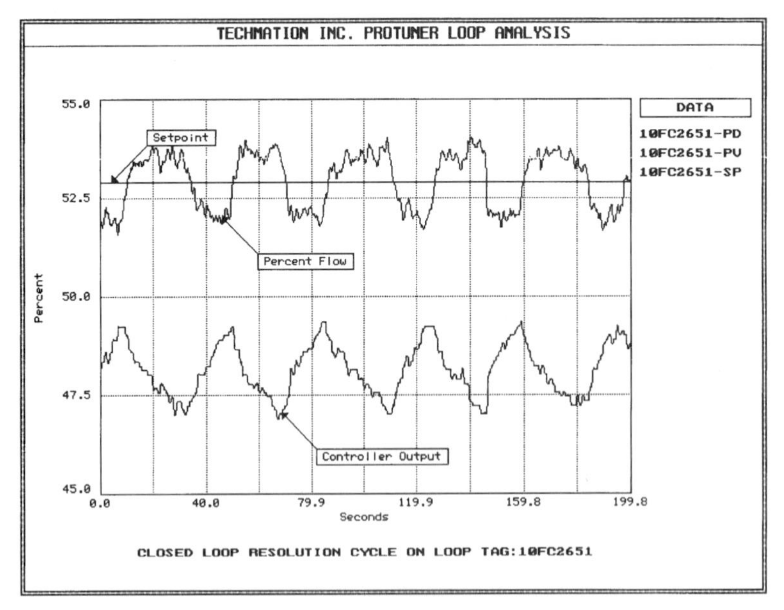

Figure 5.2 Process variable (top) and controller output (bottom) for loop in a limit cycle. (*Courtesy of Techmation, Tempe, Ariz.*)

valve jumps back the other way, and the process starts over in reverse. (See Sec. 5.13 for a discussion of controllers.)

It has been suggested that the best solution in this case might be to remove the positioner altogether or to replace it with a volume booster. In the majority of the cases, the positioner and all the advantages just mentioned can be retained. Rather than removing the positioner, try one or more of the following:

- Change the frequency response of the valve by changing the supply pressure, the spring rate, or the spring setting.
- Change the gain in the positioner if it is a model that permits gain adjustment. This feature is available on many modern positioners.
- Check the linkage to make sure it's not loose.
- Reduce the friction as a percentage of total actuator force available either by reducing friction if possible or by increasing actuator force (larger actuator or higher pressures).
- Add a volume booster.
- Check to make sure fittings are tight.
- Change the gain in the controller.
- The frequency of the limit cycle can usually be changed by changing the integral action of the controller.

If it turns out that the limit cycle is occurring because the time constant of the valve assembly is close to that for the process and none of the above helps, it may be necessary to remove the positioner and to add a volume booster, but this should be tried only as a last resort.

5.1.3 Positioner types

There are two main types of positioners, *single acting* and *double acting*. Single-acting positioners have only one output and as a result are used with spring and diaphragm actuators where the spring supplies the restoring force. Double-acting positioners have two outputs, so they can be used with actuators without springs. Most also have provisions that permit blocking off one of the outputs so that they can be used in the single-acting mode as well. With double-acting devices, one needs to be careful that the crossover pressure is properly adjusted so that the system is stiff and less sensitive to changes in process loads. It should normally be set at 75 percent of supply pressure. Follow the instruction manual to accomplish this correctly.

As far as the operating principle is concerned, there are two types here as well, force balanced and motion balanced. In the *force-balanced positioner,* the input signal acts on either a bellows or a diaphragm to create a force that is balanced by a second force from a spring connected to the valve position feedback mechanism. When these two forces are balanced, the positioner is in equilibrium.

In the *motion-balanced design,* one end of a beam is positioned by the input signal bellows, and any movement of the beam is compensated for by the correcting movement of the position feedback mechanism. In one model, the beam acts on the flapper to restrict a nozzle and change the output pressure from the positioner (**Fig. 5.3**). There is no clear-cut advantage to either approach, although most modern positioners employ the force-balance approach.

Internally, most positioners use either a *nozzle-flapper* approach that controls a single or double relay or a *spool valve.* The nozzle-flapper works as shown in **Fig. 5.4,** with the flapper moving with respect to the nozzle and changing the pressure being fed to the relay. The relay then provides an output to the actuator that moves the valve and the feedback linkage until a new equilibrium position is established that corresponds to the input signal. This is commonly referred

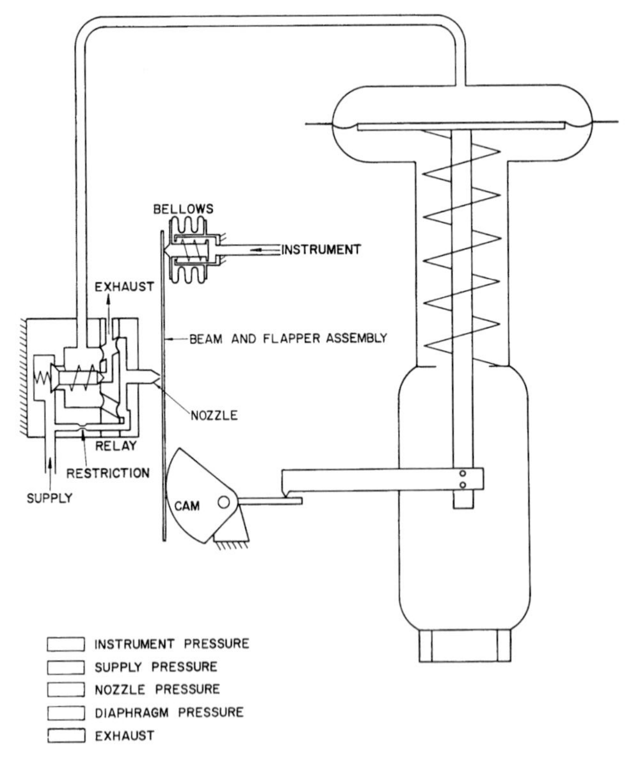

Figure 5.3 Motion-balance positioner. (*Courtesy of Fisher Controls International, Inc., Marshalltown, Iowa.*)

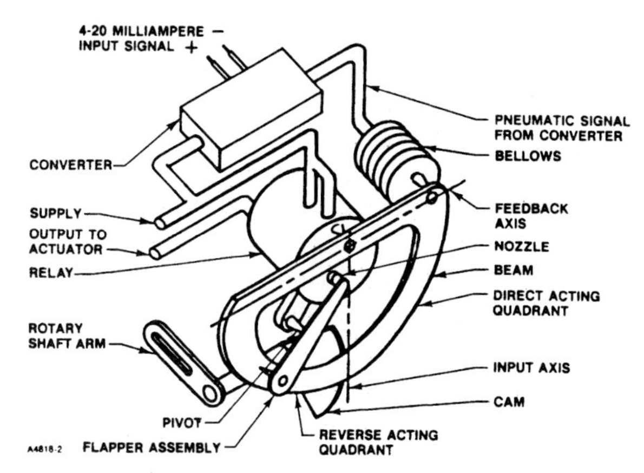

Figure 5.4 Nozzle—flapper operation. (*Courtesy of Fisher Controls International, Inc., Marshalltown, Iowa.*)

to as a **two-stage device** because of the interaction of the nozzle flapper and the relay.

The spool-valve design is shown in **Fig. 5.5** and is typified by an input signal module that positions a spool valve that feeds both sides of the actuator. In this case, as the spool valve moves downward due to a change in the input signal to the module, the spool valve opens the exhaust port on output 2 (the top of the cylinder) and opens the supply port on output 1. This tends to open the valve, and the posi-

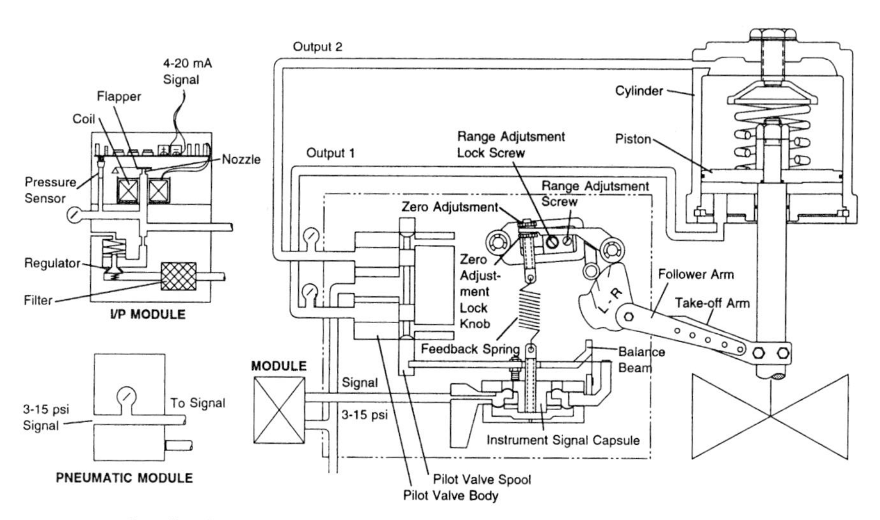

Figure 5.5 Spool-valve positioner. (*Courtesy of Valtek International, Inc., Springville, Utah.*)

tion feedback linkage reacts, pulling back up on the balance beam until the spool valve is in the equilibrium position again and the control valve is in the position corresponding to the new input signal. This is considered to be a single-stage device.

In general, the two-stage devices have better dynamic performance (faster stroke and better frequency response) than the spool valve designs, but the spool valve approach is simpler and more rugged because there are no elastomers or seats to wear out when compared to the pneumatic relay approach. Spool-valve designs do seem to be more sensitive to air quality problems.

5.1.4 Positioner selection guidelines

The following list covers items that should be considered before selecting what type of positioner might be the best choice for a given application:

1. Can the positioner be *split-ranged* and how easy is it to accomplish? Split-ranging means that the positioner only reacts to part of the input signal. For example, the zero and span might be adjusted so that the valve would fully stroke on a 3- to 9-psi input rather than the full 3- to 15-psi range. This permits using one input range to control two or more valves in tandem, depending on the application.
2. How easy is it to *adjust* the zero and span? Can it be done without removing the cover? Sometimes access to these adjustments needs to be restricted to keep them from being incorrectly adjusted.
3. How *stable* are the zero and span? If they tend to drift with temperature, vibration, time, or input pressure, the device will have to be constantly recalibrated to guarantee that the valve will properly reach the travel endpoints. This is especially important for valves that must shut off.
4. How *accurate* is the device? Ideally, the valve assembly should position itself at exactly that point which corresponds to the input signal every time, regardless of which direction it is stroking or what loads are present on the trim parts. Unfortunately, this doesn't and can't happen. All systems exhibit some error, and arriving at a standard definition for the amount of error expected is not easy. Before continuing with the positioner selection guidelines, "accuracy" needs to be covered in more detail.

Accuracy and error. While determining real error may not be easy, it is critical in determining the quality of control that can be expected, so it does need to be examined very closely. Error can change with test conditions, test methods, and environmental conditions, so in order to come up with a standard definition for error to use in comparing performance, the test parameters have to be spelled out in detail. The Instrument Society of America (ISA) has developed a standard definition for instrument error that serves this purpose very well. For a complete definition, see ISA standard S51.1. A brief review of the most commonly used terms follows.

An important point to make on accuracy claims by vendors is that they normally talk about accuracy of an individual device. What really counts is how the device works as part of the control system including the actuator and valve. "Accuracy" information on a positioner by itself means very little without considering how it will perform in concert with the loads the valve will put on it. Unfortunately, most vendors still provide only device-specific information. To really understand the performance of the device, you must determine how it will work on the valve in question. The commonly used terms are:

Deviation cycle. The standard definition of error or accuracy starts with a test where the input to a system is gradually changed from an initial point, usually 50 percent of the input range, to 100 percent of the range, down to 0 percent, and then back to 50 percent. The output from the system is measured at the same time and plotted versus the input as shown in **Fig. 5.6.** Note that this is usually done in a static fashion. In other words, the input is changed from one discrete point to the next, and the system is permitted to stabilize before the output is read. This has important implications for a control valve because the input for a valve usually changes continuously rather than discretely, and there is not always time for the

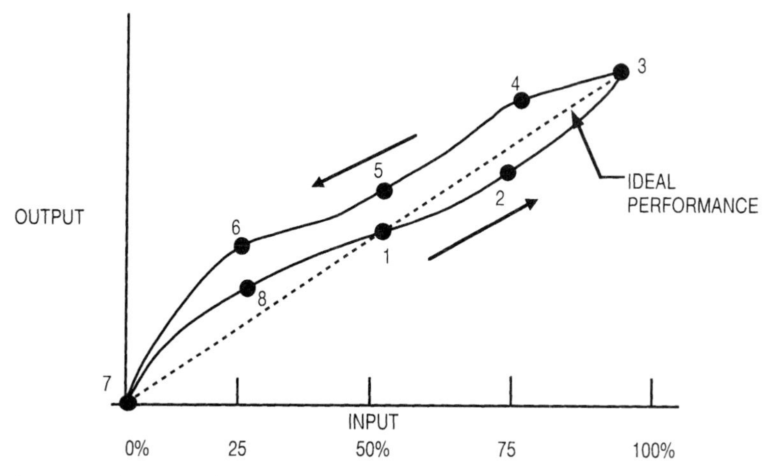

Figure 5.6 Accuracy plot.

system to stabilize before the next change is experienced. As a result, there may be more error in service than this type of test would indicate. A dynamic test has the potential to better represent real service, and a new test technique based on this type of input is covered in Chap. 13.

If **Fig. 5.6** is transformed by plotting the output deviation from the ideal line as a percent of output span versus the input in percent, a traditional deviation cycle plot results, which will be referred to in the discussion of the other common error terms (see **Fig. 5.7**). The values reflected in this curve are usually referred to as the static accuracy of the device.

Linearity. This is usually discussed as nonlinearity and is defined by taking the midpoint of the two lines shown in **Fig. 5.7** and expressing the deviation of this midpoint line from the ideal. In other words, the error due to hysteresis and deadband is subtracted out and what is left over is nonlinearity. **Figure 5.8** shows the midpoint line drawn in for a given deviation cycle. Terminal-based nonlinearity is illustrated here and is the maximum deviation between the midpoint line and a line drawn between the endpoints, points 7 and 3. It is expressed as a percentage of output span. Independent nonlinearity is the same type of deviation, but it is plotted from a best-fit straight line, so it is always less than terminal-based nonlinearity.

Hysteresis plus deadband. This type of error is reflected in the spread between the outputs as the device is stroked in two different directions. In **Fig. 5.8,** it is the difference between points 2 and 4, points 1 and 5, and points 8 and 6. It is primarily the result of

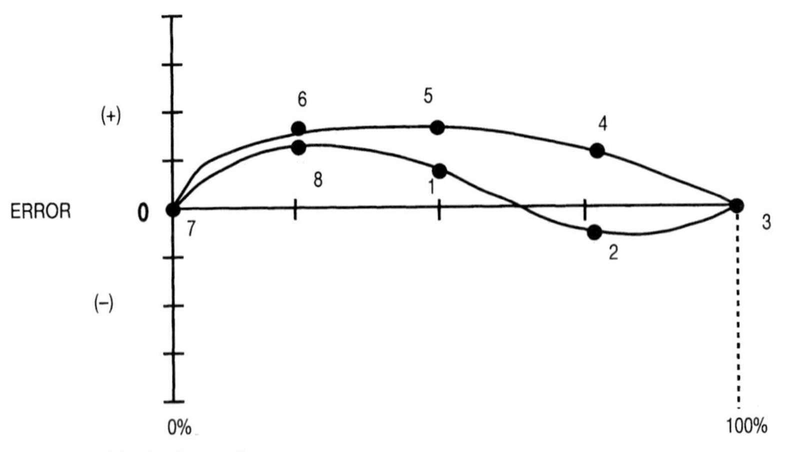

Figure 5.7 Deviation cycle.

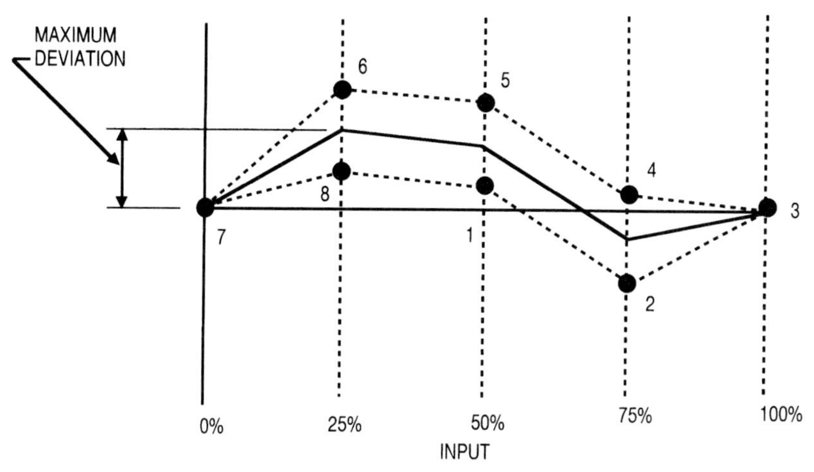

Figure 5.8 Terminal-based linearity.

"slop" in linkages and friction, and normally the reported value is the highest spread measured on the curve, expressed as a percentage of output span.

Deadband. To determine deadband, the system is allowed to reach equilibrium at some intermediate point. The input is then gradually changed in one direction until a measurable output change is detected. The input is then reversed until another change in the output is detected, and the deadband is the magnitude that the input had to change to reverse the output direction. It is usually checked at three points, and the highest level reported is expressed as a percentage of the input span. Hysteresis cannot be measured directly, so you have to measure deadband and then subtract it from the above term to determine hysterestic error.

Hysteresis. This is the maximum difference that can occur in input values for any single output value during a calibration cycle, excluding errors due to deadband, preferably expressed as a percentage of the calibration cycle amplitude.

Repeatability. This term reflects the ability of the system to provide consistent readings over a series of tests. It is particularly important for control valves since the control system can account for and partially correct for absolute errors in positioning accuracy as long as they are consistent. Nonrepeatability is defined as one-half of the standard deviation for an infinite number of tests. Statistical methods can be used to reduce the number of samples to 10, multiply the standard deviation by 50, and then divide by the output span to come up with a value for percent repeatability. Even though this is a very important element of system performance, it is not usually reported in vendor catalogs, and you will have to dig to come up with representative values.

Returning to the positioner selection guidelines:

5. How sensitive is the device to *air quality* problems? Very few air supplies provide air that meets the ISA guidelines on air quality (ISA Std. F7.3). As a result, the pneumatic device must be able to handle things like dirt, moisture, and oil if it is to survive in the real world. The bigger the internal orifices are, the better. Check to see if the instrument has a clean-out device that will permit the orifices to be occasionally cleaned without requiring complete disassembly. Regardless of which device is selected, always use a filter in combination with pneumatic control devices such as I/Ps, positioners, and controllers.
6. Are the zero and span settings *interactive* or independent? If they are interactive, it can increase the amount of time required to adjust them since you have to go back and forth between the two to gradually approach the correct settings.
7. Does the positioner have a *bypass,* allowing the input signal to be used to directly stroke the valve? This can sometimes facilitate the verification of actuator settings like benchset and seat load, as well as permitting limited adjustment or maintenance on the positioner without taking the valve off-line.
8. How *fast* is the positioner? The larger the airflow, the faster the system will be able to respond to setpoint and load changes, meaning less error and better control. Be aware, however, that too much speed can result in stability problems, as mentioned earlier in this section. Many modern positioners now include a gain adjustment so that the positioner response can be tuned to the application.
9. What is the *frequency response* of the device? In general, higher frequency response provides better control characteristics, but be sure that the response is defined using consistent test methods and that it is evaluated with the device in combination with the actuator. The response of the device alone really means nothing since the load due to the actuator will drastically modify the response.
10. What is the *maximum rated supply pressure* for the device? Some positioners are only rated for 50 psi. If the actuator is rated for a higher pressure, the positioner becomes the limiting factor in

determining actuator force output. Also note that most single-acting positioners require that the supply be set 5 psi higher than the desired output to account for pressure losses in the device.

11. What is the *positioning resolution* of the valve and positioner package? (See definition in Sec. 9.3.) This will have a very marked effect on the control performance since better resolution will permit the valve to get closer to the ideal position and will cut down on the chances of the valve overshooting and then limiting cycling.
12. Can the *action* of the device be easily reversed? Sometimes it may be necessary to change from an increasing-signal-to-close approach to an increasing-signal-to-open arrangement. Also, can the *feedback* be characterized to fit the application? How is it accomplished?
13. How complex are the internal workings of the positioner? More parts mean more training to properly maintain the equipment and more parts in stock.
14. What is the *steady-state air consumption*? It may be important to minimize this.
15. What does the *feedback linkage* look like? How close is it to giving the real plug position? Is it sturdy and protected from the elements? Will it resist corrosion? How easy is it to connect properly?

Note that the positioner is the most important valve accessory in most control valve applications, so it has been given a great deal of emphasis here. However, many of the comments regarding its selection and accuracy could apply equally as well to the accessories in the following sections.

5.2 Transducers

In engineering, the word *transducer* is a general term that applies to any device that converts from one physical property to another. For control valves it carries a more precise definition and implies a device that converts an electrical signal normally generated by the control system into a pneumatic signal that the valve can work with.

The I/P converts current to pressure, and the E/P converts voltage to pressure. Standard *inputs* are 4 to 20 mA and 0 to 10 V, respectively, and standard *outputs* for both types are 3 to 15 psi and 6 to 30 psi, but many different input and output ranges are possible.

Most transducer designs use either a torque motor or a voice coil to translate an electrical signal into a force that can be used to position a beam with respect to a pressure nozzle. The position of the beam

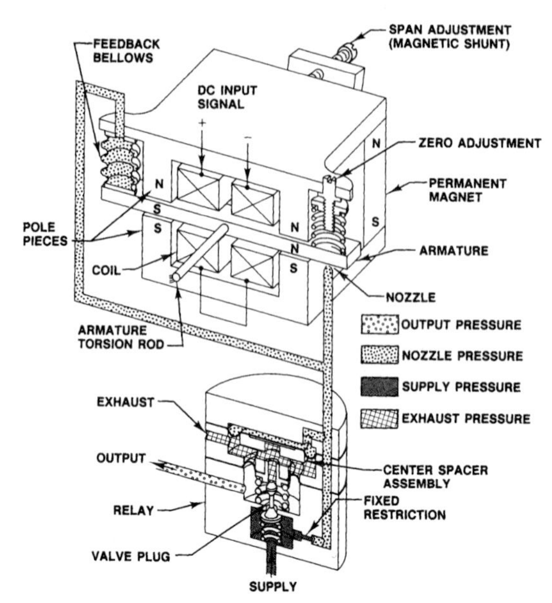

Figure 5.9 Transducer schematic. (*Courtesy of Fisher Controls International, Inc., Marshalltown, Iowa.*)

changes the pressure inside the nozzle, which in turn is fed back to a relay that provides the output from the device. Pressure feedback is provided to correct the beam position and to bring the system into equilibrium at a new output pressure (**Fig. 5.9**). In at least one model, the output pressure is fed back electronically through a P/I converter.

There is no position feedback with a transducer. The output is pressure, not travel, and the pressure is a function of the input signal. The relationship is usually linear but can be something different. Normally a transducer is used with a positioner for the reasons stated in the preceding section. If it is directly connected to the actuator, the response will be much slower because the relays in the transducer are smaller than for most positioners. You can also expect a lot of hysteresis if the friction is high. See Sec. 5.1.2 for a more detailed discussion of the pluses and minuses associated with the use of transducers alone.

Recent developments have included more accurate *pressure feedback* to reduce drift and a unique feature where the pressure output drops to zero if the input goes below 3.9 mA for a 4- to 20-mA range. What this effectively accomplishes is that the output range of the transducer is increased from 3 to 15 psi nominal to 0 to 15 psi nominal. The drop in pressure to zero psi is useful if the I/P is directly connected to the actuator (no positioner) because it can provide additional actuator load to close and shut off the valve. On a valve with a 3- to 15-psi benchset and spring-close action, for instance, the valve would throttle with a conventional I/P because the 3- to 15-psi range could overcome the spring load and position the valve correctly. However, the 3- to 15-psi output would be just enough to overcome the spring. There would be nothing left to load the seat and shut the valve off. By dropping the pressure to zero, additional spring load can come into play to help close the valve.

Transducer modules can also be supplied as an integral part of the positioner, simplifying connections and lowering the total cost. Be aware that the transducer portion can be susceptible to vibration, so there may be an advantage in a separate unit located away from the valve where the vibration is lower. Don't get too far away, however, or the volume in the transmission lines can begin to slow the transducer down, hurting control performance. Many plants get themselves in trouble by locating the I/Ps on a central panel for convenience, not realizing that the increased line length has hurt their control system response. From a selection standpoint, look for some of the new developments mentioned earlier. Otherwise, go through the same type of evaluation list that was covered for the positioners.

5.3 Boosters

Boosters, as the name implies, are used to amplify a pneumatic signal's volume or pressure so that faster stroking times can be obtained for a given control valve or to increase the effective pressure range of the output. A typical design is shown in **Fig. 5.10.** The input signal acts on the top diaphragm, which opens a port that regulates output pressure from the device. As the output pressure increases, it acts on the lower diaphragm and corrects the position of the port until the two forces on the diaphragms are balanced. If the two diaphragms are of equal area, the output pressure is the same as the input signal but with a much higher volume because of the capacity of the relay. This is called a 1:1 *volume booster* and is the typical type used on a control valve.

If the diaphragms are of different sizes, the ratio of input-to-output pressure is inversely proportional to the ratio of the top to the bottom areas. A 3-to-1 ratio is not uncommon and might be used to convert a 3- to 15-psi signal into a 9- to 45-psi output which could then be used to power an actuator and valve that required the increased range to properly stroke.

On volume boosters, in particular, the increased gain in the relay has a tendency to cause instability where the valve makes large jumps in travel that would be detrimental to control performance. For this reason most volume boosters come with a bypass adjustment that

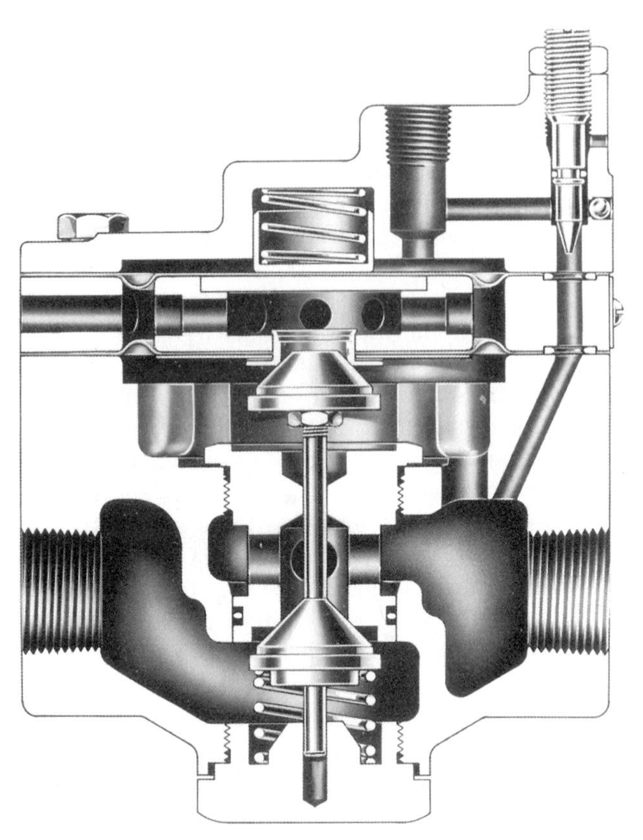

Figure 5.10 Volume booster cross section. (*Courtesy of Fisher Controls International, Inc., Marshalltown, Iowa.*)

permits some of the input signal to go around the diaphragm assembly and straight to the output. The stability is improved in this way because the pressure on the input diaphragm builds up more slowly. In fact, for small changes in input signal, the booster relay will not work at all if the bypass is opened far enough. For most control applications, this is the desired adjustment, so for small throttling-type changes in input, a stable change in position will result, but for large changes, the booster will react and get the valve to the new position in a minimum amount of time. Anytime instability is experienced and a booster is involved, the bypass should be checked. The other thing to watch for is to make sure that the booster has a *supply line* that has enough volume to drive it and that the supply is separate from that of the positioner. If the supplies aren't separate, they can interact, and the drop in supply pressure that occurs when the booster relay opens will cause variations in positioner output.

5.4 Solenoid Valves

Solenoid valves are electrical devices where an electrical signal to the valve determines the position of a plug inside the valve. Solenoid valves on control valves are used in an on-off mode to control the flow of air to and from the pneumatic devices on the valve, including the actuator, positioner, and I/P.

Solenoid valves can be *two-*, *three-*, or *four-way models,* which means that they can have two, three, or four pressure openings (**Fig. 5.11**). In the two-way models, there is an inlet and an outlet, and the valve is either *normally open* or *normally closed.* The convention adopted for these valves is that "normally" implies deenergized (i.e., a normally open valve closes when power is supplied to it). A two-way valve used at the inlet to the actuator could either permit normal flow to the actuator when open or trap air inside the actuator when closed. This type of device is useful if the valve needs to be locked in last position upon either the receipt of some signal (normally open) or on the loss of the signal (power) to the solenoid (normally closed).

Three-way valves are used where, instead of lock-in-last position, the valve must assume a position at one end of the stroke or the other. An actuator with a spring is used to provide the force necessary to put the valve where it should be, and the three-way valve is connected to provide normal airflow with the solenoid in one position and to block the input and exhaust the actuator pressure when the solenoid valve moves to the other end of its stroke. Fail action can be on receiving a signal or on loss of power, depending on whether a normally open or normally closed solenoid is selected.

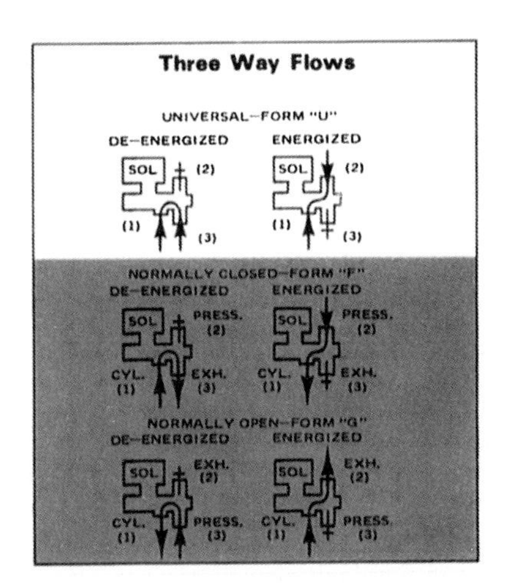

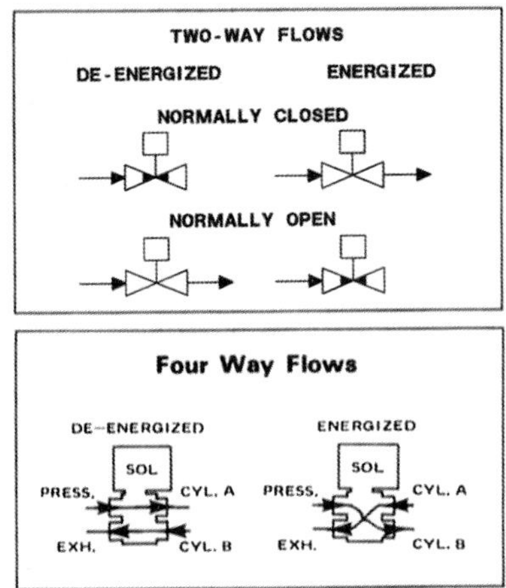

Figure 5.11 Solenoid valve schematics: two-, three-, and four-way flows. (*Courtesy of Fisher Controls International, Inc., Marshalltown, Iowa.*)

Four-way valves are also used to put a valve at one end of its stroke, but in this case, the actuator has pressure on both sides, and the solenoid supplies one side of the piston and exhausts the other. It then reverses itself when activated, driving the control valve to the other end of its stroke. The fail-safe action is determined by the pressure connections in this case.

In chemical plants, solenoids are used either to control valves used in on-off service or for *safety interlock systems* where normal throttling occurs through the valve until some type of upset occurs and an override is needed to open, close, or hold the valve in the last position. A tank-level alarm, for instance, might trip a solenoid valve that then closes the valve feeding the tank. This could be done using either a normally open or normally closed solenoid, but it's preferable to use a normally closed model so that the valve is driven open with the electrical signal and held there. That way, if something goes wrong with the power supply or connections, the control valve will close, and the operator will know about it and can take corrective action. If a normally open valve fails, it will not be detected until the control valve is asked to perform its safety override function, and then it may be too late.

As far as *installation* is concerned, the solenoid should be placed at the inlet to the actuator to be absolutely sure that nothing else can keep it from properly positioning the valve in the safety mode (**Fig. 5.12**). If it is installed at the inlet to the positioner, for example, the positioner relay could potentially fail and keep the valve at some intermediate position. Pay particular attention to the pressure connections to make sure that the ports are properly connected. Many solenoids will not work if the ports are connected incorrectly. It is also recom-

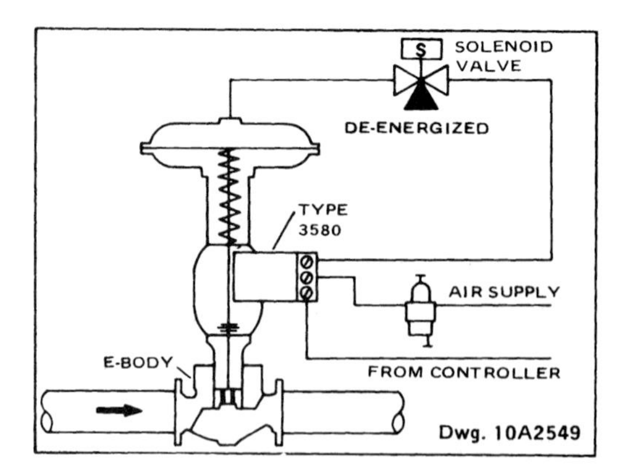

Figure 5.12 Typical installation of a solenoid valve. (*Courtesy of Fisher Controls International, Inc., Marshalltown, Iowa.*)

mended that solenoids be stroked occasionally to be sure that they don't get stuck in one position or another. Be careful with the supply pressure as well. Some piloted models will not function properly below a certain pressure. Finally, make sure that the flow-rating capacity for the solenoid valve is consistent with the stroking time required for the control valve. If the port is too small, the stroking speed of the control valve will be adversely affected during normal operation.

As far as electrical connections are concerned, solenoids can operate on either 120 V ac or 25 V dc and can be supplied to meet any of the hazardous environment specifications covered later in this chapter.

5.5 Quick Exhausts

Quick exhausts are very simple devices that are used to provide rapid stroking movement in the exhausting direction for a given pneumatic circuit. A schematic is shown in **Fig. 5.13.** As you can see, the airflow in the forward direction passes over a diaphragm to the outlet port. As long as the airflow is in that direction, the diaphragm stays pressed against the exhaust port, keeping it closed. Once the inlet pressure starts to drop, however, and the airflow begins to reverse, the diaphragm will be lifted off the exhaust port, and the air will rapidly escape through it. In control valve applications, the quick exhaust is mounted in the line supplying the actuator. It acts as a pass-through in the supply direction but provides a very large exhaust port, when the flow is reversed, to allow the valve to stroke very quickly in the exhausting direction. The quick stroke may be an application requirement or part of an interlock function.

Be sure to tube these devices up correctly, or they will not function. Also, they can be unstable if a *bypass line* is not installed around them. If instability is a problem, add a needle valve in parallel with the quick exhaust and then open the valve to the point where stable operation is established.

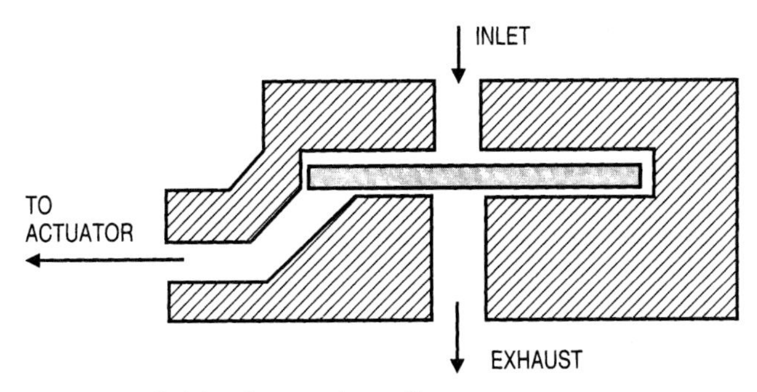

Figure 5.13 Quick exhaust schematic.

5.6 Limit Switches

Limit switches are used to provide a remote indication that a control valve has passed some point in its travel. Normally they are used to indicate that the valve is approaching the open and/or closed position, and the remote indication shows up in the control room. They are usually mounted on the valve near the opening in the actuator yoke so that they can sense valve stem movement. The simplest types are mechanically operated; a trip arm is rotated through contact with an arm or link attached to the stem (or shaft for a rotary valve) as shown in **Fig. 5.14.**

Note that the *activating force* can be quite high for some models, so the mounting method and the feedback linkage need to be sturdy. The switches can be damaged if the activating arm is rotated too far, so care must be taken to avoid this. Adjust them so that full activation occurs within one-eighth of an inch of final valve position to avoid this type of problem. There is also a significant amount of deadband in these devices, so the valve may pass back over the original trip point without the switch deactivating. Buildup of dirt, debris, or ice on the arm or the linkage can change the activation point. To avoid this problem, keep these parts clean through regular inspection or by enclosing them in a protective cover.

When a limit switch trips or activates, it closes or opens an electrical circuit. The circuit can then be used to provide position indication as noted earlier, or it can activate an alarm, an interlock solenoid, or another electrical relay. Switches can be purchased to meet any of the

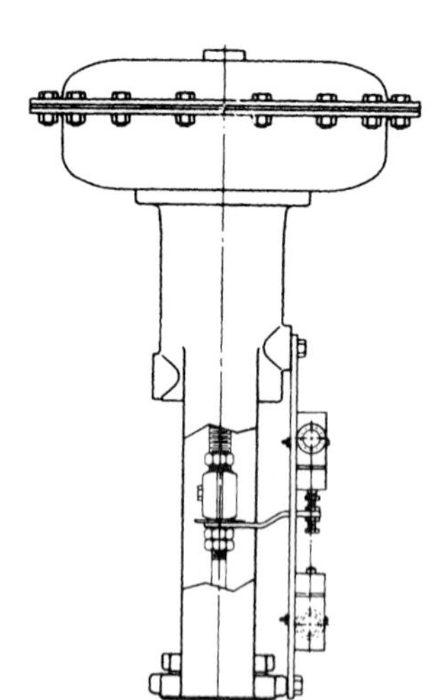

Figure 5.14 Limit switches mounted on an actuator. (*Courtesy of Fisher Controls International, Inc., Marshalltown, Iowa.*)

hazardous classifications. *Noncontact proximity-type switches* have become more popular recently because there is no actuating force, and they are easier to seal against the environment.

There are some conventions that apply for limit switches that can best be explained by **Fig. 5.15.** Limit switches can have a number of internal circuits and can be connected in several different ways. On a basic switch the terminals are labeled *common, normally open,* and *normally closed.* The common is always connected to one side of the circuit, and then you can select whether you want the circuit closed or open with no pressure on the activating arm. The convention is that the word *normally* implies the circuit state with no force applied. For each type of application, there is a preferred circuit arrangement that will guard against circuit failure not being detected. For instance, for a two-switch arrangement, the circuit in **Fig. 5.16** has one of the lights on at all times, so if neither light is on, the operator knows that there has been a circuit failure that needs to be repaired. If the circuit is arranged as in **Fig. 5.17,** no lights could be interpreted as the valve being in the intermediate position when it really is the result of a failed circuit.

Often the terminology *single-throw, double pole* is used when referring to switches. The number of poles refers to the number of individual circuits that can be controlled by a single switch. The number of throws is either single or double and determines the number of terminals that are switched in each circuit when the switch is activated.

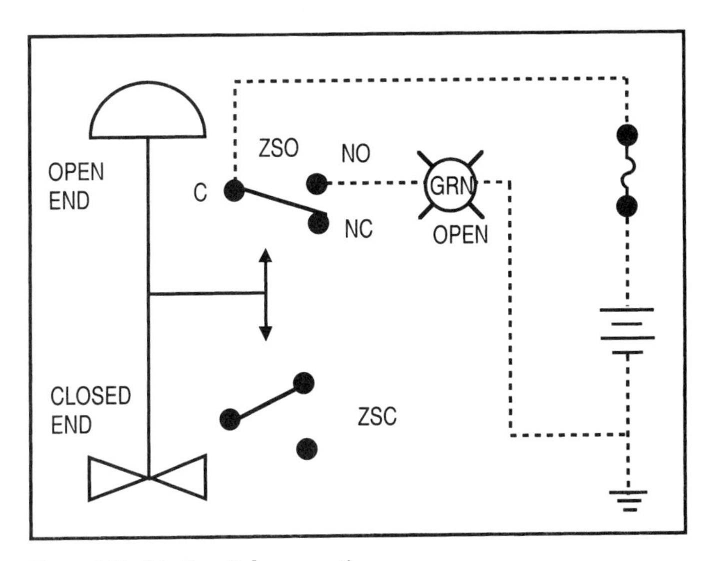

Figure 5.15 Limit switch conventions.

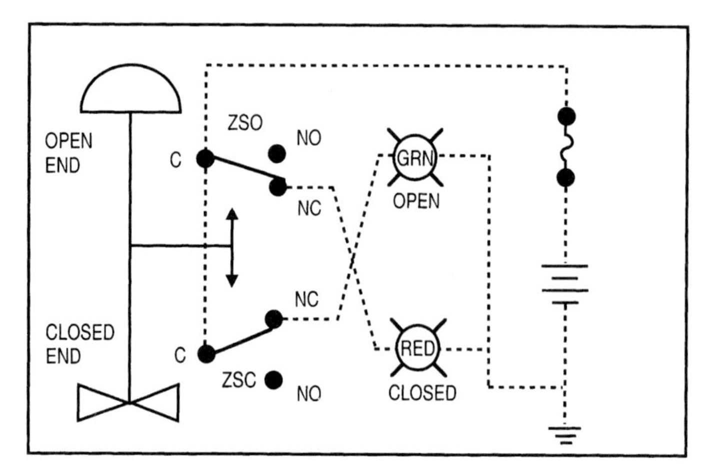

Figure 5.16 Fail-safe limit switch circuit.

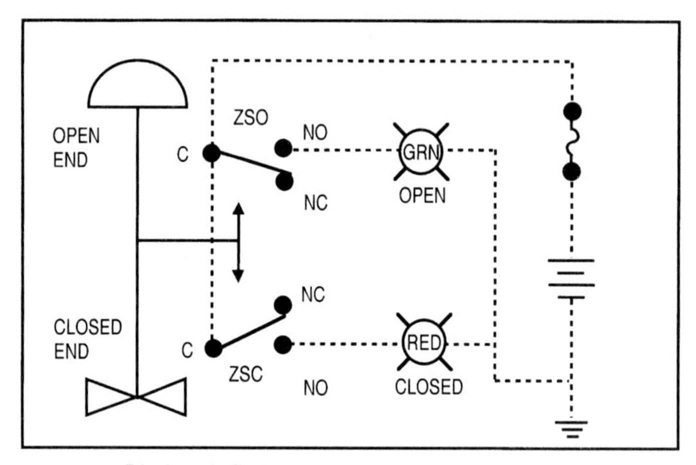

Figure 5.17 Limit switch circuit.

The switches shown in **Fig. 5.15** are single throw since the common terminal remains connected regardless of switch position. Only the normally open (NO) and normally closed (NC) connections change. In a double-throw switch, there are four terminals, not three, and activating the switch results in the action shown in **Fig. 5.18.** Double-throw switches can sometimes simplify the circuitry.

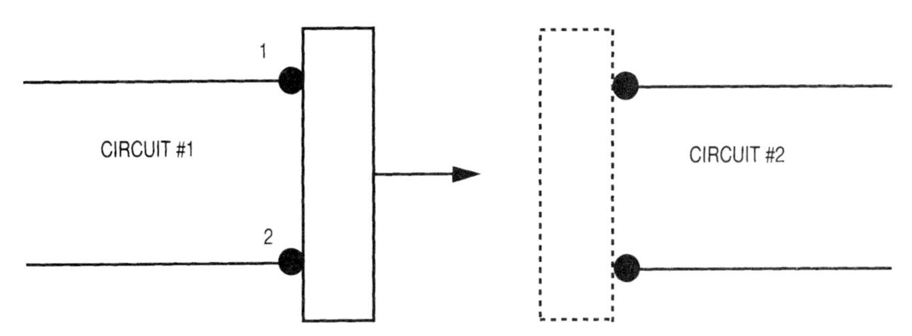

Figure 5.18 Double-throw limit switch.

5.7 Handwheels, Travel Stops

5.7.1 Sliding-stem valves

Handwheels are used to manually manipulate and position control valves. They come in two primary versions, *top mounted* and *side mounted.* Top-mounted handwheels are simpler and lighter but generally only push or pull in one direction with an actuator spring required to supply the force in the other direction. It should also be noted that most top-mounted handwheels do not have a very good mechanical advantage, so the force required to turn the handwheel can be excessive. They are actually of more use as adjustable travel stops to temporarily limit valve stroke. If the valve stroke change is going to be permanent, it's better to change the actuator stroke itself; otherwise there is a risk that the adjustable travel stop will be misadjusted. See **Figs. 5.19** and **5.20** for examples of the two different types of top-mounted handwheels.

Side-mounted handwheels are designed so that the actuating force is more reasonable, and they are usually double acting, so a spring is not required. They are also designed so that the handwheel is rotated in a vertical plane, making them easier to use. They come in two versions. One employs a *clutch* that is engaged to manipulate the valve (**Fig. 5.21**), and the other has sufficient deadband so that the valve can fully stroke without coming in contact with the handwheel mechanism when the handwheel is in the neutral position (**Fig. 5.22**). These models have the disadvantage that to reverse direction requires that the handwheel mechanism be stroked through the deadband range.

The primary caution when employing handwheels is that they not be inadvertently left in the wrong position. This can cause the valve stroke to be restricted and result in control problems or, in the worst case, can prevent the valve from assuming a safety interlock position.

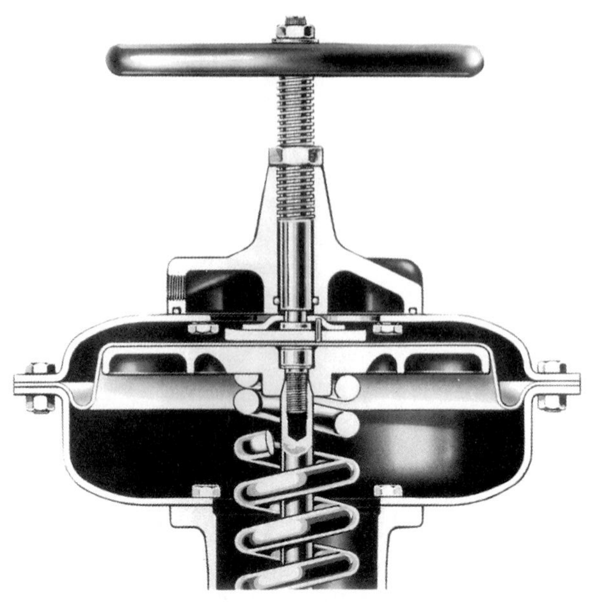

Figure 5.19 Top MTD handwheel, direct-acting actuator. (*Courtesy of Fisher Controls International, Inc., Marshalltown, Iowa.*)

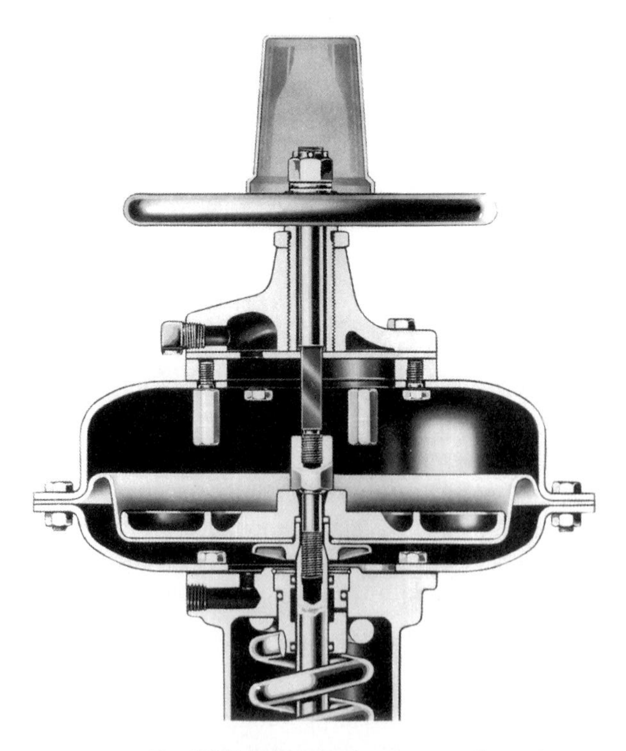

Figure 5.20 Top MTD handwheel, reverse-acting actuator. (*Courtesy of Fisher Controls International, Inc., Marshalltown, Iowa.*)

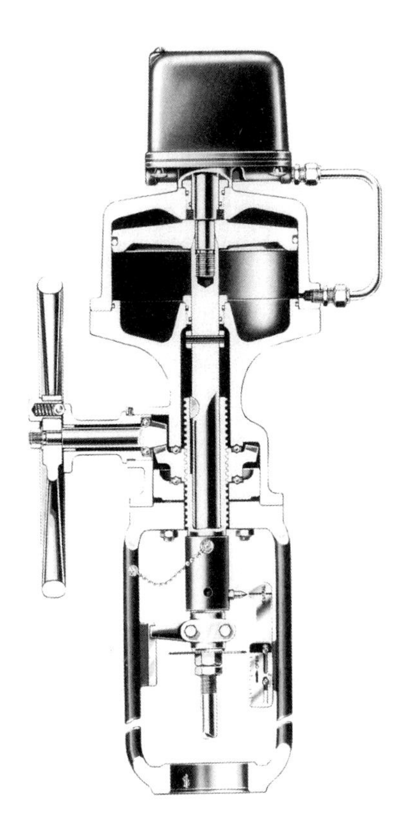

Figure 5.21 Side-mounted handwheel with clutch. (*Courtesy of Fisher Controls International, Inc., Marshalltown, Iowa.*)

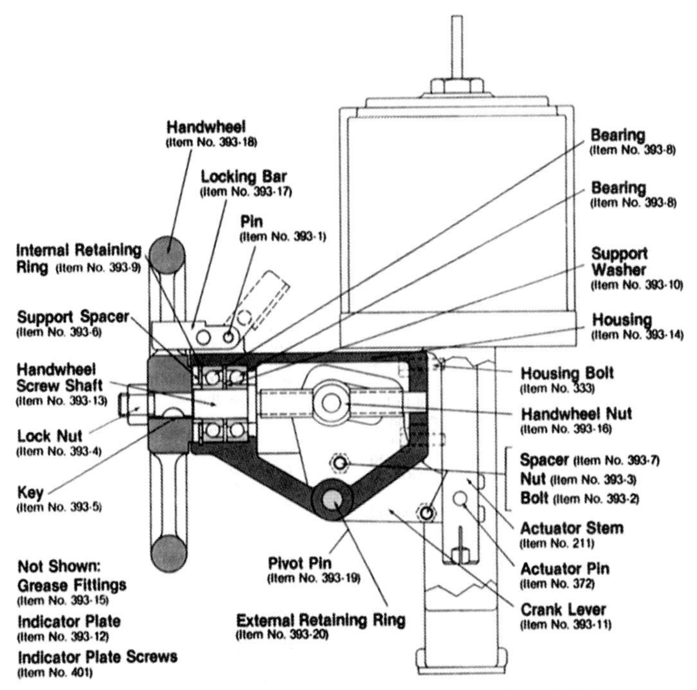

Figure 5.22 Side-mounted handwheel with deadband. (*Courtesy of Valtek International, Inc., Springville, Utah.*)

For this reason, handwheels should only be used where absolutely necessary, and even then access should be restricted to qualified personnel through the use of a lockout system. Also, make sure that any handwheel used has an effective locking mechanism to guard against it going out of adjustment due to things like vibration.

Adjustable travel stops look much like the handwheel designs already discussed, but instead of a handwheel, they employ a bolt or socket head capscrew that has to be turned with a wrench.

5.7.2 Rotary valves

For rotary valves with spring-type actuators, top-mounted handwheels and travel stops just like the ones discussed for sliding-stem valves can be used. More effective handwheel designs are mounted on the end of the valve shaft and utilize a clutch arrangement as shown in **Fig. 5.23.**

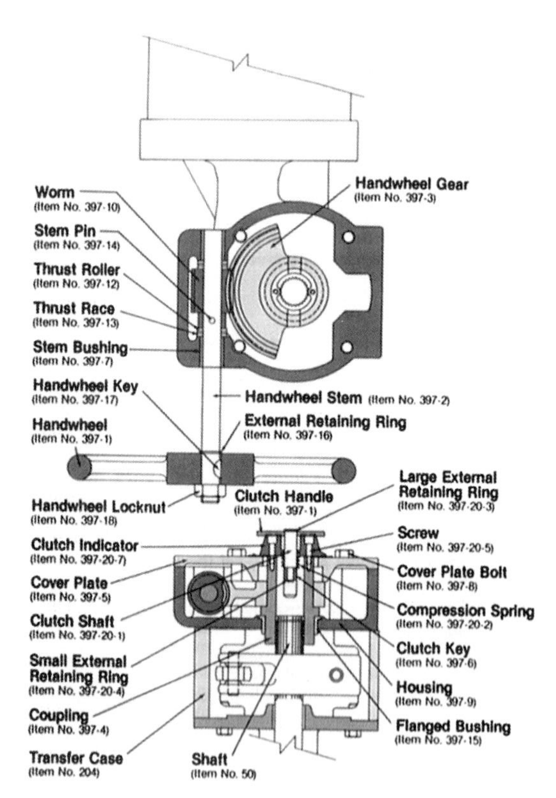

Figure 5.23 Rotary handwheel with clutch. (*Courtesy of Valtek International, Inc., Springville, Utah.*)

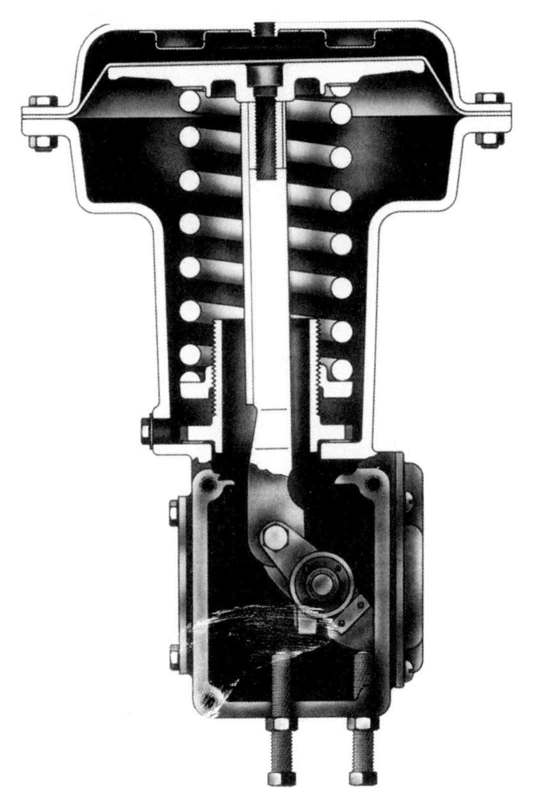

Figure 5.24 Rotary travel stops. (*Courtesy of Fisher Controls International, Inc., Marshalltown, Iowa.*)

Adjustable travel stops are particularly important for rotary valves since many rotary designs do not have hard stops inside the valve, and if they rotate too far, they can damage internal parts. A common approach is shown in **Fig. 5.24.** The same type of precautions described for sliding-stem valves apply here, as well.

5.8 Snubbers

Snubbers are devices used to increase the damping on the stem movement of control valves. A typical construction is shown in **Fig. 5.25.** In this case, the snubber is mounted on the actuator and connected to the valve stem. A piston with hydraulic fluid on both sides is used to resist the rapid stem movement usually associated with instability due to fluid forces on the internal parts. The snubber can be adjusted using a bypass to increase or decrease the damping action, depending on the magnitude of the instability.

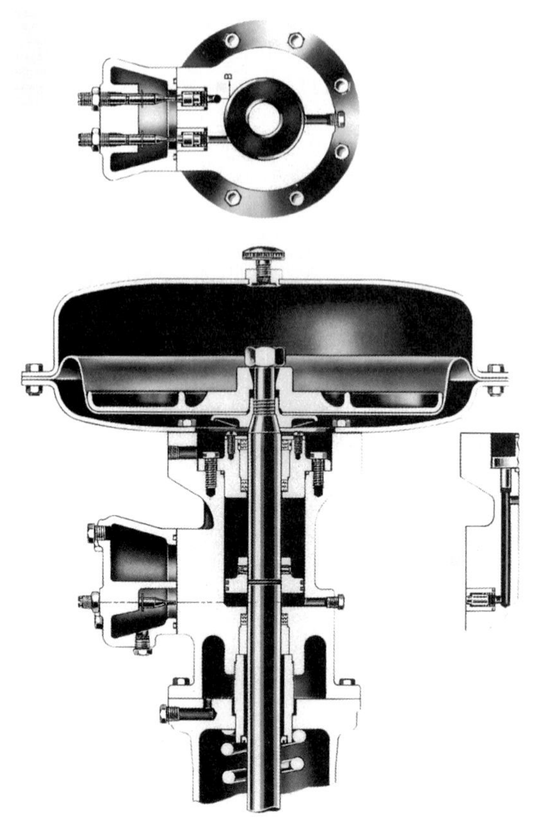

Figure 5.25 Hydraulic snubber. (*Courtesy of Fisher Controls International, Inc., Marshalltown, Iowa.*)

Adding a snubber will help to reduce the movement, but it carries a price in that rapid movement in response to control signals is also adversely affected. It is also treating a symptom, not the root cause of the problem. The fluid forces on the plug are still there and are being transmitted through the actuator and valve where they can cause eventual mechanical failure. In most cases, it is preferable to work on the flow forces by changing the flow direction or contours inside the valve to the point where the instability problem is solved without the use of a snubber.

5.9 Regulators

There are a number of different types of devices referred to as regulators. In the context of control valves, what we're talking about is a device that is used to reduce the available "shop air" or instrument air pressure, which is usually set at 100 to 110 psi, down to a pressure that the pneumatic devices on the valve can withstand. When used in this way, the device is sometimes referred to as the airset. Many valve vendors are now making actuators and pneumatic accessories that can handle shop air directly, eliminating the need for a supply pressure regulator.

On valves where they are used, they should include a built-in filter, a outlet pressure gage, and a chamber to help collect moisture or oil that might be carried along in the air supply system. They should also have some type of relief device to allow pressure to bleed off as necessary and a locking device to ensure that the setting will not change due to vibration. Most airsets, as they are sometimes called, are good for up to 150-psi inlet pressure, but the outlet pressure range is limited for each of the springs used inside the regulator. As a result, you must have a ballpark idea of the outlet pressure required so that the proper spring can be selected initially.

Determining the *required supply pressure* can be a little tricky. Some plants use a rule of thumb that says that the supply pressure should be set at 5 psi over the benchset (the benchset is the stroking pressure range for the actuator). Unfortunately, on valves with high friction and/or shutoff requirements, this rule doesn't work and will result in leakage or even a reduction in valve stroke. What should be determined is the maximum pressure required to either fully open or close and/seat the valve with friction, and then the supply pressure should be set 5 psi higher than this value to account for pressure losses in the accessories. At the same time, the pressure setting should take into account the maximum allowable pressure limits of the accessories and be set no higher than the values published in the vendor literature.

Flow capacity should also be an overriding concern in the selection of the regulator. Many valves are limited in the way they respond to control signal changes by the lack of flow capacity through the airset. Make sure that the rated capacity of the regulator is well above the capacity of the other accessories, particularly the positioner. Also routinely check to make sure that the regulator is clean.

5.10 Position Feedback Devices

On valves in critical services it is sometimes desirable to know the actual position of the valve at any given point in time. A local indication is given by the travel scale that is normally mounted near the connection between the actuator and valve stems for a sliding-stem valve and on the end of the shaft for rotary valves. For a remote indication, a device has to be mounted on the valve that can convert the valve position into a signal that can be transmitted to another location such as the control room. The signal itself is normally an electrical signal of 0 to 10 V dc or 4 to 20 mA, but there are devices that provide a pneumatic signal of 3 to 15 psi or 6 to 30 psi. Currently, the pneumatic devices are nearly all being replaced by electrical devices that can interface directly with the control system. Nearly all position feedback devices require a separate power source of something like 24 V dc, which can be a problem for some valves in remote locations.

5.11 Pneumatic Lock-Ups and Trips

Essentially, *lock-ups* and *trips* are pneumatic on-off switches that change position depending on the signal that they receive at their input port. In a lock-up system, the switches are used to hold a valve in the last position if a designated signal is sensed or received. In most cases, supply pressure is the signal being monitored, and if it falls below a certain adjustable threshold level, the system triggers, locking air in the actuator to keep the valve from moving. Once adequate supply pressure is reestablished, the system opens up again and normal throttling control resumes. A schematic is shown in **Fig. 5.26.** For single-sided actuators such as spring and diaphragm types,

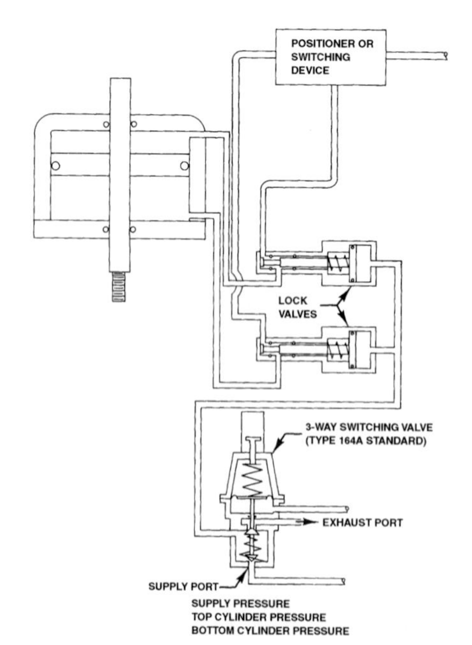

Figure 5.26 Lock-up system. (*Courtesy of Fisher Controls International, Inc., Marshalltown, Iowa.*)

only one lock valve is required. Devices are now available that combine the functions of the three-way valve and the lock valves. Note that all these systems have inherent deadband, which means that the supply pressure has to recover above the original trip point before the system resets. Deadband is usually about 20 percent of the trip point pressure. Also note that "lock-up" is never perfect. The control valve may move slightly due to forces on the valve stem, and leakage will eventually result in stem movement as well.

Similar systems can be set up to drive the valve to either the open or closed position rather than locking it in the last position. In this case, a trip valve is used to sense supply pressure, and it switches the input from a volume tank to one side or the other of the piston to put the valve in the desired position on loss of supply (**Fig. 5.27**). Of course, a spring in the actuator is a much simpler and more reliable way of accomplishing this. Note that these are not simple systems to install and maintain. Care must be taken to ensure that all the ports are properly connected and that no leakage is present.

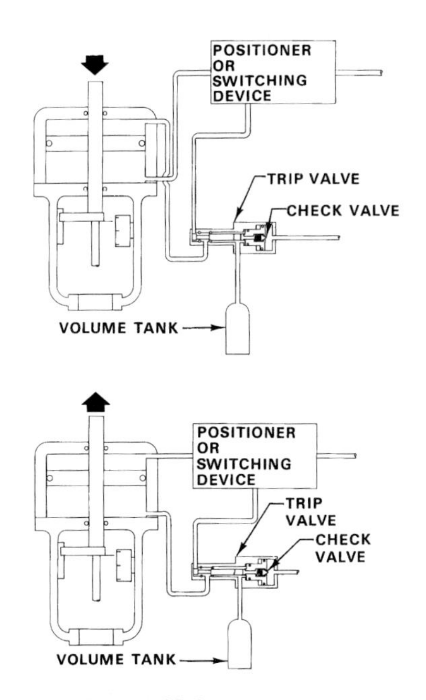

Figure 5.27 Fail-safe systems. (*Courtesy of Fisher Controls International, Inc., Marshalltown, Iowa.*)

5.12 Transmission Lines

One element that does not always receive the attention it deserves is the *tubing connections* between the various pneumatic devices on a control valve. They need to be airtight, and more importantly, they need to be of sufficient size to permit the accessories to position the valve in a timely and efficient manner; $\frac{1}{4}$-in tubing is a bare minimum and $\frac{3}{8}$ in is probably better. Practical experience has shown that in many cases the ability of the valve to track the input signal is limited by the size of the pneumatic lines used on and to feed the supply for the valve. Overall control could be improved, in these cases, by installing larger lines.

At the same time, the length of the lines needs to be taken into account from a dynamic response standpoint. If the flow capacity of the accessory in question is low in comparison to the line volume to be filled, the response of the system to a change in input from the control system will be adversely affected. The common practice of mounting I/Ps on a single rack, serving a number of valves at various distances is a case in point. While it is more convenient from a maintenance and calibration standpoint to place the I/Ps in a central location, the loss in performance due to the lag associated with filling the long transmission lines can far outweigh any perceived savings in time. It is better to keep the line length below 30 ft for I/Ps feeding positioners and less than 10 ft for I/Ps that are feeding actuators directly. This may dictate an alternate approach for things like split-ranging two valves off of one signal, but it really is justified by optimized control performance.

The last subject with regard to transmission lines is that of supply line capacity. In many cases, a large actuator is being used to position a valve where fast response is desired. Unfortunately, the accessories may have been selected with speed in mind, but the supply line is too small or too long to be able to keep up with the demands from the actuator. Not only do you end up with slow response to setpoint or load changes, the response is different depending on stroking direction since only the supply direction is affected, not the exhaust. This can make control performance erratic and the loop difficult to tune.

5.13 Pneumatic Controllers

A *controller,* in the context of the process industries, is a device with one or more inputs that are measurements of process properties (or process variables) and a variable output that can be changed by the internal workings of the controller to help bring the process properties closer to their desired values, called the setpoints. In essence the controller constantly compares the process variables to the setpoint, measures the error, and then modifies the output depending on the magnitude of the error. A schematic is shown in **Fig. 5.28.**

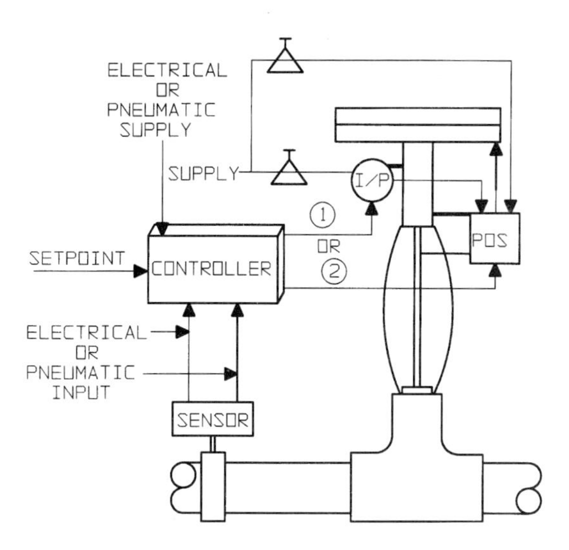

Figure 5.28 Controller in a typical loop (shows electrical or pneumatic versions).

If a controller looks at a single variable, it is referred to as a single-element controller. For simplicity's sake, the balance of this section will deal only with single-element controllers. There are both pneumatic and electronic controllers. The pneumatics can work directly with the control valve and process variable (PV), and the electronic versions need some type of transducer to convert the PV into an electronic signal and the output into a pneumatic signal that can work with the valve. Pneumatic controllers will be discussed here since they are occasionally found on or near the valve. Electronic controllers are outside the scope of this book. Even the coverage of the pneumatic controller will be relatively brief because of the complexity of the subject. For a more detailed discussion of process control and controllers, see one of the first three references at the end of the chapter.

Before getting into the mechanics of a pneumatic controller, a short review of basic control theory is probably in order. First of all, any controller can react to an error between the process variable and the setpoint in a number of different ways. This control reaction is called the *control mode* for the controller. There are three traditional modes of control: *proportional, integral,* and *derivative.* Industry practice is to refer to each control mode by the first letter of the name: P for proportional, I for integral, and D for derivative.

Any traditional process control text will provide you with a very detailed mathematical derivation for each of the modes mentioned above. This is impressive, but it doesn't really provide any useful information or any insight into how the adjustments of each of the modes will affect a process. To keep things simple, the formula for a proportional mode controller is:

$$CO(t) = K[(SP - PV(t)] + Y$$

where CO(t) = the controller output as a function of time
K = a constant called proportional gain
SP = the setpoint (the desired value)
PV(t) = the process variable as a function of time
Y = a constant that reflects a baseline output of the controller

If we move Y to the left-hand side of the equation, we can think of $[CO(t) - Y]$ as the change in controller output, and it becomes clear that the change in controller output is proportional to the error between SP and PV. So much for the math. Practically speaking, nearly every controller encountered on a control valve will utilize the proportional mode for control, so it's important to understand what's going on. This mode is called proportional because, as the equation indicates, the change in controller output is proportional to the error, or the difference between the setpoint and process variable.

It's really very simple if we look at an example. Consider your car as the process and assume that you have a cruise control installed that is unusual in that it has a K adjustment on it. This is the controller for the process. Think of the speed setting as the setpoint, and the process variable that we want to control is the actual speed of the car as measured by the speedometer. The output of the controller controls the position of the gas pedal, which, in turn, adjusts the speed of the car. In the simplest of controllers with no proportional action, if we set the speed (SP) at a nominal 30 mph and then put the car in motion on the highway, the controller will act such that the gas pedal is going to be either full on or full off. Anyone who's ever taken a cab in New York City has experienced this type of process control and can testify that it's not a very pleasant experience, nor is it an efficient way of controlling the process.

Anytime the speed is below the setpoint, it's "pedal to the metal" until the car reaches the setpoint speed, and then the pedal is completely released until the speed drops below setpoint again. What happens is that the car is very seldom at the setpoint since the process is continually overshooting in both directions. The solution: proportional control.

With proportional control, the correction (the change in controller output) is now a linear function of the error between setpoint and process variable. In other words, for a given setting of K on our speed control, the controller will look at the initial error between actual speed and 30 mph and depress the pedal to speed the car up (assuming we are below 30 mph to begin with). As the speed of the car approaches the setpoint, the error decreases, and the pedal moves back in the other direction because the correction has to decrease as a linear function of the error. What happens is we get much smoother control, and the process variable begins to stabilize near the setpoint.

The optimum setting for K will depend on a number of variables. The higher the setting, the bigger the correction so, in the case of the car, we would want to choose a setting that will keep the average speed close to the setpoint, given the changing loads on our system due to things like wind resistance, tire condition, and hills. It also needs to be selected so that the corrections are not too large. If we "floor it" every time the error is 5 mph or more, we'll end up using a lot more gas than we need to and will put unnecessary wear and tear on the car without a significant improvement in accuracy. We also risk putting the car into a state of constant speed swings due to overshooting the setpoint in both directions (instability). And finally, it needs to take into account the characteristics of the car itself. If the engine is very powerful and can change the speed of the car fairly quickly, the K, or gain setting, on the controller needs to be lower to keep from overshooting the setpoint. The whole process of choosing controller settings is referred to as tuning and can get to be very complicated, depending on some of the process characteristics already identified above.

The operation of a typical process loop is really very similar to the example just covered. The process variable might be something like pressure or temperature, and the goal of the system is to keep it as close as possible to the setpoint, just like our speed control. We also see load changes that tend to push our PV away from the setpoint, like the hills or wind in our example. And finally, the controller proportional gain (K) needs to be adjusted in light of the gain in the process and the dynamic response of the control valve. This is analogous to the response of the car to changes in the position of the gas pedal.

In keeping with the "simple is better" philosophy, we would expect to find a gain setting on a controller equivalent to the K term just described. Unfortunately, this is not the case. What is used is a proportional band that is the inverse of the gain multiplied by 100 and expressed as a percent. It is defined as the change of the measured input, as a percentage of total input span, that results in a 100 percent change in the controller output. The range of possible adjustments for proportional band is large but might vary from 0.1 to 300 percent on a typical controller. This corresponds to a gain range of 1000 to 0.33. The important thing to remember here is that raising the proportional band reduces the gain and makes the controller react more slowly to errors between the setpoint and PV.

Integral action is also a very common control mode and is normally used along with the proportional mode. It is sometimes required because of an inherent problem with proportional control. In a proportional-only system there will always be a steady-state offset between the setpoint and the process variable. The offset can be reduced by increasing the gain setting, but we reach a point where further gain increases are no longer advisable because of the risk of unstable operation. Integral action allows us to set the gain at manageable levels and still reduce the steady-state offset to a minimum. The equation for integral action is:

$$\mathrm{CO(t)} - Y = \frac{1}{T_i \int (\mathrm{SP} - \mathrm{PV})}$$

where CO, SP, PV, and Y are all as defined above
T_i is the integral time

Again, trying to put this in plain English, it means that the change in the controller output (the correction) is a linear function of the integral of the error, not of the error itself. What this means is that the tendency for a steady-state error to exist will diminish with integral action because the integral of even a small error will grow over time, resulting in an increasing correction until the error is eliminated or reduced.

The relative strength of the correction is inversely proportional to the variable T_i, meaning that increasing T_i (integral time) will reduce the overall effect of the integral mode of the controller. Once again, most controllers will not have an adjustment called integral. Look for the *reset* knob marked in repeats per minute, the inverse of T_i. Increasing the reset rate will reduce the integral time, which increases the integral portion of the total correction. Increasing the integral action tends to reduce the steady-state error but can also result in loop instability, so it should be adjusted with care, taking into account the gain setting, since these two settings interact.

The third mode referred to above is called *derivative action*. The equation is as follows:

$$\mathrm{CO(t)} - Y = \frac{T_d d(\mathrm{SP} - \mathrm{PV})}{\mathrm{dt}}$$

where CO, PV, Y, and SP are all as defined above
T_d is the derivative time

In this case, the controller correction is a linear function of the derivative of the error. The derivative is like an early warning system that supplies a large correction relatively soon after a load change. This anticipating action is possible because the derivative of an error leads the error in terms of absolute magnitude for most error functions. Looking at the car again should help to clarify how the three modes act and interact. Let's assume that it is traveling at a constant 55 mph on a level road. Now, let's see how it reacts to a load change depending on whether we have proportional-only, proportional plus integral, or proportional, integral, and derivative speed control.

In **Fig. 5.29,** a hill represents a load change (a step) for the car. The proportional mode reacts to the load change but ends up with a steady-state error, as mentioned above. If the reduction of the error is deemed to be important, reset action can be added, and it continues to correct the process after the proportional action has subsided. Note that this improves the steady-state error near the end of the curve, but the maximum deviation from setpoint is still roughly the same. This is where derivative action is of interest.

The correction from the proportional mode is a function of the absolute error between the setpoint and the process variable. This permits the error immediately after the upset to grow to be relatively large before the proportional correction becomes large enough to

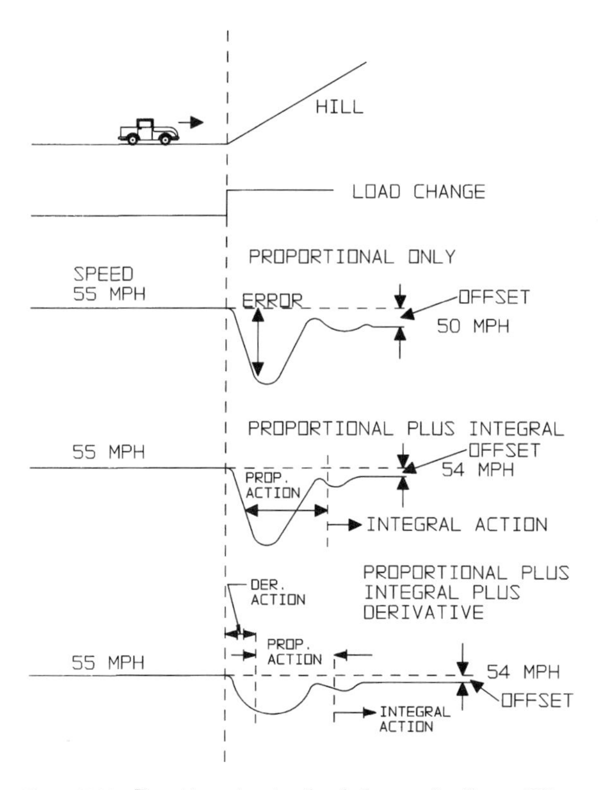

Figure 5.29 Reactions to step-load change for three different controller modes.

reverse the trend. Derivative action looks at the slope of the error line, not the absolute error, so it reacts faster and provides a large correction immediately after the upset, reducing the maximum deviation. Derivative action is commonly called *rate* and, like gain and reset, can also be adjusted on most pneumatic controllers. The higher the rate, the stronger the derivative action, but once again, be careful adjusting it because all three control modes interact. Tuning a three-mode controller can be relatively complicated and is outside the scope of this book. For more information on this subject, see reference 4.

A standard pressure controller is shown in **Fig. 5.30.** In this case, the controller is mounted on the control valve and its output is directly connected to the actuator of the valve. In many cases, the pneumatic output of the controller is compatible with the actuator benchset, so the positioner shown in **Fig. 5.28** is not required. **Figure 5.31** shows the internals for the same controller where one can see the adjustments for setpoint, gain, rate, and the like. In this case a bourdon tube is used to sense pressure directly, so no external sensing unit is required. The schematic for this device is shown in **Fig. 5.32,** which shows how the control mode adjustments actually work.

Figure 5.30 Pneumatic pressure controller. (*Courtesy of Fisher Controls International, Inc., Marshalltown, Iowa.*)

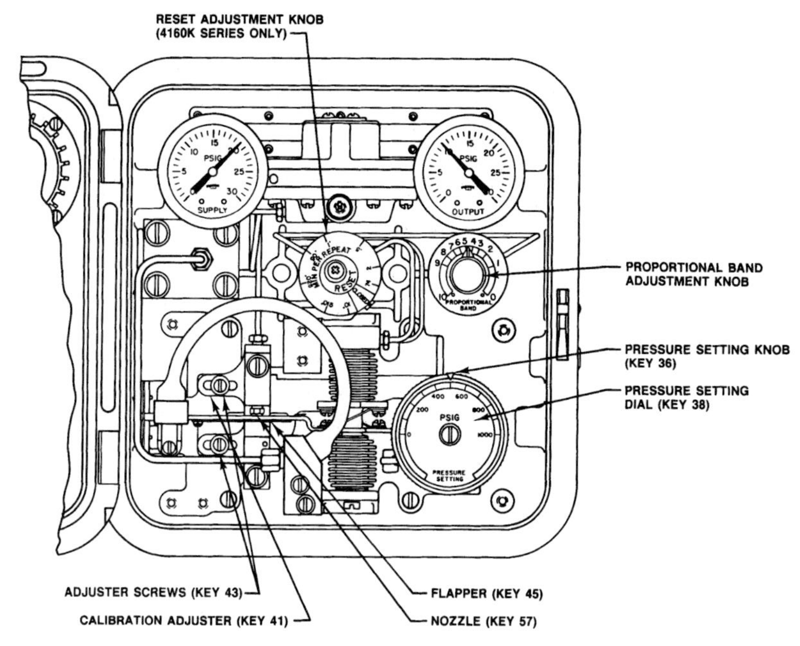

Figure 5.31 Pressure controller internals. (*Courtesy of Fisher Controls International, Inc., Marshalltown, Iowa.*)

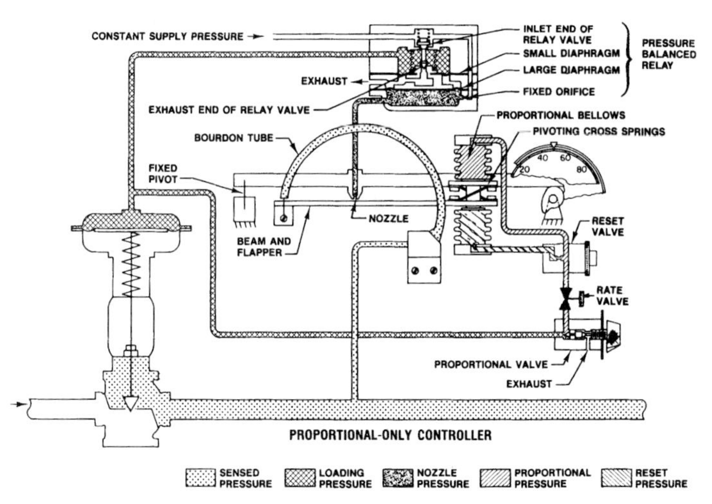

Figure 5.32 Controller schematic. (*Courtesy of Fisher Controls International, Inc., Marshalltown, Iowa.*)

5.14 Hazardous Area Classifications

The following is a general discussion of hazardous area classifications and the measures that must be taken to ensure the safety of personnel and equipment. For more detailed information, consult one of the following standards:

Common standards—North America

ANSI/ISA RP12.6, 1987—Installation of Intrinsically Safe Instrument Systems in Hazardous (Classified) Locations.

ANSI/ISA S12.12, 1984—Electrical Equipment for Use in Class I, Division 2 Hazardous (Classified) Locations. Provides requirements for the design, construction, and marking of electrical equipment for use in class I, division 2 locations.

Canadian Electrical Code (CEC), 1993, Part I, Section 18—Defines hazardous locations by class, division, and group and provides installation criteria.

Canadian Standards Association (CSA) Standard C22.2, No. 94.1991—Special Purpose Enclosures.

National Electric Code (NEC), Article 500, 1993—Defines the hazardous area classes, divisions, and groups and provides installation criteria.

National Electrical Manufacturer's Association (NEMA) Standard 250, 1985—Enclosure for Electrical Equipment (1000 Volts Maximum). Describes definitions and standards for electrical enclosures for U.S. installations.

NFPA 497M. 1991—Classification of Gasses, Vapors and Dusts for Electrical Equipment in Hazardous (Classified) Locations.

Common Standards—Europe and Asia Pacific

Standards Association of Australia (SAA) as 2380.2, Flameproof Enclosure "d"

SAA as 2380.7, Intrinsic Safety "i."

SAA 2380.9, Type of Protection "n"—Non Sparking.

British Standards Institute (BSI) BS 6941, Type of Protection "N."

CENELEC EM 50014, General Requirements.

CENELEC EN 50018, Flameproof Enclosures "d."

CENELEC EM 50019, Increased Safety "e."

CENELEC EN 50020, Intrinsic Safety "i."

International Electrotechnical Commission (IEC) 79-1, Construction and Test of Flameproof Enclosures of Electrical Apparatus.

IEC 79-4, Method of Test for Ignition Temperature.

IEC 79-10, Classification of Hazardous Areas.

IEC 79-11, Construction and Test of Intrinsically-Safe and Associated Apparatus.

IEC 79-15, Electrical Apparatus with Type of Protection "n."

IEC 529, Enclosure Ratings.

5.14.1 North America

Any electrical device that is to be used in a hazardous area needs to be certified for use according to the requirements specified for the type of area in question. Certification and/or approval is provided by three principal groups in North America: Factory Mutual (FM), Underwriters Laboratories (UL), and Canadian Standards Association (CSA).

Types of protection. The types of protection required depend on the risk involved. The types that are normally encountered on equipment used on control valves in chemical plants are:

Dust-ignitionproof (DIP). A protection concept that prevents dust from entering an enclosure and will not allow arcs, sparks, or heat generated inside of the enclosure to cause ignition of exterior accumulations or atmospheric suspensions of a specified dust on or near the enclosure.

Explosionproof (XP). A protection concept that requires electrical equipment to be capable of containing an internal explosion of a specific flammable vapor-air mixture, thereby not allowing the release of burning or hot gases to the external environment which may be potentially explosive. Also, the equipment must operate at a safe temperature with respect to the surrounding atmosphere.

Intrinsically safe (IS). A protection concept that requires electrical equipment to be incapable of releasing sufficient electrical or thermal energy to cause ignition of a specific hazardous substance under "normal" or "fault" operating conditions.

Nonincendive (NI). A type of protection which requires electrical equipment to be nonsparking and incapable of releasing sufficient electrical or thermal energy to cause ignition of a specific hazardous substance under "normal" operating conditions.

Hazardous location classification. Each type of protection can be used to reduce the risk in one or more hazardous location classifications. The classifications are defined using four designations: class, division, group, and temperature code. The designations are defined as follows:

Class. The class defines the general nature of the hazardous material in the surrounding atmosphere. Class I locations are those in which flammable gases or vapors are, or may be, present in the air in quantities sufficient to produce explosive or ignitable mixtures. Class II locations are those that are hazardous because of the presence of combustible dusts. Class III locations are those in which easily ignitable fibers or flyings may be present but are not likely to be in suspension in sufficient quantities to produce ignitable mixtures.

Division. The division defines the probability of hazardous material being present in an ignitable concentration in the surrounding atmosphere. Division 1 locations are those in which the probability of the atmosphere being hazardous is high due to flammable material being present continuously, intermittently, or periodically. Division 2 locations are those which are presumed to be hazardous only in an abnormal situation.

Group. The group defines the hazardous material in the surrounding atmosphere. The specific hazardous materials within each group and their automatic ignition temperatures can be found in Article 500 of the National Electrical Code (NEC) and in NFPA 497M. Groups A, B, C, and D apply to class I locations; Groups E, F, and G apply to class II locations. Group A atmospheres contain acetylene. Group B atmospheres contain hydrogen, fuel, and combustible process gases containing more than 30 percent hydrogen by volume or gases or vapors of equivalent hazard such as butadiene, ethylene oxide, propylene oxide, and acrolein. Group C atmospheres include such gases as ethyl ether and ethylene or gases or vapors of equivalent hazard. Group D atmospheres contain gases such as acetone, ammonia, benzene, butane, cyclopropane, ethanol, gasoline, hexane, methanol, methane, natural gas, naphtha, and propane or gases or vapors of equivalent hazard. Group E atmospheres contain combustible metal dusts, including aluminum, magnesium, and their commercial alloys, or other combustible dusts whose particle size, abrasiveness, and conductivity present similar hazards in the use of electrical equipment. Group F atmospheres contain combustible carbonaceous dusts, including carbon black, charcoal, coal, or coke dusts that have more than 8 percent total entrapped volatiles or dusts that have been sensitized by other material so that they present an explosion hazard. Group G atmospheres contain combustible dusts not included in Groups E or F, including flour, grain, wood, plastic, and chemicals.

Temperature code. A mixture of hazardous gases and air may be ignited by coming into contact with a hot surface. The conditions under which a hot surface will ignite a gas depends on surface area, temperature, and the concentration of the gas. The approvals agencies test and establish maximum temperature ratings for the different equipment submitted for certification. Equipment that has been tested receives a temperature code that indicates the maximum surface temperature attained by the equipment. The following table shows the different temperature codes.

	Maximum surface temperature	
Temperature code	°F	°C
T1	842	450
T2	572	300
T2A	536	280
T2B	500	260
T2C	446	230
T2D	419	215
T3	392	200
T3A	356	180
T3B	329	165
T3C	320	160
T4	275	135
T4A	248	120
T5	212	100
T6	185	85

The NEC states that any equipment that does not exceed a maximum surface temperature of 212°F (based on 104°F ambient temperature) is not required to be marked with the temperature code. When a temperature code is not specified on the approved apparatus, it is assumed to be T5. Most devices mounted on control valves fall into this category.

Detailed discussion of protection techniques. There are four different protection techniques used to address hazardous area concerns:

explosion proof, intrinsically safe, dust ignition proof, and nonincendive. Each is defined below.

Explosionproof technique. This technique is implemented by enclosing all electrical circuits in housing and conduits strong enough to contain any explosion or fires that may take place inside the instrument. All electrical wiring leading to the field instrument must be installed using threaded rigid metal conduit, threaded steel intermediate metal conduit, or Type MI cable. The advantages of this technique are:

1. Users are familiar with this technique and understand its principles and applications.
2. Sturdy housing design provides protection to the internal components of the instruments and allows their application in hazardous environments.
3. An explosionproof housing is usually weatherproof, as well.

The disadvantages of this technique are:

1. Circuits must be deenergized or the location rendered nonhazardous before housing covers may be removed.
2. Opening of the housing in a hazardous area voids all protection.
3. Generally requires use of heavy bolted or screwed enclosures.

Intrinsically safe technique. This technique operates by limiting the electrical energy available in circuits and equipment to levels that are too low to ignite the most easily ignitable mixtures of a hazardous area. This technique requires the use of intrinsically safe barriers between the hazardous and safe areas to limit the current and voltage, thereby avoiding the development of sparks or hot spots in the circuitry of the instrument under fault conditions. The advantages of this technique are:

1. Lower cost. No rigid metal conduit or armored cable is required for field wiring of the instrument.
2. Allows greater flexibility since it permits simple components such as switches, contact closures, thermocouples, RIDs, and other non-energy-storing instruments to be used without special certification but with appropriate barriers.
3. Ease of field maintenance and repair. There is no need to remove power before adjustments or calibration are performed on the field instrument. The system remains safe even if the instrument

is damaged because the energy level is too low to ignite most easily ignitable mixtures. Note that test and calibration instruments must have the appropriate approvals for hazardous areas or be used with a hot work permit.

The disadvantage of this technique is that high-energy consumption applications are not applicable to this technique because the energy is limited at the source (or the barrier). This technique is limited to low-energy applications such as dc circuits, solenoid valves, electropneumatic converters, etc.

Dust-ignitionproof technique. This technique results in an enclosure that will exclude ignitable amounts of dusts and will not permit arcs, sparks, or heat otherwise generated inside the enclosure to cause ignition of exterior accumulations or atmospheric suspensions of a specified dust on or near the enclosure.

Nonincendive technique. This technique is limited to division 2 applications only. It allows for the incorporation of circuits in electrical instruments which are not capable of igniting specific flammable gases or vapor-in-air mixtures under normal operating conditions. Each instrument must have specific approval from a testing agency in order to qualify as nonincendive. The advantages of this technique are:

1. It uses electronic equipment which normally do not develop high temperatures or produce sparks strong enough to ignite the hazardous environment.
2. Its cost is lower than other hazardous environment protection techniques because there is no need for explosionproof housings or energy limiting barriers.
3. For nonincendive circuits, wiring is permitted using any of the methods suitable for wiring in ordinary locations (as allowed by the exception of the NEC).

The disadvantages of this technique are:

1. This technique is applicable to division 2 locations only.
2. It places constraint on control room to limit energy to field wiring (normal operation is open, short, or grounding of field wiring) so that arcs or sparks under normal operation will not have enough energy to cause ignition.
3. Both the field instrument and control room device may require more stringent labeling.

The following table summarizes the locations where the various types of protection can be used:

Type of protection	Class I division 1	Class I division 2	Class II division 1	Class II division 2
Explosion-proof (XP)	X	X		
Dust ignition-proof (DIP)			X	X
Intrinsically safe (IS)*	X	X	X	X
Nonincendive (NI)		X		

*Intrinsically safe apparatus used in division 2 must still use the same barriers as it would if installed in division 1.

5.14.2 Other world areas

Again, use of electrical devices in hazardous areas requires that the design be approved by an authorized agency. Outside of North America these include:

Location	Abbreviation	Approval agency
United Kingdom	BASEEFA	British Approvals Service for Electrical Equipment in Flammable Atmospheres
Germany	PTB	Physikalische-Technische Bundesanstalt
France	LCIE	Laboratorie Central des Industries Électriques
Australia	SAA	Standard Association of Australia

Types of protection. There are four types of protection typically encountered on control valves outside of North America:

Flameproof. A type of protection that requires electrical equipment to be capable of containing an internal explosion of a specific flammable vapor-air mixture, thereby not allowing the release of burning or hot gases to the external environment, which may be potentially explosive. Also, the equipment must operate at a safe temperature with respect to the surrounding atmosphere. This is similar to explosionproofing used in North America. This type of protection is referred to by IEC as Ex d.

Increased safety. A type of protection in which various measures are applied so as to reduce the probability of excessive temperatures and the occurrence of arcs or sparks in the interior and on the external parts of electrical apparatus that do not produce them in normal service. These measures typically involve "robust" construction features such as special cable glands, specially spaced and secured terminal connections, double insulation of windings, rugged enclosure design, or combinations of such features. This type of protection is referred to by IEC as Ex e. *Note:* Increased safety may be used along with the flameproof type of protection.

Intrinsically safe. A type of protection that requires electrical equipment to be incapable of releasing sufficient electrical or thermal energy to cause ignition of a specific hazardous substance under "normal" or "fault" operating conditions. This type of protection is referred to by IEC as Ex i.

Hazardous area classifications. Hazardous areas outside North America are classified by gas group and zone.

Group. Electrical equipment is divided into two groups. Group I covers electrical equipment used in mines, and Group II covers all other electrical equipment. Group II is further subdivided into three subgroups: A, B, and C. The specific hazardous materials within each group can be found in CENELEC EN 50014, and the automatic ignition temperatures for some of these materials can be found in IEC 79-4. Group I (Mining) covers atmospheres containing methane or gases or vapors of equivalent hazard. Group IIA covers atmospheres containing propane or gases or vapors of equivalent hazard. Group IIB covers atmospheres containing ethylene or gases or vapors of equivalent hazard. Group IIC covers atmospheres containing acetylene or hydrogen or gases or vapors of equivalent hazard.

Zone. The zone defines the probability of hazardous material being present in an ignitable concentration in the surrounding atmosphere. *Zone 0* is locations where an explosive concentration of gas or vapor is present continuously or is present for long periods of time. *Zone 1* is locations where an explosive concentration of gas or vapor is likely to be present for short periods of time under normal operating conditions. *Zone 2* is locations where an explosive concentration of gas or vapor is likely to be present for very short periods of time due to an abnormal condition.

Temperature code. A mixture of hazardous gases and air may be ignited by coming into contact with a hot surface. The conditions under which a hot surface will ignite a gas depend on surface area, temperature, and the concentration of the gas. The approval agencies test and establish maximum temperature ratings for the different equipment submitted for certification. Group II equipment that has been tested receives a temperature code that indicates the maximum surface temperature attained by the equipment. It is based on a 104°F (40°C) ambient temperature:

	Maximum surface temperature	
Temperature code	°F	°C
T1	842	450
T2	572	300
T3	392	280
T4	275	135
T5	212	100
T6	185	85

Detailed discussion of protection techniques. There are four different protection techniques used to address the hazardous zones just defined. They are flameproof, increased safety, intrinsically safe, and Type N. They are defined as follows.

Flameproof technique. This technique is implemented by enclosing all electrical circuits in housings and conduits strong enough to contain any explosion or fires that may take place inside the instrument. The advantages of this technique are:

1. Users are familiar with this technique and understand its principles and applications.
2. Sturdy housing designs provide protection to the internal components of the instruments and allows their application in hazardous environments.
3. Flameproof housing is usually weatherproof as well.

The disadvantages of this technique are:

1. Circuits must be deenergized or the location rendered nonhazardous before housing covers may be removed.

2. Opening of the housing in a hazardous area voids all protection.
3. Generally requires use of heavy bolts or screwed enclosures.

Increased safety technique. The increased safety technique incorporates special measures to reduce the probability of excessive temperatures and the occurrence of arcs or sparks in normal service. The advantages of this technique are:

1. Increased safety enclosures provide at least 1P54 enclosure protection.
2. Installation and maintenance are easier than for flameproof enclosures.
3. The wiring costs are significantly reduced over flameproof installations.

The disadvantage of this technique is that it is limited in the apparatus for which it may be used. Normally used for apparatus such as terminal boxes and compartments.

Intrinsically safe technique. This technique operates by limiting the electrical energy available in circuits and equipment to levels that are too low to ignite the most easily ignitable mixtures in a hazardous area. This technique requires the use of intrinsically safe barriers to limit the current and voltage between the hazardous and safe areas to avoid the development of sparks or hot spots in the circuitry of the instrument under fault conditions. The advantages of this technique are:

1. Lower cost. Less stringent rules for field wiring of the instrument.
2. Allows greater flexibility since it permits simple components such as switches, contact closures, thermocouples, RTDs, and other non-energy-storing instruments to be used without special certification but with appropriate barriers.
3. Ease of field maintenance and repair. There is no need to remove power before adjustments or calibration are performed on the field instrument. The system remains safe even if the instrument is damaged because the energy level is too low to ignite most easily ignitable mixtures. Testing and calibration instruments must have the appropriate approvals for hazardous areas or must be used with appropriate hot work permits.

The disadvantage of this technique is that high-energy applications are not applicable to this technique because the energy is limited at the source (or the barrier). This technique is limited to low-energy applications such as dc circuits, solenoid valves, electropneumatic converters, etc.

The various types of protection are typically used in the locations designated in the table below. Refer to the appropriate national regulations for specifics.

Type of protection	Zone 0	Zone 1	Zone 2
Flameproof, Ex d		X	X
Increased safety, Ex e		X	X
Intrinsically safe, Ex ia	X	X	X
Intrinsically safe, Ex ib		X	X
Nonincendive, Ex n or Ex N			X

References

1. Lloyd, Sheldon G., and Anderson, Gary, *Industrial Process Control,* 1st ed., Fisher Controls, Marshalltown, Iowa, 1971.
2. Eckman, D. P., *Automatic Process Control,* John Wiley, New York, 1958.
3. Murrill, Paul W., *Fundamentals of Process Control Theory,* ISA, Research Triangle Park, N.C., 1981.
4. Jury, Floyd D., *TM-28 Fundamentals of 3-Mode Controllers,* Fisher Controls, Marshalltown, Iowa, 1973.
5. ASCO Cat. No. 32: 2, 3, & 4—Way Solenoid Valves, ASCO Inc., N.J., 1988.
6. *ANSI/ISA S7.3-1975 Quality Std. for Instrument Air* (*R1981*), ISA, Research Triangle Park, N.C., 1981.
7. Jury, Floyd D., *Fundamentals of Closed Loop Control,* Fisher Controls, Marshalltown, Iowa, 1975.
8. Lloyd, Sheldon G., *Guidelines for the Application of Valve Positioners,* Fisher Controls, Marshalltown, Iowa, 1969.
9. Driedges, Walter, "Limit Switches Key to Valve Reliability," *Intech,* vol. 40, no. 1, p. 35, ISA, Research Triangle Park, N.C., 1993.
10. ISA Std. S51.1, ISA, Research Triangle Park, N.C., 1979.
11. "Valtek #5, Auxiliary Handwheels and Limit Stops," Valtek No. 49015, Valtek, Springville, Utah, May 1987.
12. *Fisher Controls Instruction Manual: 4150K & 4160K,* Fisher Controls, Marshalltown, Iowa, March 1982.

Chapter

6

Typical Applications

Once again, it should be noted that it is difficult to characterize or describe typical applications in the chemical industry because there are so many different processes. However, there are a few applications that are seen more often than others and that can be used to illustrate the considerations made when selecting a valve to meet the demands of a given application. Before talking about specific applications, however, a general background will be provided in the area of what is called *severe service* since so many of the applications that will be covered later fall into this area.

6.1 Severe Service

Severe service, as the name implies, is a general term that applies to any application that requires a valve to perform above and beyond the call of normal duty. It includes the following categories:

Noise

Cavitation

Flashing

Erosion

Corrosion

High temperature

High pressure

Low temperature (i.e. cryogenic)

Fire safety

6.1.1 Noise

Anytime a fluid passes through a control valve, an energy transformation takes place where a portion of the energy contained in the fluid is converted into other forms such as heat and sound. This energy transformation is the mechanism that permits a pressure drop to be created across the valve since there is no mechanical means of extracting energy from the flowstream as there would be with something like a turbine. The production of sound energy becomes a concern if it exceeds certain levels that are either dangerous for personnel in the area or could be detrimental to the condition of the trim inside the valve. *Noise,* in this case, can be defined as any sound that is considered to be undesirable.

As far as personnel exposure is concerned, OSHA has set limits that spell out both the instantaneous and prolonged exposures that are considered to be safe for industrial workers exposed to noise. These limits carry many qualifiers, and to properly understand them, we need to better understand certain characteristics of sound.

First of all, sound can be characterized by intensity and frequency. The intensity is a general measure of the strength and is expressed in a unit of sound pressure called the decibel (dB). Frequency is measured in cycles/second or hertz (Hz). The intensity is the best indicator of the potential for a given sound to do damage to the human ear and as a result, a special frequency-weighted intensity unit has been developed that takes into account the way the human ear responds to different frequencies. This is called the dBA, and when sound intensity is expressed in the context of the human ear, it is the unit that is normally employed or implied. The decibel is a logarithmic unit, and for each increase of 3 dB, the sound intensity doubles.

With this in mind, we can now return to the government regulations that state that the instantaneous exposure to noise should be less than 115 dBA and that the 8-h exposure should be less than 90 dBA. Note that these are general rules and that if the frequency content of the noise is centered in the 1000- to 5000-Hz region where the ear is most sensitive, the limits are slightly lower. Practically speaking, for control valves, these limits have been redefined, so in most cases the noise from control valves is limited to 85 dBA if the valve is to be used in an area where personnel exposure occurs on a regular basis.

With regard to the equipment itself, studies by a number of manufacturers have indicated that prolonged operation at very high noise levels will result in physical damage due to vibration that is normally associated with high noise levels. A good rule of thumb is to keep noise levels below 110 dBA to ensure that trim damage will not result.

Knowing that noise can be detrimental to personnel and equipment, we need to know how it can be controlled. To control it, we need to know how it is produced. There are three possible sources for noise in a control valve: mechanical vibration, hydrodynamic noise, and aerodynamic noise. *Mechanical vibration* is the relative motion of internal or external parts caused by the fluid flowing through the valve. While mechanical vibration can provide clues as to the physical deterioration of the valve parts, the noise generated by the vibration is usually low in frequency and intensity, so it does not pose an exposure problem for personnel in the area. Its effect can be minimized through the tight control of clearances between the moving pieces and by emphasizing large bearing areas in valve design. It is for this reason that cage-guided sliding-stem valves are popular in severe service, since they feature very effective bearing designs. Vibration will be covered again in Part 3, but for now it can be discounted as a source of significant noise problems.

Hydrodynamic noise is generated in liquid flows. It is primarily due to the formation and/or collapse of vapor bubbles in the fluid. The formation of the bubbles is called *flashing,* and the formation and subsequent collapse of the bubbles is referred to as *cavitation.* Neither of these phenomena produce noise in the "dangerous" range but can result in significant damage to the trim parts, nevertheless. With this potential for damage in mind, the cause for both flashing and cavitation, along with potential solutions, will be covered in more detail later in this chapter.

Aerodynamic noise is the primary source that one has to worry about when working with control valves. It is generated by the turbulent expansion or compression of gases and results from the shear forces created as the gas hits obstructions in the flowstream, decelerates, expands, or changes direction. The generation of noise in a control valve is concentrated in the recovery region immediately downstream of the vena contracta (the vena contracta is the area in the flowstream where the flow cross section is at a minimum). Noise levels generated depend on the amount of flow, the ratio of pressure drop to the inlet pressure (pressure ratio), and the pressure drop for a valve. Given the right conditions, flow through the valve can result in noise levels of greater than 120 dBA, which is like standing next to a 747 while it's taking off. Now, assuming a noise problem exists, how can it be corrected?

There are two basic approaches to reduction of aerodynamic noise: *source treatment* and *path treatment.* Source treatment is an attempt to limit the generation of noise at its source, while path treatment tries to attenuate the level of noise after it's produced. In general, source treatment is considered to be more effective and will be discussed first.

Source treatment can take many forms. Since the level of noise generated depends on such things as pressure drop, the amount of flow, and the pressure ratio, any change in these properties can help to reduce the noise levels. It's not always possible, but don't neglect the possibility of changing the service conditions or system layout so that one of these critical properties is reduced, thereby reducing the noise. If the application will not permit these changes, the next step in source treatment is to look at the trim inside the valve. Source treatment at the trim level can be done many different ways and each vendor has its own approach to addressing this problem. However, there are some common threads. As far as trim design is concerned, nearly all the designs try to break the flowstream into multiple paths as a first approach. As the need for noise reduction increases, the second step involves taking the pressure drop in multiple steps, which also tends to reduce the velocity of the fluid passing through the valve. Examples of several noise-reduction trims for sliding-stem valves are shown in **Figs. 6.1** through **6.4.**

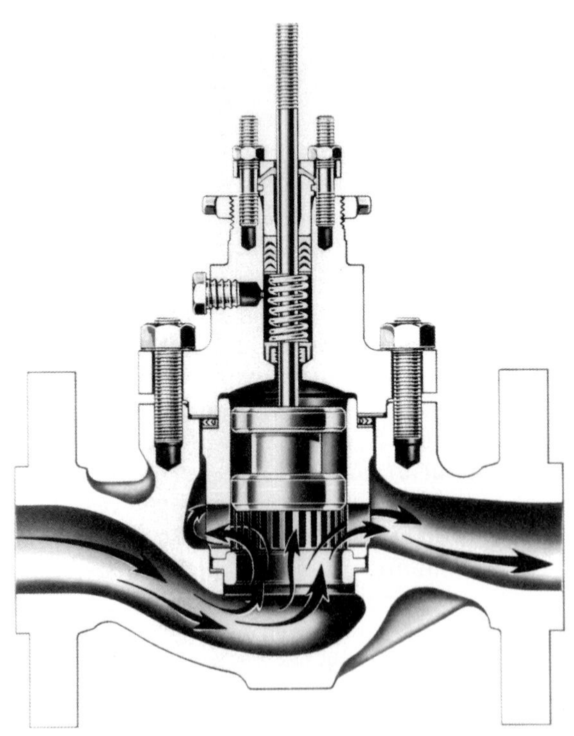

Figure 6.1 Single-stage, multipath, noise reduction. (*Courtesy of Fisher Controls International, Inc., Marshalltown, Iowa.*)

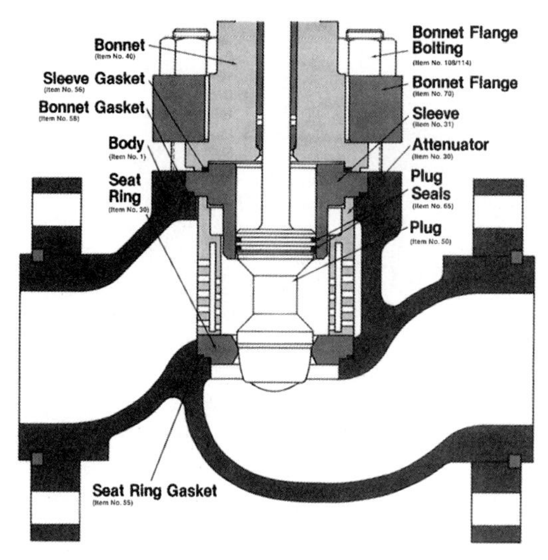

Figure 6.2 Multistage, Multipath, noise reduction. (*Courtesy of Valtek International, Inc., Springville, Utah.*)

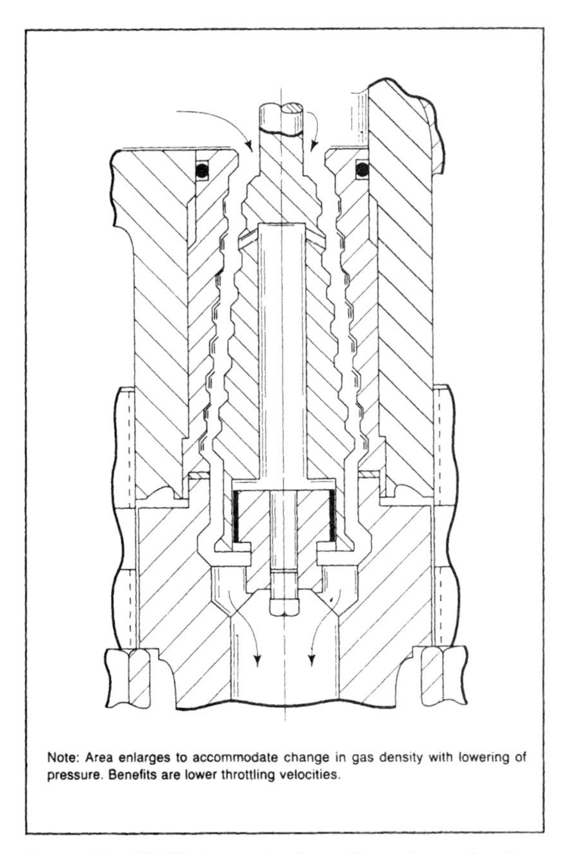

Figure 6.3 Multistage, single-path, noise reduction. (*Courtesy of Masoneilan-Dresser, Inc., Houston, Tex.*)

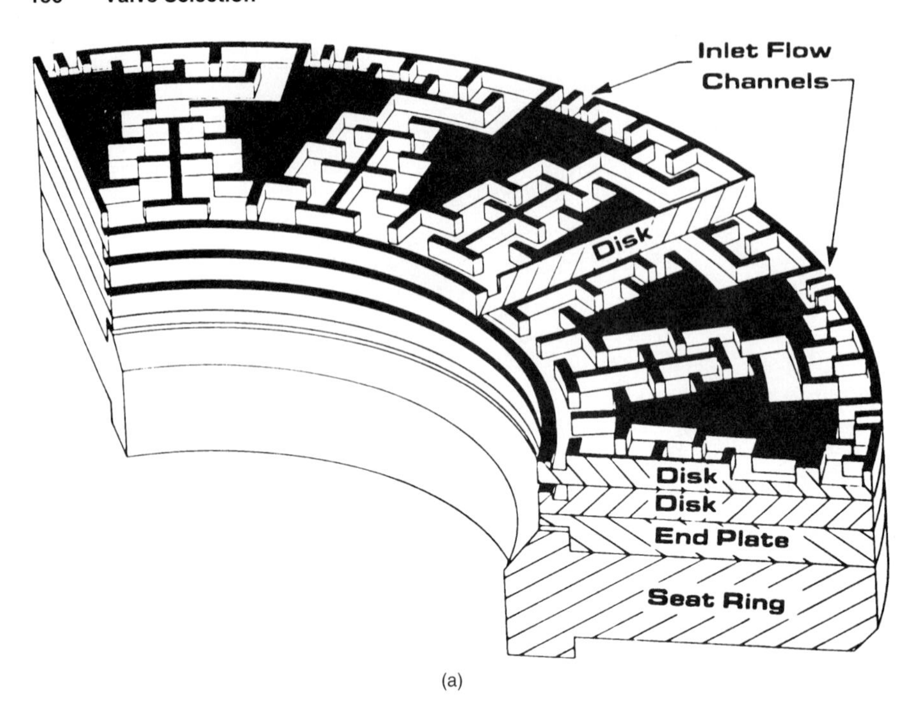

(a)

(b)

Figure 6.4 Multistage, multipath, stacked plate, noise reduction. (*Courtesy of Control Components, Inc., Rancho Sante Marguerita, Calif.*)

A number of observations can be made about these figures. Note that in breaking the flow into several paths to reduce noise, the cross section of each individual path is smaller, which can lead to plugging of the holes if the fluid has any entrained solids. Plugging is one of the big drawbacks in most antinoise trims and needs to be addressed by thoroughly flushing the lines before trim installation or by the judicious use of upstream filters that can be routinely cleaned. Plugging reduces the capacity of the trim and also tends to cause sticking and jumping because the debris tends to get trapped between the plug and the cage. If plugging is projected to be a problem, the trims shown in **Figs. 6.1** and **6.3** are generally a better choice because of their larger flow paths. However, they do not provide the same levels of noise attenuation as the designs with more flow paths (smaller holes).

As far as the *rotary designs* are concerned (**Figs. 6.5, 6.6,** and **6.7**), they generally are not as effective as the sliding-stem trims, with possible attenuation limited to about 10-dBA range. They will also be more prone to damage due to vibration because of the inherent limitations in bearing surfaces and guiding. For the fluted butterfly design shown in **Fig. 6.5,** one side benefit is a significant reduction in dynamic torque on the disc.

It should be noted that most antinoise trims are designed to flow up over the plug and out through the cage. This approach limits the interaction of the individual jets, which helps to keep the noise low. However, on an unbalanced valve this means bigger actuators

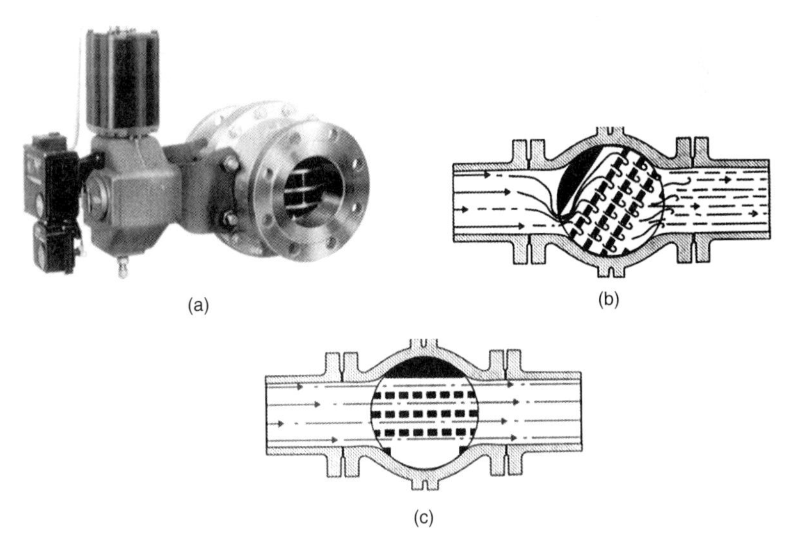

Figure 6.5 Rotary noise attenuation—full ball valve. (*Courtesy of Neles-Jamesbury, Inc., Worcester, Mass.*)

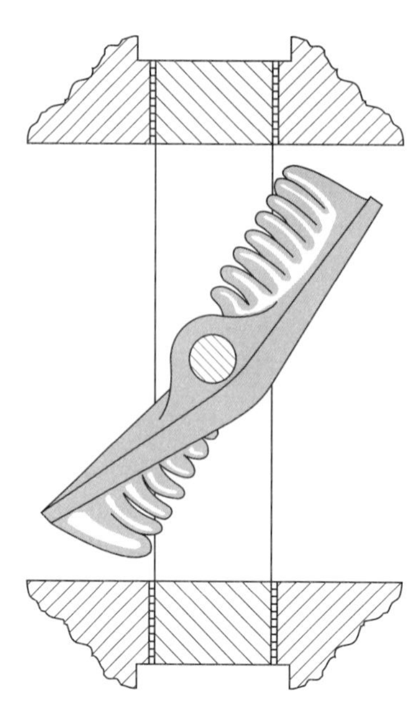

Figure 6.6 Rotary noise attenuation—butterfly valve. (*Courtesy of Valtek International, Inc., Springville, Utah.*)

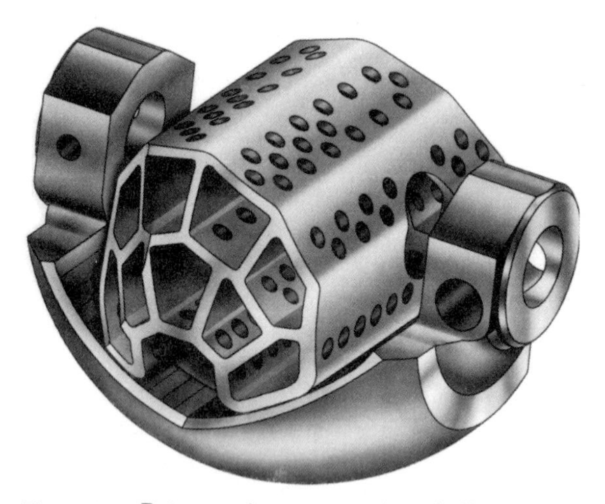

Figure 6.7 Rotary noise attenuation—ball segment valve. (*Courtesy of Fisher Controls International, Inc., Marshall-town, Iowa.*)

because the flow force tends to push the plug off the seat, and on a balanced design this can cause instability and jumping near the seat since the net force with the plug on the seat is pushing down, but this disappears very quickly when the valve is opened. See Sec. 2.6 for a more detailed description of this phenomena. An option that can help mitigate the consequences of this instability near the seat is to move the openings away from the seat so that throttling will not occur as often in this range of travel.

Another feature to look for in antinoise trim is an expanding flow path, which is best illustrated by the trim in **Fig. 6.3.** Without going into the theory in depth, the pressure of the fluid is dropping as it flows through the trim. As the pressure drops, the fluid expands, and unless the flowing cross-sectional area is increased, the velocity of the fluid will increase, producing more noise.

One precaution that should be taken when using antinoise trim is that the *outlet velocity* needs to be within certain limits if the full benefits of the trim design are to be realized. Each different type of trim has its own set of limits that need to be respected, which can sometimes mean that the solution to a noise problem might be as simple as using an expanded length of pipe on the outlet to limit velocity. In some cases, the internal valve cavity is oversized to address this concern.

Capacity is also a consideration when using noise-control trim. Both pressure-staging and multipath approaches tend to reduce the potential flow through the trim; therefore take a close look at the required capacity before selecting one of the options. Many valve vendors alleviate the inherent capacity problems by using special extended stroke designs. The plugging problems mentioned earlier tend to make this capacity problem even worse.

The final comment on antinoise trim relates to drilled-hole trim and the shape of the holes. Most antinoise trims utilize a straight-through hole because it's easier to manufacture. However, utilizing a stepped hole, in effect, provides another pressure-reducing stage that further improves noise abatement, and it has been found that the shape and depth of the "step" has a profound effect on the shape of the resultant fluid jet and, if optimized, can further improve antinoise performance.

If the utilization of antinoise trim in the valve itself does not provide sufficient noise attenuation, a device that takes a portion of the pressure drop and reduces the noise generated in the valve can be added in line with the valve. These devices can take several forms. One option is to add a "basket" in the bottom of the valve, as shown in **Fig. 6.8.** A more traditional approach is to insert an in-line diffuser downstream of the valve, as shown in **Fig. 6.9.** Ideally, the pressure drop should be shared between the valve and the diffuser so that the noise generated by the two devices is essentially the same. In combining the two sources into one noise level at any particular point, we can consider the valve to be a point source and the pipe around the diffuser to be a line source.

Because the sound intensity drops very rapidly as the distance from the source is increased, one possible solution to personnel exposure concerns would be to restrict access inside the zone where the SPL

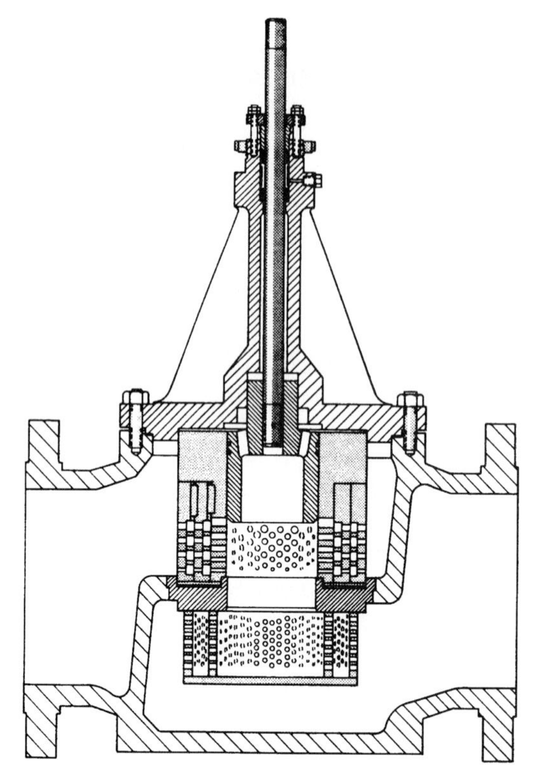

Figure 6.8 Integral diffuser "basket." (*Courtesy of Kent-Introl, Ltd., London, Eng.*)

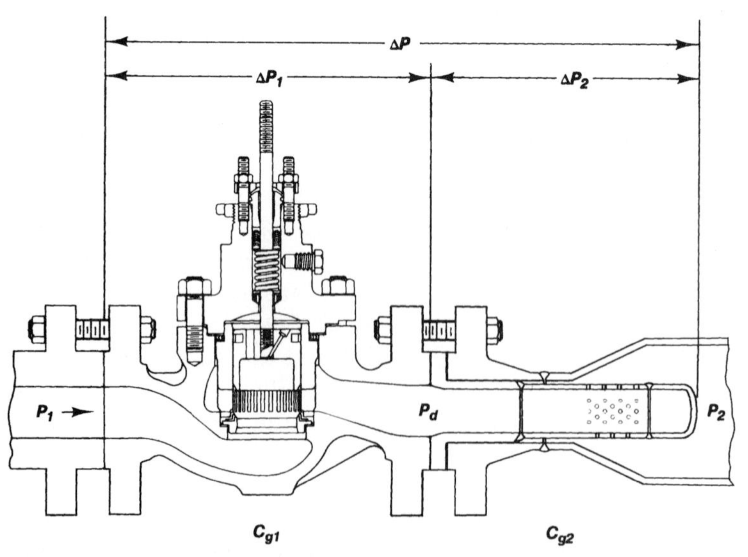

Figure 6.9 Control valve and in-line diffuser. (*Courtesy of Fisher Controls International, Inc., Marshalltown, Iowa.*)

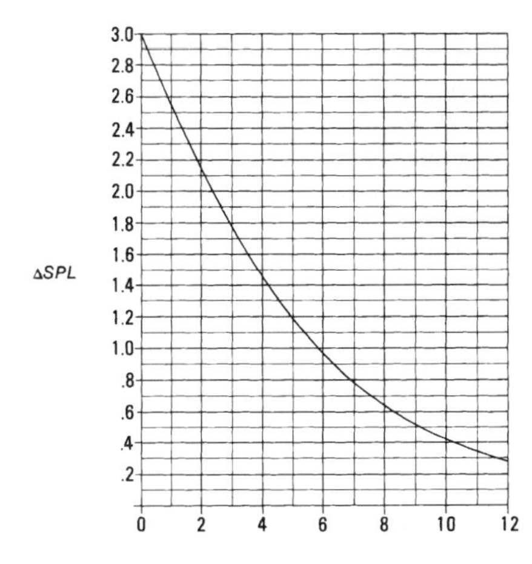

Figure 6.10 Noise source combination curve. (*Courtesy of Fisher Controls International, Inc., Marshalltown, Iowa.*)

limit is exceeded. Be aware that other sound sources can add to the total sound level at any given point; therefore, be careful to include all sources when setting up restricted zones.

When looking at how sound sources combine, there is a simplified method of adding two sources together. Use **Fig. 6.10** to calculate the noise level of each source at the desired point. Take the difference between the two values and enter this difference on the horizontal axis of **Fig. 6.10.** Determine the vertical intercept of this point and add this value to the larger of the two values found for the individual sources at the given point. In many cases, the proper design of a valve/diffuser combination can result in a noise attenuation of 25 dBA or more.

A cheaper, simpler version of the diffuser principle, called a flat plate diffuser, is illustrated in **Fig. 6.11.** The downside with this approach is that the capacity of the diffuser plate may be limited to the point that the pressure drop cannot be properly shared between the valve and the diffuser, and as a result, the noise levels might be higher than with a traditional diffuser design. Note that all diffusers are fixed restrictions whose effectiveness varies with the throttling position of the valve. They should be sized taking into account the condition that results in the most noise and the condition at which the valve will spend most of its time. They are not a good choice for effective noise control over a broad range of flows.

Path treatment consists of either limiting the sound transmitted to the environment or actually attenuating the sound after it has been generated. Limiting transmission can be accomplished by using *heavier schedule pipe* at the valve outlet or installing acoustical insulation on the outside of the pipe wall. Various tables exist for the effects of

Figure 6.11 Flat plate diffuser. (*Courtesy of Fisher Controls International, Inc., Marshalltown, Iowa.*)

pipe wall thickness, and any valve vendor can provide an estimate of the attenuation due to this effect. Be aware that increasing pipe thickness is an expensive way of increasing attenuation and that it has only a local effect. Once the pipe schedule returns to normal, the noise level returns as well. *Acoustical insulation* works in much the same way in that it is very effective where it is applied, but the noise level on exposed, downstream pipe is affected very little. Typical local reductions in sound levels due to pipe schedule and insulation are 5 and 20 dBA, respectively.

In-line silencers are like large mufflers (**Fig. 6.12**) and tend to be more effective in path treatment because they actually reduce the noise level, not just the noise transmitted to the environment. A comparison of the various path treatment methods is illustrated in **Fig. 6.13.**

One final note on noise prediction and treatment. Be aware that noise predictions are just estimates and that typical accuracy is ± 5 dBA. In addition, vendors sometimes stretch the rules somewhat in claims regarding attenuation for a particular product. Ask for documented test results that support the vendor's claims rather than simply taking them at face value.

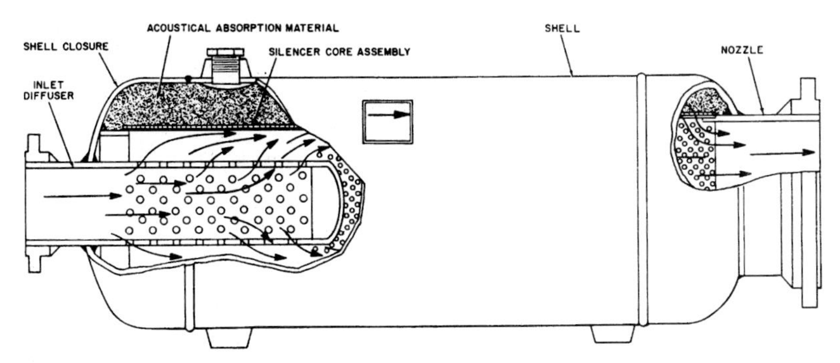

Figure 6.12 Typical silencer. *(Courtesy of Fisher Controls International, Inc., Marshalltown, Iowa.)*

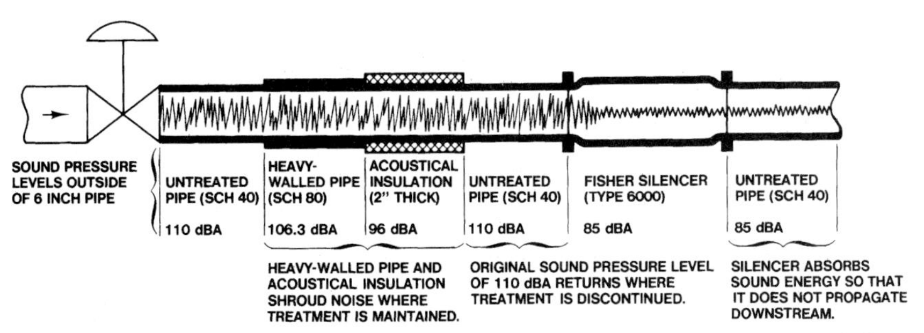

Figure 6.13 Path treatment methods. *(Courtesy of Fisher Controls International, Inc., Marshalltown, Iowa.)*

6.1.2 Cavitation and flashing

Flashing and cavitation are examined together here because flashing is actually the first stage of cavitation. Flashing occurs in liquid flows and is defined as the change in state from a liquid to a vapor. It occurs because of the relationship between the pressure in a fluid and its velocity (Bernoulli's equation), which indicates that as the velocity of the fluid changes, the pressure of the fluid varies inversely. If we treat a control valve as a restriction in the flow path where the cross-sectional area of flow decreases, given that the flow rate has to remain constant through the valve, we know that the velocity of the fluid has to increase as it goes through the restriction. As the velocity increases, the pressure has to decrease, and if the pressure falls below the vapor pressure for the fluid at the given temperature, a portion of the fluid will begin to "boil," changing into a vapor. This boiling is called flashing, and the potential for it depends on the vapor pressure of the fluid in relation to the inlet pressure at the valve and on the velocity of the fluid inside the valve.

Flashing causes two problems: reduced capacity and erosion of the trim and body. The capacity problem stems from the fact that as the bubbles form in the fluid, they take up more space than the equivalent liquid. Holding P_1 constant and increasing the pressure drop across the valve, increases the velocity of the fluid and more bubbles form. What this means is that the normal flow equation for the valve that says that the flow increases with the square root of the pressure drop no longer applies, and we gradually move toward a condition where no matter how much the pressure drop is increased (assuming constant P_1), the flow through the valve remains the same. This is called *choked flow* and is addressed in more detail in Chap. 8. Choked flow does not occur at the same conditions for different types of valves or even for different travels for the same valve. It varies with flow geometry and has to be determined experimentally for each valve and for each point in travel. To check for choked flow, a test is run where the valve is held at a given point in travel and the pressure drop is gradually increased using water as the test medium. As the pressure drop increases, a point is reached where the flow no longer varies directly with the square root of the pressure drop (**Fig. 6.14**). The point at which fully choked flow is reached is defined by the dimensionless factor K_m, called the valve recovery coefficient. It is published by most valve manufacturers for each valve at 10 percent intervals of stroke and is normally lowest at full travel.

Erosion of the valve parts occurs with flashing because of the two-phase flow that is present. As the degree of flashing increases, portions of the flowstream end up being more vapor than liquid, and the remaining liquid is carried along in the vapor at relatively high velocities. These high-velocity liquid droplets can act like solid particles if they impinge on the internal parts of the valve and tend to erode them, leaving their surfaces with a very shiny, pitted appearance. Increasing the size of the valve outlet will reduce the velocity of the

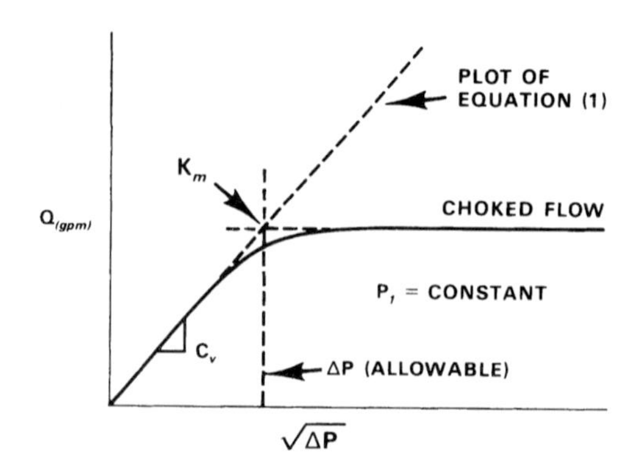

Figure 6.14 K_M defined. (*Courtesy of Fisher Controls International, Inc., Marshalltown, Iowa.*)

fluid and help to reduce damage. Hardened trim materials and high-chrome body materials have also shown good resistance to flashing conditions. Another potential solution is to keep the flashing from occurring near the expensive trim parts by using an angle valve that discharges straight into the pipeline or by dumping the outlet directly into a vessel.

Cavitation occurs for the same reasons as flashing, but a second step is necessary. Like flashing, fluid velocity increases as it passes through the valve, causing a drop in pressure and bubble formation, but in cavitation the pressure recovers sufficiently so that the bubbles formed collapse upon themselves. A schematic of flow through the valve showing pressure is illustrated in **Fig. 6.15.** Flashing occurs as the pressure drops below the vapor pressure just downstream of the vena contracta, and then the bubbles implode when the pressure recovers above the vapor pressure. It is this bubble implosion that causes the serious damage seen with cavitation. In fact, cavitation is without a doubt the most severe of all the services that a control valve can be exposed to. The mechanics of how the damage is caused are still being debated, but from a practical standpoint all you need to know is that when it occurs, it is extremely destructive. An example of the potential damage is shown in **Fig. 6.16.** Cavitation is noisy and is often described as sounding as if gravel were flowing through the valve. However, as mentioned earlier, the noise generated is not the primary concern. It is usually of low intensity and low frequency, so it does not pose a problem for personnel.

The big concern is trim and body damage. If allowed to continue unchecked, it can severely damage valve parts in a very short period of time. Utilization of very hard, resilient materials like Stellite,

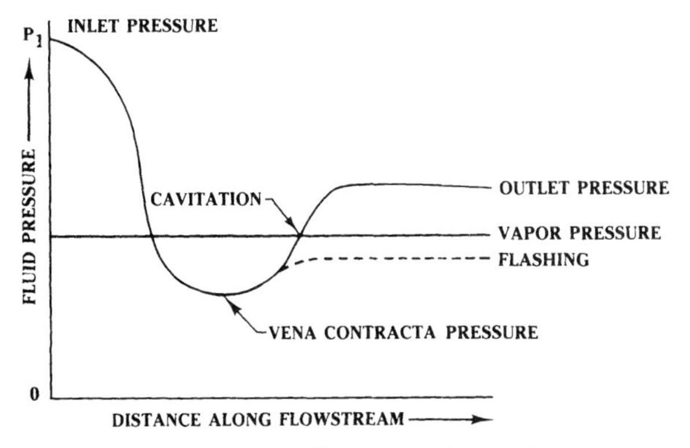

Figure 6.15 Pressure profile in a valve with cavitation. (*Courtesy of Fisher Controls International, Inc., Marshalltown, Iowa.*)

Figure 6.16 Typical appearance of cavitation damage. (*Courtesy of Fisher Controls International, Inc., Marshalltown, Iowa.*)

440C, Colomonoy, and 17-4PH will buy some time, but the only real solution is to either prevent it from happening or to make sure it happens in an area of the flowstream that is not in direct contact with the internal parts.

Before going into too much detail on possible solutions, it needs to be pointed out that there are different degrees or levels of cavitation and that the solution to be selected depends on just how intense the cavitation really is. Over time, a number of general guidelines have been developed that will help you understand the risks associated with cavitation and how best to avoid cavitation-related damage. It should be noted that these guidelines are broad in nature and that exceptions will surface from time to time.

1. Although cavitation can and will occur in any compressible fluid, experience has shown that significant damage is usually only associated with water. Even though cavitation may be theoretically occurring with another fluid based on the flowing conditions and vapor pressure, the damage will not normally be of concern and no corrective action is normally required.

2. Cavitation damage varies directly with the velocity of the fluid through the valve, flow capacity, pressure drop, and difference between the downstream pressure and the vena contracta pressure. Anything that can be done to reduce any of these characteristics will be of benefit.

3. Below certain pressure drops, cavitation, even though it occurs, does not result in measurable damage. These limits depend on valve type and flow capacity and are lower for rotary valves than for sliding-stem valves. A typical limit for a 4-in globe body would be a pressure drop of 200 psi.

4. An easy way of characterizing the potential for damage for a set of service conditions is the use of a factor called the *application ratio,* defined as $(P_1 - P_2)/(P_1 - P_v)$ where P_v is the vapor pressure, and P_1 and P_2 are the upstream and downstream pressures, respectively. This ratio must be less than 1 or the service is flashing, not cavitation. The closer the ratio gets to 1, the higher the potential for damage.

5. The other factor that is critical in looking at damage potential is the *recovery characteristic* for the valve employed. The recovery characteristic is a measure of how much the pressure "recovers" at the exit of the valve. It is illustrated in **Fig. 6.17** where the pressure profiles for high- and low-recovery valves are shown. In reviewing these profiles, it is apparent that the high-recovery valve will be more prone to cavitate since its downstream pressure

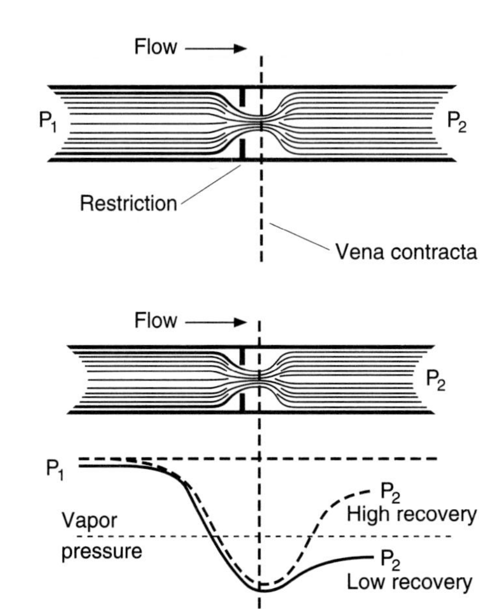

Figure 6.17 Valve recovery characteristics. (*Courtesy of Fisher Controls International, Inc., Marshalltown, Iowa.*)

is more likely to exceed the vapor pressure, which is the differentiating characteristic between flashing and cavitation. Rotary valves, in general, are inherently high-recovery values because of their line-of-sight flow path. It is for this reason that they are not usually recommended for applications involving the potential for cavitation. There are some staged-pressure-drop trims available now for rotary valves that have been shown to be effective as long as the pressure drops are not too high. Examples are shown in **Figs. 6.5, 6.6,** and **6.7.**

Recovery characteristics have been quantified through the use of the K_m factor defined above. K_m indicates the conditions under which the flow is choked, and if we have choked flow and the application ratio is less than 1, full-fledged cavitation will occur. It is important to note that significant levels of cavitation can and will occur before flow is fully choked. It is for this reason that a second term is used called K_c that is usually less than K_m and is a more accurate measure of when damage-producing cavitation will begin to occur. Like K_m, it is determined experimentally.

6. In summary, to select a valve for cavitation service:

   Determine the pressure drop.

   Calculate the application ratio. If it is equal to or greater than 1, the service is flashing, and cavitation is not an issue.

   Pick a valve that can handle the pressure drop and that has a K_c higher than the application ratio.

With the above as background, we can now look at various approaches taken to improve the cavitation resistance of valve designs. One method employed is to keep it away from the metal parts, as shown in **Fig. 6.18.** Like antinoise trim, drilled-hole technology is used to separate the flow into numerous paths, but in this case the flow is outside-in, so the flow from opposing holes meets in the middle of the cage where pressure recovery occurs. Since it is the pressure recovery that results in the damaging phase of cavitation, the bubble implosion takes place in the middle of the cage and, at least in theory, does not result in trim damage. This is a cost-effective approach for many applications but is limited to pressure drops of up to 600 psi. It should also be noted that it controls cavitation rather than eliminating it, so the noise and vibration can still be present.

True prevention of cavitation is achieved by either reducing the amount of fluid that flashes or by preventing the pressure from recovering to above the vapor pressure. The standard approach here is to separate the pressure drop into multiple stages. To understand why

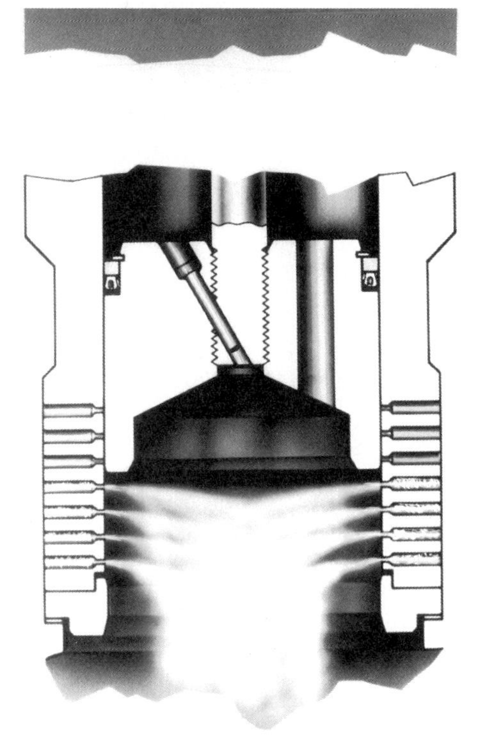

Figure 6.18 Controlling cavitation by keeping it away from the trim. (*Courtesy of Fisher Controls International, Inc., Marshalltown, Iowa.*)

this is effective, let's return to the pressure profile for flow through a valve (**Fig. 6.14**). Cavitation is caused when the "dip" in pressure at the vena contracta drops below the vapor pressure and then recovers to above it. The farther the pressure drops below the vapor pressure, the more intense the cavitation. It has been determined that the larger the pressure drop is for a given restriction, the farther the vena contracta pressure will drop below the downstream pressure. It follows that if the pressure drop is reduced, the vena contracta "dip" will also be reduced along with the intensity of the cavitation. In practice, this can be achieved by using several valves in series or by using a breakdown orifice in conjunction with the valve. The orifice approach is fine if the flowing conditions don't change too drastically. If they do, the orifice will only be effective at one set of conditions since it is a fixed restriction. The multiple valve approach also works but is an expensive way of addressing the problem.

What most valve manufacturers have done is to package multiple stages within one valve. The staging is usually accomplished in the cage, as illustrated in **Fig. 6.19,** or by using a stepped plug, as shown in **Fig. 6.20.** The principle behind multiple stages is illustrated in

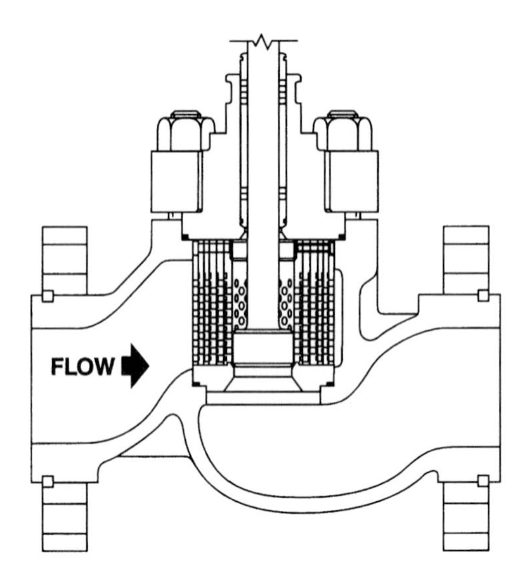

Figure 6.19 Anticavitation trim-staging in cage. (*Courtesy of Valtek International, Inc., Springville, Utah.*)

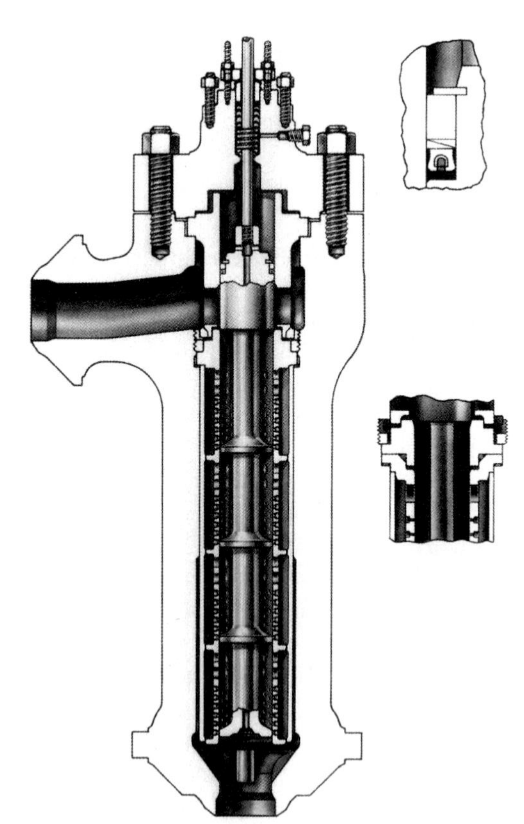

Figure 6.20 Anticavitation trim-stepped plug. (*Courtesy of Fisher Controls International, Inc., Marshalltown, Iowa.*)

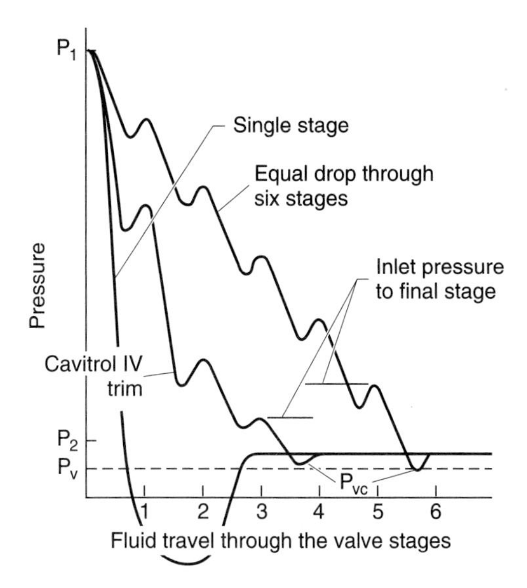

Figure 6.21 Characterized pressure staging. (*Courtesy of Fisher Controls International, Inc., Marshalltown, Iowa.*)

Fig. 6.21, where it can be seen that for identical overall pressure drops, the vena contracta pressure is much lower for a single stage than for a multiple stage. This figure also illustrates another important feature referred to as *characterizing*. If we size the various stages so that the largest pressure drop occurs in the earlier stages, we can further improve the performance of the package since it is the later stages that begin to approach the vapor pressure and determine the degree of cavitation. This also tends to protect the seat in the valve since, in most constructions, it is located near the last stage.

6.1.3 Erosion and corrosion

These two phenomena are probably the most common causes for valve failure in the chemical process industry. They are treated together here because they are often related. Erosion action can wear away the protective coatings that some materials count on for their corrosion protection. Once these coatings are depleted, corrosion occurs at a very rapid rate until failure occurs.

Looking at *erosion* first, it is defined as the wearing away of material and can have a number of causes, including solid particle impingement, cavitation, and high-velocity liquid impingement. Damage can occur on the trim parts, resulting in poor guiding or shutoff, and it can also affect the body wall, eventually resulting in through-body leaks that can be very dangerous for personnel in the area. Given the serious consequences and the relatively high incidence of failure, erosion problems have received a lot of attention from both end users and manufacturers. As a result, a number of solutions are available, depending on the cause of the erosion.

Looking at solid particle impingement first, the potential for damage depends on the velocity of the particles, the concentration of the particles in the flowstream, the angle of attack on the metal parts, and the hardness of the particle in relation to the valve parts. A common misconception is that the particles must be harder than the metal valve parts for damage to occur. This is not true. The rate and degree of damage *are* a function of particle hardness, but damage can and will occur even when the particles are softer than the parts which they affect.

Given these damage factors, it follows that one way to reduce damage is to reduce the *fluid velocity.* This can be done by staging the pressure drop as described in Sec. 6.1.2 or by increasing the flow cross section inside the valve. When adopting the staging approach, be aware that the staging design should be selected with the particle size in mind to avoid plugging. Also, the higher the concentration of particles, the greater the damage. Anything that can be done to reduce the concentration will be beneficial. Unfortunately, this is normally defined by the application and is not something you can affect.

Proper design of the flowstream to limit the direct impingement of particles is a very effective technique. This is called *streamlining,* and one of the features of the approach is to make sure that the flow path runs parallel to critical surfaces such as the seat, plug, and body wall. Angle valves flowing down are a good option here, and just recently, eccentric plug valves have been found to be very effective in controlling erosion due to their use of massive trim pieces and naturally streamlined flow path (**Fig. 6.22**). In general, the valve should be designed so that erosion, if it does take place, occurs on parts that are easily replaceable rather than on the valve body.

One of the common techniques employed for erosive service is to select hardened materials for the valve parts. Using hardened material overlays in the guiding and seating areas is called *hardfacing* and commonly employs a grade of Stellite with very high hardness, although Colmonoy is also becoming a popular choice for this. If solid parts are employed rather than an overlay, the choices increase and include 17-4PH, Stellite, Colmonoy, 440C, tungsten carbide, and ceramics, listed in increasing order of performance, with ceramics being the best. The choice depends on the severity of the problem, cost, and material compatibility with the fluid. Ceramics, in particular, have improved in recent years to the point where they can provide the best combination of hardness and resilience if they can be fabricated in the shape required. Note that even with the hardest materials, some erosion will always occur. Material selection is essen-

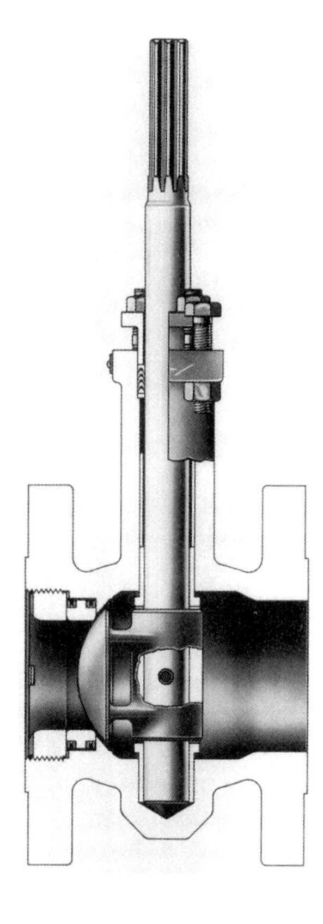

Figure 6.22 Eroision-resistant trim. (*Courtesy of Fisher Controls International, Inc., Marshalltown, Iowa.*)

tially a delaying tactic, but the right choice can result in a service life of several years.

As far as the body is concerned, stainless and chrome-moly steels resist better than carbon steels. In some cases, it may be cheaper or easier to determine the areas of damage in a carbon steel body after some time in service and then repair the area by overlaying it with a more resistant material.

Cavitation, another cause of erosion, was already covered in a preceding section. Prevention is the best bet here because no material, no matter how hard, can resist cavitation for very long. However, if material selection is to be considered as a stopgap solution, the materials listed above offer the best protection.

High-velocity liquid impingement is the third and last cause of erosion to be covered here. There are two different types. The first involves two-phase flow where liquid droplets are being carried along in a gaseous flowstream. If velocities are high enough, the droplets can act like solid particles and erode the trim in much the same way. A flashing flaw is an example of this type of erosion.

The other type of liquid erosion is where a *liquid* jet is directed against a trim or body part. The velocities have to be very high for this to result in damage, and the damage is usually very localized when compared to the liquid droplets erosion just covered. One example of this type of damage is called *wire drawing,* and it occurs across the seat and plug when the valve is closed. Entrained solids or defects in the seating areas cause a microscopic leak where the fluid velocity is very high. If left uncorrected, it will develop into a leak path running across the face of the seating surface.

Although the cause of liquid impingement erosion is different from those previously discussed, the solutions are the same: streamlining, material selection, and velocity control through pressure staging.

One other solution that might be considered for any type of erosion is a boot valve. It has a very streamlined flowpath and stands up well in erosive service but is somewhat limited in terms of pressure. See the example shown in **Fig. 2.6.**

The other major problem with control valves in the chemical process industries is *corrosion.* This is because many different types of fluids may be handled and material selection with regard to compatibility can become a very complicated issue. Since all the possible options cannot be covered here, you should work with your local valve vendors to be sure that the fluid specified for the application is an accurate reflection of reality. Most vendors, if supplied with the right information, can come up with an acceptable option. As a general guide, see **Table 6.1,** which is a very handy table from Fisher Control's *Control Valve Handbook.* Of the exotic alloys listed in this table, the Hastelloys probably see the most use in the chemical industry, particularly for chlorine and acid service. In severely corrosive applications, the body, along with the trim can be made from a Hastelloy alloy.

Lined valves are also an option that shouldn't be forgotten for corrosive service. The boot valves already mentioned can provide a good combination of erosion/corrosion resistance. For severely corrosive applications, a PTFE-lined valve like that shown in **Fig. 6.23** can be the best choice. They are limited to about 350°F and have relatively poor erosion resistance. Also be aware that some fluids can permeate through the liner and bellows material, permitting the fluid to attack the base material or to leak to atmosphere.

One final note on corrosion. Many times its root cause is actually related to erosion. Many materials rely on a protective layer of surface corrosion to prevent damage to the substrate. If erosion is causing the continuous wearing away of the protective layer, the material underneath can be attacked and corrosion can continue unchecked until failure occurs. In this case, if the erosion problem is addressed, the corrosion problem is solved at the same time.

TABLE 6.1

Fluid	Material													
	Carbon steel	Cast iron	302 or 304 stain-less steel	316 stain-less steel	Bronze	Monel*	Hastellot† B	Hastelloy† C	Durimet‡ 20	Titanium	Cobalt-base alloy 6	416 stain less steel	440C Hard stain less steel	17-4PH hard stain-less steel
Acetaldehyde	A	A	A	A	A	A	I.L.	A	A	I.L.	I.L.	A	A	A
Acetic acid, air free	C	C	B	B	B	B	A	A	A	A	A	C	C	B
Acetic acid, aerated	C	C	A	A	A	A	A	A	A	A	A	C	C	B
Acetic acid vapors	C	C	A	A	B	B	I.L.	A	B	A	A	C	C	B
Acetone	A	A	A	A	A	A	A	A	A	A	A	A	A	A
Acetylene	A	A	A	A	I.L.	A	A	A	A	I.L.	A	A	A	A
Alcohols	A	A	A	A	A	A	A	A	A	A	A	A	A	A
Aluminum sulfate	C	C	A	A	B	B	A	A	A	A	I.L.	C	C	I.L.
Ammonia	A	A	A	A	C	A	A	A	A	A	A	A	A	I.L.
Ammonium chloride	C	C	B	B	B	B	A	A	A	A	B	C	C	I.L.
Ammonium nitrate	A	C	A	A	C	C	A	A	A	A	A	C	B	I.L.
Ammonium phosphate (mono-basic)	D	C	A	A	B	B	A	A	B	A	A	B	B	I.L.
Ammonium sulfate	C	C	B	A	B	A	A	A	A	A	A	C	C	I.L.
Ammonium sulfite	C	C	A	A	C	C	I.L.	A	A	A	A	B	B	I.L.
Aniline	C	C	A	A	C	B	A	A	A	A	A	C	C	I.L.
Asphalt	A	A	A	A	A	A	A	A	A	I.L.	A	A	A	A
Beer	B	B	A	A	B	A	A	A	A	A	A	B	B	A
Benzene (benzol)	A	A	A	A	A	A	A	A	A	A	A	A	A	A
Benzoic acid	C	C	A	A	A	A	I.L.	A	A	A	I.L.	A	A	A
Boric acid	C	C	A	A	A	A	A	A	A	A	A	B	B	I.L.
Butane	A	A	A	A	A	A	A	A	A	I.L.	A	A	A	A
Calcium chloride (alkaline)	B	B	C	B	C	A	A	A	A	A	I.L.	C	C	I.L.
Calcium hypochlorite	C	C	B	B	B	B	C	A	A	A	I.L.	C	C	I.L.
Carbolic acid	B	B	A	A	A	A	A	A	A	A	A	I.L.	I.L.	I.L.
Carbon dioxide, dry	A	A	A	A	A	A	A	A	A	A	A	A	A	A
Carbon dioxide, wet	C	C	A	A	B	A	A	A	A	A	A	A	A	A
Carbon disulfide	A	A	A	A	C	B	A	A	A	A	A	B	B	I.L.
Carbon tetrachloride	B	B	B	B	A	A	B	A	A	A	I.L .	C	A	I.L.
Carbonic acid	C	C	B	B	B	A	A	A	A	I.L.	I.L.	A	A	A
Chlorine gas, dry	A	A	B	B	B	A	A	A	A	C	B	C	C	C

TABLE 6.1 (*Continued*)

Fluid	Material													
	Carbon steel	Cast iron	302 or 304 stain-less steel	316 stain-less steel	Bronze	Monel*	Hastellot† B	Hastelloy† C	Durimet‡ 20	Titanium	Cobalt-base alloy 6	416 stain less steel	440C Hard stain less steel	17-4PH hard stain-less steel
Chlorine gas, wet	C	C	C	C	C	C	C	B	C	A	B	C	C	C
Chlorine, liquid	C	C	C	C	B	C	C	A	B	C	B	C	C	C
Chromic acid	C	C	C	B	C	A	C	A	C	A	B	C	C	C
Citric acid	I.L.	C	B	A	A	B	A	A	A	A	I.L.	B	B	B
Coke oven gas	A	A	A	A	B	B	A	A	A	A	A	A	A	A
Copper sulfate	C	C	B	B	B	C	I.L.	A	A	A	I.L.	A	A	A
Cottonseed oil	A	A	A	A	A	A	A	A	A	A	A	A	A	A
Creosote	A	A	A	A	C	A	A	A	A	I.L.	A	A	A	A
Ethane	A	A	A	A	A	A	A	A	A	A	A	A	A	A
Ether	B	B	A	A	A	A	A	A	A	A	A	A	A	A
Ethyl chloride	C	C	A	A	A	A	A	A	A	A	A	B	B	I.L.
Ethylene	A	A	A	A	A	A	A	A	A	A	A	A	A	A
Ethylene glycol	A	A	A	A	A	A	I.L.	I.L.	A	I.L.	A	A	A	A
Ferric chloride	C	C	C	C	C	C	C	B	C	A	B	C	C	I.L.
Formaldehyde	B	B	A	A	A	A	A	A	A	A	A	A	A	A
Formic acid	I.L.	C	B	B	A	A	A	A	A	C	B	C	C	B
Freon, wet	B	B	B	A	A	A	A	A	A	A	A	I.L.	I.L.	I.L.
Freon, dry	B	B	A	A	A	A	A	A	A	A	A	I.L.	I.L.	I.L.
Furfural	A	A	A	A	A	A	A	A	A	A	A	B	B	I.L.
Gasoline, refined	A	A	A	A	A	A	A	A	A	A	A	A	A	A
Glucose	A	A	A	A	A	A	A	A	A	A	A	A	A	A
Hydrochloric acid (aerated)	C	C	C	C	C	C	A	B	C	C	B	C	C	C
Hydrochloric acid (air free)	C	C	C	C	C	C	A	B	C	C	B	C	C	C
Hydrofluoric acid (aerated)	B	C	C	B	C	C	A	A	B	C	B	C	C	C
Hydrofluoric acid (air free)	A	C	C	B	C	A	A	A	B	C	I.L.	C	C	I.L.
Hydrogen	A	A	A	A	A	A	A	A	A	A	A	A	A	A
Hydrogen peroxide	I.L.	A	A	A	C	A	B	B	A	A	I.L.	B	B	I.L.
Hydrogen sulfide, liquid	C	C	A	A	C	C	A	A	B	A	A	C	C	I.L.
Magnesium hydroxide	A	A	A	A	B	A	A	A	A	A	A	A	A	I.L.
Mercury	A	A	A	A	C	B	A	A	A	A	A	A	A	B

TABLE 6.1 (*Continued*)

Fluid	Material													
	Carbon steel	Cast iron	302 or 304 stain-less steel	316 stain-less steel	Bronze	Monel*	Hastellot† B	Hastelloy† C	Durimet‡ 20	Titanium	Cobalt-base alloy 6	416 stain less steel	440C Hard stain less steel	17-4PH hard stain-less steel
Methanol	A	A	A	A	A	A	A	A	A	A	A	A	B	A
Methyl ethyl ketone	A	A	A	A	A	A	A	A	A	I.L.	A	A	A	A
Milk	C	C	A	A	A	A	A	A	A	A	A	C	C	C
Natural gas	A	A	A	A	A	A	A	A	A	A	A	A	A	A
Nitric acid	C	C	A	B	C	C	C	B	A	A	C	C	C	B
Oleic acid	C	C	A	A	B	A	A	A	A	A	A	A	A	I.L.
Oxalic acid	C	C	B	B	B	B	A	A	A	B	B	B	B	I.L.
Oxygen	A	A	A	A	A	A	A	A	A	A	A	A	A	A
Petroleum oils, refined	A	A	A	A	A	A	A	A	A	A	A	A	A	A
Phosphoric acid (aerated)	C	C	A	A	C	C	A	A	A	B	A	C	C	I.L.
Phosphoric acid (air free)	C	C	A	A	C	B	A	A	A	B	A	C	C	I.L.
Phosphoric acid vapors	C	C	B	B	C	C	A	I.L.	A	B	C	C	C	I.L.
Picric acid	C	C	A	A	C	C	A	A	A	I.L.	I.L.	B	B	I.L
Potassium chloride	B	B	A	A	B	B	A	A	A	A	I.L.	C	C	I.L.
Potassium hydroxide	B	B	A	A	B	A	A	A	A	A	I.L.	B	B	I.L.
Propane	A	A	A	A	A	A	A	A	A	A	A	A	A	A
Rosin	B	B	A	A	A	A	A	A	A	I.L.	A	A	A	A
Silver nitrate	C	C	A	A	C	C	A	A	A	A	B	B	B	I.L.
Sodium acetate	A	A	B	A	A	A	A	A	A	A	A	A	A	A
Sodium carbonate	A	A	A	A	A	A	A	A	A	A	A	B	B	A
Sodium chloride	C	C	B	B	A	A	A	A	A	A	A	B	B	B
Sodium chromate	A	A	A	A	A	A	A	A	A	A	A	A	A	A
Sodium hydroxide	A	A	A	A	C	A	A	A	A	A	A	B	B	A
Sodium hypochloride	C	C	C	C	B-C	B-C	C	A	B	A	I.L.	C	C	I.L.
Sodium thiosulfate	C	C	A	A	C	C	A	A	A	A	I.L.	B	B	I.L.
Stannous chloride	B	B	C	A	C	B	A	A	A	A	I.L.	C	C	I.L.
Stearic acid	A	C	A	A	B	B	A	A	A	A	B	B	B	I.L.
Sulfate liquor (black)	A	A	A	A	C	A	A	A	A	A	A	I.L.	I.L.	I.L.
Sulfur	A	A	A	A	C	A	A	A	A	A	A	A	A	A
Sulfur dioxide, dry	A	A	A	A	A	A	B	A	A	A	A	B	B	I.L.

TABLE 6.1 (*Continued*)

Fluid	Material													
	Carbon steel	Cast iron	302 or 304 stain-less steel	316 stain-less steel	Bronze	Monel*	Hastellot† B	Hastelloy† C	Durimet‡ 20	Titanium	Cobalt-base alloy 6	416 stain less steel	440C Hard stain less steel	17-4PH hard stain-less steel
Sulfur trioxide, dry	A	A	A	A	A	A	B	A	A	A	A	B	B	I.L.
Sulfuric acid (aerated)	C	C	C	C	C	C	A	A	A	B	B	C	C	C
Sulfuric acid (air free)	C	C	C	C	B	B	A	A	A	B	B	C	C	C
Sulfurous acid	C	C	B	B	B	C	A	A	A	A	B	C	C	I.L.
Tar	A	A	A	A	A	A	A	A	A	A	A	A	A	A
Trichloroethylene	B	B	B	A	A	A	A	A	A	A	A	B	B	I.L.
Turpentine	B	B	A	A	A	B	A	A	A	A	A	A	A	A
Vinegar	C	C	A	A	B	A	A	A	A	I.L.	A	C	C	A
Water, boiler feed	B	C	A	A	C	A	A	A	A	A	A	B	A	A
Water, distilled	A	A	A	A	A	A	A	A	A	A	A	B	B	I.L.
Water, sea	B	B	B	B	A	A	A	A	A	A	A	C	C	A
Whiskey and wines	C	C	A	A	A	B	A	A	A	A	A	C	C	I.L.
Zinc chloride	C	C	C	C	C	C	A	A	A	A	B	C	C	I.L.
Zinc sulfate	C	C	A	A	B	A	A	A	A	A	A	B	B	I.L.

KEY: A = Good choice
B = Proceed with caution
C = Not recommended
I.L. = Information lacking

*Trademark of International Nickel Co.

†Trademark of Stellite Division, Cabot Corp.

‡Trademark of Duriron Co.

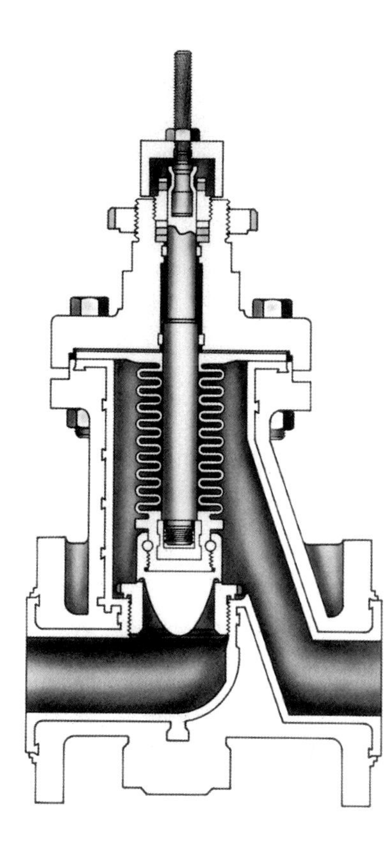

Figure 6.23 Typical lined valve construction. (*Courtesy of Fisher Controls International, Inc., Marshalltown, Iowa.*)

6.1.4 High and low temperatures

High-temperature applications pose two different types of problems. The first is that the materials selected must function properly at the given elevated temperature conditions. The second involves the fact that the materials used in the valve assembly will grow at different rates as the valve passes from ambient to elevated temperatures. The valve must be able to be assembled and sometimes function at ambient conditions but still function properly from a guiding, clearance, and shutoff standpoint at high temperatures. This is not as easy as it sounds.

First of all, as far as material compatibility is concerned, there are several things to keep track of. As the temperature increases, the choice for body material changes. Below about 800°F, WCB is the best choice. From 800 to about 1100°F, the chrome-moly alloys such as WC9 and C5 make sense, although they are more expensive and require heat treatment if welded. Stainless steels such as 304 and 316 will also work in this range and offer better corrosion and erosion resistance but are even more expensive than the chrome-moly alloys.

TABLE 6.2 Temperature Limits for Common Trim Materials

Material	Temperature limit, °F	
	Upper	Lower
300 series stainless	1100	−425
Hastelloy B†	700	−450
416	800	−20
440C	800	−20
17-4PH (CB-7CU)	800	−350
K-Monel*	900	−400
Monel	900	−400
Nitronic 50‡	1000	−325
Hastelloy C†	1000	−300
Chrome plate	1100	−450
Inconel*	1200	−400
Alloy 6	1500	−460
Alloy 20	1000	−400

*Trademark of International Nickel Co.
†Trademark of Stellite Division, Cabot Corp.
‡Trademark of Armco Steel Corp.

For the trim materials, the choices for high temperature service are shown in **Table 6.2.**

For the soft parts such as gaskets, seals, and packing, the typical choices are listed in **Table 6.3.**

Once the proper selection of body, trim, and soft parts given the service temperature has been made, you need to make sure that the parts will work together as described earlier. The details of the design techniques required to assure that this will happen are not covered here, but be aware that *differential thermal expansion* is the root

TABLE 6.3 Temperature Limits for Common Soft Parts

Material	Temperature limit, °F	
	Upper	Lower
Neoprene	180	−40
Nylon	200	−100
Polyethylene	200	−100
Nitrile	200	−40
Composition	300	−325
Carbon/PTFE	500	−250
PTFE	450	−350
Graphite	800 (standard)	−325
	1100 (special)	−325
Kel-F	300	−350
PEEK	550	−40

cause of many problems seen with valves operating at elevated temperatures. Anytime leaks or vibration occur for this type of valve, a review should be carried out on the materials employed to make sure that differential thermal expansion is not causing the parts to separate and allowing the above conditions to exist.

One other construction that may be encountered occasionally in high-temperature applications is the *extension bonnet.* As the name implies, it is a valve bonnet that is longer than normal, permitting the temperature in the packing box area to be lower than that of the valve body. This is sometimes necessary to permit the packing to work properly. For an extension bonnet to do its job, it should never be insulated. You may also occasionally encounter a finned bonnet, which was popular in the past for high-temperature applications. In recent years, it has been determined that the finned bonnet is no better at lowering packing box temperature than a bonnet with no fins, so they are no longer used.

Low-temperature service is generally understood to mean temperatures below freezing down to −150°F, while the term *cryogenic* applies to service temperatures of −150 down to −450°F. Low-temperature service requires the use of special grades of carbon steel such as LCB or LC3 for the valve body. The extension bonnet just described is also used here to limit the heat transfer between the fluid and the surrounding environment and to raise the temperature of the packing box so that the packing seals better and to prevent the formation of ice on the stem. If ice does form, it can be dragged through the packing on sliding-stem valves and can hurt sealing performance. This is less of a risk with rotary valves, but the concerns regarding heat transfer are the same whether a sliding stem or rotary design is used. Trim and soft-part material selection needs to take the limits listed in **Tables 6.2** and **6.3** into account. The change from ambient to service temperature in this case is usually not enough to warrant any special attention with respect to thermal differential expansion.

Cryogenic service is much more complicated and usually involves the handling of liquefied gases such as oxygen, nitrogen, hydrogen, fluorine, helium, and methane (properties of these fluids are shown in **Table 6.4**). Operating at −450°F requires that somc special measures be taken. The material choices become severely limited and are reflected in **Table 6.5.** There also has to be some serious consideration given to thermal expansion rates of the materials used since the service temperature can be more than 500°F below the ambient temperature.

Most valves in this service also use a very special extension bonnet like that shown in **Fig. 6.24.** It is specially designed to reduce the

TABLE 6.4 Properties of Cryogenic Fluids

Gas	He	H_2	CH_4	N_2	O_2	F_2
Density, NTP* lb/ft^3	.0103	.0052	.0415	.0724	.0827	.0982
Boiling point 1 atm. °F	−452.1	−423.2	−258.7	−320.4	−29 7.4	−306.6
Vapor density at B.P. lb/ft^3	1.06	.084	.111	.288	.296	
Liquid density at B.P. lb/ft^3	7.62	4.37	26.46	50.41	71.27	94.2
Heat of vaporization at B.P. Btu/lb	8.8	193	219.2	85.2	91.7	74.1
Critical temperature °F	−450.3	−400.3	−116.5	−232.8	−18 1.1	−200.2
Critical pressure, psia	33.2	187.7	673.1	492.3	736.9	808.3

*NTP = normal temperature and pressure, 70°F. and 14.696 psia.

TABLE 6.5 Materials for Cryogenic Service*

Name of part	Suitable materials
Valve body	Copper and all copper alloys
	Aluminum and all aluminum alloys
	Austenitic (300 series) stainless steel alloys
	Nickel and nickel alloys
	Monels
	Hastelloys
Valve trim	Austenitic stainless steel (with or without Stellite)
	K-Monel (for hardened trim)
	Hastelloys
Guide bushings	Hard aluminum bronze
	K-Monel
	Alloy G
Valve packing	Teflon v-ring
	Graphite
Valve bolting	Austenitic stainless steel, strain hardened
Valve gasket	Graphite
	Spiral-wound graphite and stainless steel
	Teflon (in trapped gasketed joints) or Kel-F

*When gaskets, valve seats, and bushings are made from nonmetallic materials, use either Teflon or Kel-F.

heat exchange between the fluid and the environment and to keep the packing box temperature at a maximum. The very long extension is normally made of 300 series stainless to keep conductive heat transfer to a minimum and features a dead space in which boiled-off vapors collect, further insulating the top of the valve from the fluid. Normally a Kel-F seal is used on the stem near the bottom of this dead space to keep the liquid from collecting inside it and potentially causing a pressure buildup when the system is shut down and the temperature begins to climb.

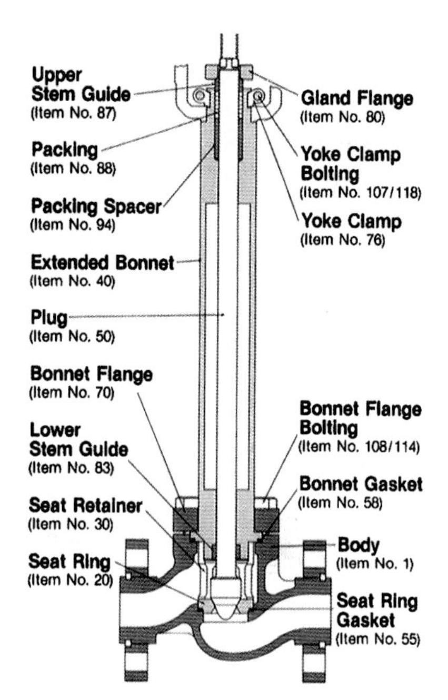

Figure 6.24 One-piece cryogenic bonnet. (*Courtesy of Valtek International, Inc., Springville, Utah.*)

A slightly different design is shown in **Fig. 6.25** where the extension is welded to the body and the bolted joint is moved outside the "cold" area. This can simplify maintenance if the valve is enclosed inside an insulated box since the valve can be opened and the trim removed without removing the box. Note also that in this figure the

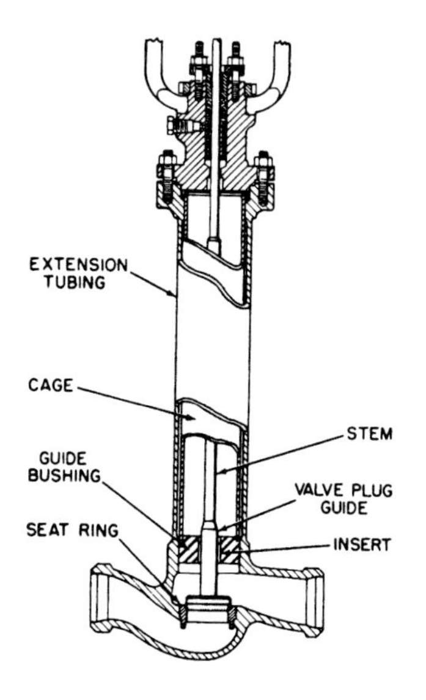

Figure 6.25 Control valve for cryogenic service. (*Courtesy of Fisher Controls International, Inc., Marshalltown, Iowa.*)

valve is a welded design. This is sometimes done to reduce the chance of leakage and to eliminate the mass of the flanges since they add to the metal that has to be cooled down during start-up. Where heat transfer has to be absolutely minimized, vacuum jacketing like that shown in **Fig. 6.26** is used. Other valve features are unchanged.

For liquid oxygen and liquid fluorine, the situation is further complicated by the need for degreasing of all wetted parts. This is required due to the heavy risk of ignition with both these fluids if they come in contact with any hydrocarbon-based residue left on the parts during fabrication.

6.1.5 High pressure

In this case, severe service high-pressure applications are somewhat arbitrarily defined as those that are above 10,000 psi. At these levels, special design considerations come into play. In the chemical process industry pressures of up to 50,000 psi can be reached, primarily in the production of polyethylene. Other high-pressure applications include urea control and ammonia production.

When a high-pressure application is encountered, there are a number of design features that need to be considered. First of all, the valve body material is normally forged to eliminate any chance of voids or cavities, and the material is selected to provide very high strength with good ductility. Type 4140 or 4340 steels are good choices because they inherently have high strength and can be heat treat-

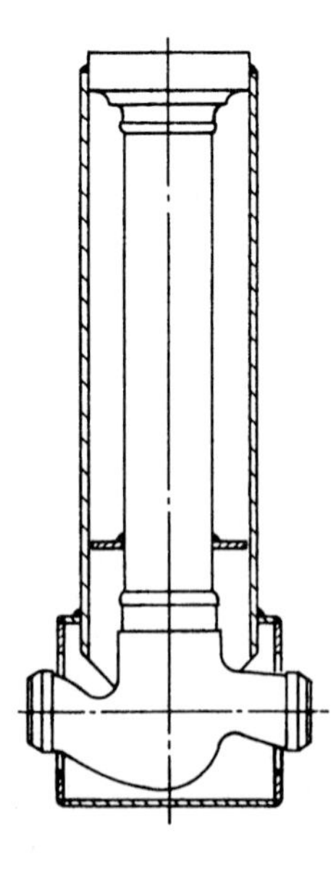

Figure 6.26 Vacuum jacketing. *(Courtesy of Fisher Controls International, Inc., Marshalltown, Iowa.)*

ed to give the desired overall characteristics. In some cases steam-tracing may be used where steam is injected into the body cavities to increase the temperature and further improve the ductility before service pressure is introduced. Line size is limited to 1 or 2 in to keep stresses low, and line connections are usually a self-energizing metal-type ring seal like that shown in **Fig. 6.27.** The metal ring seals are used because they are resistant to blowout, and their seal becomes tighter as the pressure increases.

Finishes, fits, and tolerances become very important at these pressure levels because of the tendency of seals to extrude if gaps become too large. Machining has to be very tightly controlled to keep contours and finishes very smooth to avoid stress concentration within the pressure-retaining parts. Teflon packing is still usable but must be reinforced with materials such as graphite or fiberglass to guard against extrusion. The clearances of all the packing-related parts must be tightly controlled for the same reason. All other seals should be metal to guard against blowout and need to be pressure assisted to assure that they stay tight. One option for this is a silver-plated Inconel metal O-ring that is perforated so that the high pressure can get inside the ring and help to seal against the mating parts. Material selection should also be made so that clearances don't grow with changing temperature, permitting extrusion to occur.

As far as the trim is concerned, if pressure drops are within normal ranges, standard trim can be used. If the pressure drops are high as well, special hardened trims like tungsten-carbide should be used to avoid erosion. One manufacturer prestresses a tungsten-carbide seat ring with an outside liner to keep the carbide in compression, which improves service life. Because of the high stem loads normally encountered, the stem should be made of very high-strength material and should be relatively short to keep it from failing by buckling. It

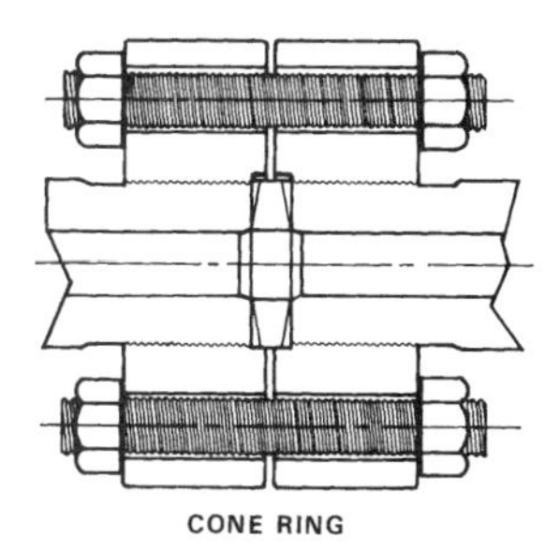

Figure 6.27 Cone ring line connection. (*Courtesy of Fisher Controls International, Inc., Marshalltown, Iowa.*)

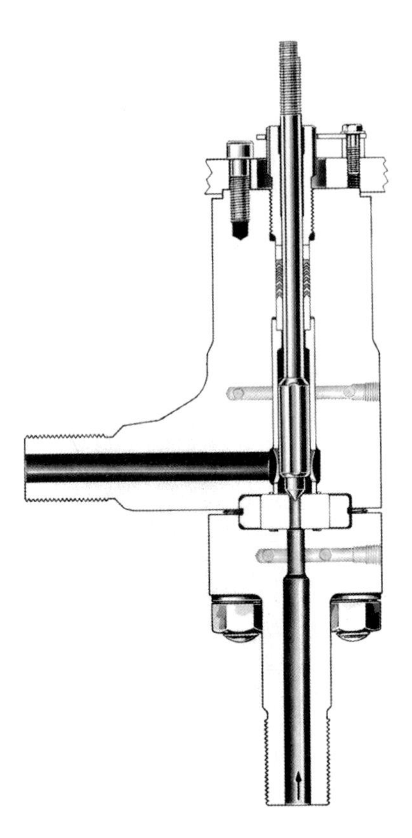

Figure 6.28 High-pressure valve body construction. (*Courtesy of Fisher Controls International, Inc., Marshalltown, Iowa.*)

should also be chrome plated to be able to stand up to the high contact stresses from the packing. A typical valve construction is shown in **Fig. 6.28.**

These valves are normally unbalanced since it would be very difficult to get dynamic seals to function at these pressure levels. Unbalanced designs with high pressure levels usually mean high actuator loads and some unique designs have been utilized to develop the loads required. One is shown in **Fig. 2.34** where dual-opposed piston actuators work through a lever-power screw system to provide very high thrust with extremely tight positioning. This type of requirement is common in the polyethylene service mentioned earlier.

6.1.6 Fire safety

Fire safety qualifications for control valves relate to their ability to perform a function that reduces the risk of damage related to a fire when and if it occurs at a chemical processing facility. Normally this means either preventing the fluid in the pipeline from flowing past the valve where it could further feed a fire or preventing the fluid from leaking to the surrounding environment with the same negative consequences. Determining whether a valve will successfully perform this function can be very complicated, particularly for a chemical plant where the nature of the fluid being handled and the characteristics of the fire can be very difficult to characterize. A number of standards exist, including:

OCMA Specification FSV.1

British Standard BS5146

Exxon BP3-14-4

API 607

FM 6033

It should be noted that these were developed for the refining industry and have been borrowed by the chemical industry, so some of the elements of the testing are not entirely appropriate, but they are used, nevertheless. Basically, these testing standards specify the conditions under which a valve is to be tested to gain the designation "fire safe." This does not mean that it will properly perform its fire-safe function under any or all conditions that might exist in a fire. This is nearly impossible. The tests really just give a relative idea of the fire-safety performance of the valves tested. Better performance under the test conditions should translate into better performance during a fire.

Most fire-safe designs employ a soft-seat seal that provides very tight shutoff until it burns away, and then a metal backup seal comes into play. To minimize through-valve leakage after the fire, special attention also needs to be paid to the body seals and the packing to limit leakage to the atmosphere.

The details of each of the test methods listed above are summarized in **Table 6.6.** If a particular application is judged to be important to fire safety within a plant, warranting a fire-safe design, be sure to review with the vendor the test method employed and how well it fits the valve construction and the possible scenarios that could exist during a fire, given the fluid handled and the type of fire that might result.

6.2 Common Applications

This next section is not meant to cover every application that might be encountered in the chemical processing plant, but it does provide some guidelines on some of the more common problems and provides several good examples of the thought process necessary when dealing with tough applications.

TABLE 6.6 Summary of Various Test Standards

Test specification	OCMAFSV.1 & BS 5146	Exxon 5P3-14-4	API 507— 2d Edition	FM 5033
Stem position	Vertical	Vertical	Horizontal	Not specified
Bore position	Horizontal	Horizontal	Horizontal	Not specified
Valve open or shut	Open	Open	Shut	Shut
Test pressure during burn	30 psi	25 psi	1000°F class pressure rating	125 psi
Test medium	Kerosene or diesel fuel	Liquid hydrocarbon	Water	Not specified
Valve body temperature	Not specified	1200°F minimum	1100°F minimum 30 min	Not specified 15 min
Burn duration	Sufficient to destroy soft seat, 15 min minimum	Sufficient to destroy soft seat, 15 min minimum		
When seat leakage is measured	After test	After test	During test	During test
Maximum external leakage	No appreciable leakage	Shall be negligible	20 ml/min/in dia	Individual drops
Maximum seat leakage	10 ml/min/in dia†	10 ml/min/in dia†	40 ml/min/in dia†	0.1 qt/min (94.6 cc/min)
Operability	3 cycles open to shut	3 cycles open to shut	1 cycle open to shut	Must be operable

*Temperature should be maintained for 5 min.
†In no case shall leakage rate exceed 100 ml/min.

6.2.1 Oxygen service

Oxygen service is unique in that it is normally low-temperature or cryogenic service and that it is very easily ignited, causing a fire. The low-temperature considerations were covered earlier in this chapter, but the subject of ignition deserves some additional discussion.

All organic and inorganic materials will react with gaseous or liquid oxygen at certain pressure and temperature conditions. The reaction that occurs can cause a fire or an explosion. Because of these inherent dangers, control valve material selection is extremely important for this application.

Many materials used in control valves have ignition temperatures that are above the normal flowing temperature of gaseous oxygen; therefore, ignition of the materials by the fluid is generally not the danger. The danger is in the ignition of these materials by abnormal, localized high temperatures. Listed below are just a few examples of what could cause localized high temperatures:

Flow velocity. Velocity criteria for oxygen service is outlined in the *Compressed Gas Association Pamphlet G-4.4.* In general, if the velocity through the port of a valve can exceed 200 ft/s, only copper-base alloy materials should be used for parts coming in contact with the fluid.

Foreign particle impingement. A foreign particle, such as weld slag, that is being carried along in the flowstream may collide with the valve body or trim. This collision could transform the particle's kinetic energy into sufficient heat to cause either the impinging particle or the surface it strikes to ignite.

Vibration. A part that is caused to vibrate may generate enough heat from the internal friction to raise its temperature to its ignition point. As a result, clearances and guiding are critical for oxygen-service valves.

Most vendors follow strict cleaning guidelines with oxygen service valves that minimize the possibility of foreign matter being present within the assembly. The end user needs to take special precautions during installation to ensure that no foreign particles are introduced at that stage.

The selection of metals should be based on their resistance to ignition and rate of reaction. Following is a comparison of these two properties for some commonly used valve materials:

Resistance to ignition in oxygen (listed in order from hardest to easiest to ignite):

Copper and copper alloys such as Monel (Monel is rated as one of the best materials for oxygen service.)

300 series stainless steel

Carbon steel

Aluminum

Rate of reaction (listed in order from slowest to most rapid rate of combustion):

Copper and copper alloys such as Monel

Carbon steel

300 series stainless steel

Aluminum; burns very rapidly

Note that stainless steel, once ignited, burns more rapidly than carbon steel. Nevertheless, the austenitic grades (300 Series) of stainless steel are considered to be much better in oxygen service than carbon steel due to their high resistance to ignition.

Organic materials have ignition temperatures below those of metals. Use of organic materials in contact with oxygen should be avoided, particularly when the material is directly in the flowstream. When an organic material must be used for parts such as valve seats, diaphragms, or packing, it is preferable to select a material with the highest ignition temperature, the lowest specific heat, and the necessary mechanical properties. For example, Viton is often used for diaphragms and O-rings, while PTFE is acceptable for packing.

6.2.2 Methyl chloride

Methyl chloride is manufactured by the direct reaction of methane and chlorine. A catalyst or the presence of light triggers the reaction, which predominantly forms methyl chloride and smaller amounts of methylene dichloride, chloroform, and carbon tetrachloride. Methyl chloride can also be manufactured by the action of hydrogen chloride on methanol with the aid of a catalyst.

Originally, methyl chloride was used only as a refrigerant, but chlorofluoromethanes have since eliminated this need for methyl chloride. It is now mainly used in silicone production, which requires methyl chloride as a raw material. The end products that are produced are antiknock additives, polymers, and silicone inorganics.

Methyl chloride reacts violently with copper, aluminum, bronze, zinc, and magnesium-bearing alloys. For this reason, 17-4PH is not acceptable for trim parts. WCB bodies should be used with 316 SST trim on all applications. O-rings should be Teflon or Viton A, and gaskets should be made from Teflon and ferrous metallic materials. Cage-guided, sliding-stem valves are mainly used in this application. If cavitations is an issue, an Alloy 6 overlay offers added protection under those conditions. Occasionally, methyl chloride service can be above 500°F, which also requires hardfaced trim.

6.2.3 Ammonia let-down

Ammonia is the principle source of nitrogen fertilizer (via urea). Natural gas or naphtha is used as the feedstock for the processing of ammonia. The steps are desulphurization, catalytic steam reformation, catalytic shift conversion, carbon dioxide removal and regeneration, methanation, and synthesis and refrigeration. In the first step, carbon monoxide is added to the natural gas to remove any traces of sulphur that are normally present. Sulphur must be thoroughly removed since it will interfere with the subsequent catalytic reaction. The catalyst steps reform the hydrocarbons into a carbon dioxide-rich solution. The carbon dioxide is then removed by scrubbing the feed gas. (The carbon dioxide released is often used for further processing of ammonia into urea.) The methanation step converts the residual carbon oxides in the feed gas into methane. The gas, after this step, is sufficiently pure for the production of ammonia since it contains basically hydrogen and nitrogen in a 3:1 ratio.

The synthesis and refrigeration step is the final stage in ammonia processing, and it also includes the ammonia let-down control valve. The gas is compressed and introduced into the synthesis converter for conversion of the hydrogen-nitrogen mixture to ammonia over a promoted iron oxide catalyst. The ammonia contained in the converter exit gas is then condensed and flashed. The ammonia let-down valve's typical process conditions are 4800-psi inlet to 300-psi outlet. This is both erosive, due to high velocities and any catalyst fines still in the process, and corrosive, due to the presence of hydrogen and ammonium carbamate produced in the synthesis step. To make a proper material selection, you must know the pressure, temperature, process concentration, and exact fluid composition.

Materials in this system must be able to handle hydrogen at elevated pressures and temperatures, as well as steam (up to 1500 psig). Tungsten carbide, NTX 40, 316/Stellite, and 440C can be used for trim materials, but the results vary depending on the above conditions. Drilled-hole cages are not recommended due to catalytic fines in the process. Caution should also be used in applying Stellited trim since any moisture acts to attack the cobalt-based Stellite. This is a very tough application given the high pressure drop and erosive/corrosive service, but some success has been seen with small flow-down angle valves with massive, hardened trim.

6.2.4 Caustic service

Common caustics include potassium carbonate, sodium carbonate, potassium hydroxide, and sodium hydroxide, and each is known by another common name. Potash is potassium carbonate ($KHCO_3$), soda ash is sodium carbonate (Na_2CO_3), caustic potash is potassium hydroxide (KOH), and caustic soda is sodium hydroxide (NaOH).

Sodium hydroxide is extremely corrosive, particularly above 150°F. Potassium hydroxide is not as corrosive as sodium hydroxide, but the same valve materials are generally used for both. Potassium carbonate and sodium carbonate are not corrosive at low temperatures. Steel and SST can be used up to 200°F for these latter two materials; above that temperature, they should be treated like NaOH.

The primary risk with all caustics is corrosion, especially as concentration and temperature increase. As a result, copper alloys like Monel are a very popular choice for this application. Inconel and K-Monel are also good. In extreme cases, even these trim materials may need to be stress relieved to guard against the threat of stress corrosion. Once again, sliding-stem cage-style valves are the common choice here, although Teflon-lined valves are also a very cost-effective option.

6.2.5 Urea let-down

Urea is produced from ammonia and carbon dioxide as follows: $2NH_3+CO_2 \rightarrow NH_2 \bullet CONH_4 \rightarrow NH_2 \bullet CONH_2+H_2O$. This reaction requires high temperature and pressure to cause the intermediate, ammonium carbamate, to break down into urea and water. The ammonia and carbon dioxide form ammonium carbamate in a zirconium-lined reactor at approximately 3200 psig. The crucial control valve in the entire urea process is the let-down valve from the reactor to the first separator. The final output of the process is urea prills or crystals, which are used in nitrogen fertilizers or urea/formaldehyde resins. Typical service conditions for the highly corrosive urea let-down service are as follows:

P_1 = 3200 psig

P_2 = 1500 psig

$\Delta P_{s.o.}$ = 3200 psid (shutoff pressure drop)

Temperature = 365°F

P_v = 225 psia (vapor pressure)

S.G. = 1.02–1.1 (specific gravity)

As far as material choices are concerned, bodies cast from 316L are typically specified, and trim material can range from Nitronic-50 to Zirconium, Zircoloy, Stellite, 316L, and Ferralium. Downstream liners made from silicon carbide are also popular. The typical valve used in this service is a small-bore angle valve used in a flow-down direction, with hardened trim made from one of the above materials.

6.2.6 Titanium dioxide (TiO_2)

TiO_2 is the white pigment in coatings, plastics, and paper, with the largest sole user of the product being the automotive industry. A handful of producers worldwide make up the majority of production. Due to environmental issues, producers have been shifting away from the older sulfate process. Most new plants today are utilizing the chloride process, and many of the older sulfate plants are either being converted or closed down.

Figure 6.29 is a simplified schematic that shows how the product is processed. **Table 6.7** contains notes on the process. The key "prob-

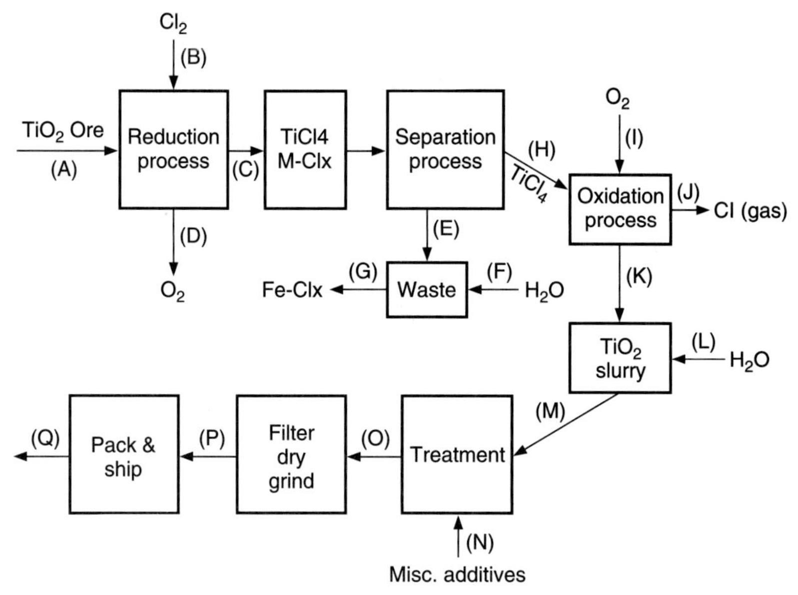

Figure 6.29 Titanium dioxide process.

TABLE 6.7 Flow Chart Analysis for Figure 6.34

Line	Explanation
Line A	Solids transported by conveyers—TiO_2 ore.
Line B	Cl_2 addition line.
Line C	Very abrasive slurry. Corrosive due to chlorides.
Line D	O_2 extracts.
Line E	Metal chlorides being extracted; corrosive slurry.
Line G	Extremely corrosive slurry; use lined or pinch valves mostly; Titanium or Tantalum required for metals.
Line H	Less abrasive after separation; liquid tickle.
Line I	Oxygen addition lines.
Line J	8- to 10-in lines of impure chlorine gas; $P \leq 50\#$; $T \leq 400°F$.
Line K	TiO_2 slurry.
Line N	Additives for chemical treatment; small 1-in lines generally.
Line P	Bag house—dry powder state uses gravity flow; 12- or 16-in high-performance butterfly valve style on new units.

lem" applications are the $TiCl_4$ ("tickle") and chlorine services. Typically, $TiCl_4$ is very erosive and potentially corrosive. The chlorine applications tend to be nonerosive but very corrosive. Service conditions for both applications are moderate with process temperatures below 300°F and pressures below 300 psi. Valve body sizes range from 1 to 6 in.

One of the greatest problems with these applications is packing integrity. Once a packing system begins to leak, atmospheric air combines with the chlorine in the system and hydrochloric acid is formed, causing corrosion to be greatly intensified. After a period of time, the valves tend to seize up and HCl continues to leak past the packing area. Live-loaded Teflon packing is a good option to use in this case to help minimize the chances of the initial leaks developing. Atmospheric corrosion at these plants is also a concern and should be discussed in detail. Epoxy painting is a good option to apply to both the valve and actuator to guard against this.

For $TiCl_4$ service, most customers follow one of two philosophies in specifying materials for a control valve. The most common is for the customer to conservatively order carbon steel valves with Stellited trim but include Hastelloy C shafts and packing parts. Hastelloy C is required to help protect the parts from corrosion damage, assuming that the packing will leak.

The second philosophy, and the one that makes more sense and is cheaper in the long run, is to utilize standard carbon steel valves with Stellited (or ceramic) trims and standard shafts and packing parts with the addition of live-loaded packing. The key here is to eliminate packing leakage altogether. Once this is accomplished, the risk of corrosive damage to the stem and packing parts is minimized.

Ceramic trim with eccentric plug valves has been used on some of the extremely erosive $TiCl_4$ applications with excellent success. The second trim material choice is Stellite. Stellited trims have worked very well, but will wear after a period of time. To make the body a little more "rugged" from an erosion standpoint, ceramic coating is also something to be considered. However, the ceramic coating available for most rotary valves gives only *erosion, not corrosion,* protection.

6.2.7 Sulfuric acid

Material and valve selection for sulfuric acid service greatly depends on concentration. As a result, two different sets of recommendations will be covered here: one for concentrations above 90 percent, and one for below.

The term *concentrated sulfuric acid* broadly refers to the concentration range from 90 to 100 percent. Concentrated sulfuric acid is a powerful oxidizing agent and desiccant because of its affinity for water. The major problems in its handling and storage relate to its hygroscopic nature (reacting with atmospheric humidity), the exothermic reactivity with water, and velocity effects which erode otherwise protective films of anticorrosive metals (see Sec. 6.1.3).

As far as body and trim materials are concerned, cast iron, gray or ductile, is more corrosion resistant than carbon steel. Gray cast iron and ductile iron have useful resistance to 99 percent sulfuric acid at temperatures greater than 212°F. Velocities up to 5 ft/s can be tolerated in cast iron pipe. The conventional 18Cr-8Ni stainless steels (i.e., S30400, CF8, S31600, CF8M, S31700, and CG8M) are inherently resistant to cold concentrated acid in the 93 to 100 percent range since they already have a naturally formed passive film. The low-carbon grades [i.e., S31603 (316L SST and CF3M)] are occasionally specified as insurance against the possibility of intergranular attack in weld repair areas. Intergranular corrosion of sensitized stainless steel is not, however, an inherent problem in concentrated acid, and the regular stainless steel grades are used in all cases possible.

S31600 and (316 and CF8M), S31603 (316L and CF3M), S30400 (304 and CF8), S30403 (304L and CF3), and S31700 (317 and CG8M) are acceptable for concentrated sulfuric acid above 93 percent and for temperatures below 85°F. S31600 (316 and CF8M), S31603 (316L and CF3M), S31700 (317 and CG8M), and 520910 (Nitronic 50) are acceptable for concentrations above 93.5 percent acid up to 212°F. Any surface contamination by chlorides, as by seawater from transportation or storage under marine conditions, can cause rapid localized pitting of stainless steel from hydrochloric acid formation.

Alloy 20 and other nickel-chromium-copper-molybdenum type of alloys will resist concentrated acid under high-velocity conditions and at temperatures higher than the lower alloy austenitic grades. The maximum temperature limit is a function of concentration but generally exceeds 160°F. With respect to trim, it should be noted that Alloy 6, CoCr-A, ENC, hard chrome, and chromium coating are not acceptable for concentrated sulfuric acid service at any temperature.

The fluorinated plastics listed below are fully resistant to concentrated sulfuric acid and can be used for soft parts including packing.

Material	Upper temperature limits, °F
PTFE (polytetrafluoroethylene)	450
PFA (perfluoralkoxy)	450
ECTFE (ethylene-chlorotri-fluorethylene)	250
FEP (fluorinated ethylene-propylene)	350
E-TFE (ethylene-tetra-fluoroethylene)	300

Summing all this up, it should be clear that *concentrated sulfuric acid* is one of the toughest applications in the chemical industry from a corrosion standpoint. Nothing's going to work very well or very long, but 316 SST bodies and trim, without hardfacing or chrome plating, are normally the best choice for concentrations up to 93 percent and temperatures up to 85°F. Beyond those limits, Alloy 20 is normally used. For the soft parts, either Teflon or graphite is acceptable, but Teflon is preferred.

As far as *dilute solutions* are concerned (<90 percent concentration), the choices are actually more limited. Alloy 20, again, sees a lot of use for bodies and trim. Exotic alloys like Zirconium, tantalum, and platinum also show good resistance to the corrosive attack but are very expensive and sometimes difficult to fabricate into the parts required in a control valve assembly. There *is* a little more flexibility in the use of hard overlays with the lower concentrations, with Alloy 6 being acceptable up to 150°F for 20 percent solutions and up to 85°F for 50 percent solutions. Soft parts, once again, can be Teflon or graphite with Teflon being the preferred choice.

6.2.8 Pressure swing absorption (PSA) skids

PSA is one of the several methods used to separate the components of air or other gases. Some of the more common PSA systems include hydrogen, nitrogen, oxygen, and methane recovery units. A skid consists of at least two tanks which are filled with an absorbent solid. As the feedstock is pressurized in the tank, the gas to be recovered is absorbed (the binding of molecules from a gaseous or liquid phase to the surface of solids) by the solids until they are saturated. The tank is then depressurized and purged with expansion gas. The purging process regenerates the tank, removing the collected gas and preparing the solids to accept more gas on the next pressurization. When the purging process is complete, the tank is again pressurized with the feedstock to full process pressure, at which time the gas begins to be absorbed by the solids again.

Three issues typically need to be addressed when selecting valves for PSA skid applications: *cycle life, shutoff,* and *pressure drop*. First, the pressure swings of this system occur in a matter of minutes, dependent upon the size of the skid and the gas to be recovered. The valves for this skid are therefore required to cycle as often as once a minute, and therefore skid manufacturers typically require valves, actuators, and positioners to have been cycle tested for up to 300,000 full stroke cycles for these applications.

Secondly, plants typically specify bidirectional Class VI shutoff to attain the maximum achievable purity from their process. Lastly, the pressure drops in these services occasionally are high enough to require piston operators, oversized yoke bosses, and strengthened stems in globe valves. They also require a spring-fail mode from the actuator.

The above requirements can be addressed by using a well-designed post-guided sliding-stem globe valve with a soft seat. The bearings must be able to withstand the high cycling, in particular. Experience has shown that a spool-valve positioner generally lasts longer than other designs and should be used in this type of application. High-performance butterfly valves are also seeing increased service in PSA skids. As a general note, typical carbon steel valves are acceptable since the processes are usually not corrosive.

6.2.9 Chlorine service

Chlorine can be made in two main ways. The first is the electrolysis of brine (sodium chloride) or potassium chloride in the following reactions:

$$2NaCl + 2H_2O \xrightarrow{\text{electricity}} Cl_2 + 2NaOH + H_2$$

$$2KCl + 2H_2O \xrightarrow{\text{electricity}} Cl_2 + 2KOH + H_2$$

Chlorine, hydrogen, and caustic soda (NaOH) or caustic potash (KOH) are produced.

The second way is by a proprietary process named Kel-Chlor (a trademark of M.W. Kellogg Co.). This process takes hydrogen chloride and dilutes it with nitrosyl sulfuric acid in dilute sulfuric acid. This is oxidized to form chlorine and water. Chlorine is further purified and used in the production of vinyl chloride, pulp stock, and other organic chemicals. Typical products are PVC pipe, solvents, antiknock ingredients for gasoline, pesticides, and refrigerants.

Chlorine is very reactive, chemically, and is never found in nature as an isolated element. Moisture in either gaseous or liquid chlorine makes it extremely corrosive to common metals. However, dry chlorine (with less than 200 ppm solubility water) can be used with common metals with no damage. Chlorine has a high affinity for moisture, including moist air in the packing box.

With no moisture present, carbon steels, WC9, C5, Monel, and Hastelloy C are good selections for dry chlorine. Dry hydrogen chloride can also use the same materials. With any moisture present, chlorine becomes extremely corrosive, and Monel, nickel, and Hastelloy B-2 can be used if no oxidizing agents are present. Hastelloy C, titanium, and tantalum can be used with or without oxidizing agents. Wet hydrogen chloride behaves the same as wet chlorine for material suggestions. All valve styles can be used on this process since material selection is the key issue. Packing should be live-loaded, if possible, to keep the chlorine from combining with moisture from the surroundings.

6.2.10 Heat-transfer fluids

Heat-transfer fluids are used to transfer heat between locations at much lower pressures than steam. Many companies manufacture heat-transfer fluids; some common trade names are Dowtherm (Dow), Syltherm (Dow Corning), Mobiltherm (Mobil), and Therminol (Monsanto). Cold-transfer fluids are also available; one example is Dowfrost (Dow Chemical).

Heat-transfer fluids range in temperature from −100 to 750°F. They are either synthetic, organic, silicone polymer, or glycol-based. They are used in:

- Indirect heating of liquids or polymers
- Batch processing heating and cooling
- Pipeline tracing
- Energy recovery
- Low-pressure cogeneration systems
- Drying and heating bulk materials
- Solar energy collection and storage
- Gas processing
- Refrigeration coil defrosting
- Fermentation cooling
- Ice skating rinks

Concerns that arise in applying heat-transfer fluids are that they are largely nonlubricating and viscous (i.e., *erosive*). Galling, buildup, binding, and accelerated trim wear are signs of incorrectly specified valves. The key is to reduce the amount of guiding surfaces by using post-guided instead of cage-guided valves. Also, the vapor phase may be difficult to contain in the packing box, so valves may need to be installed with stems in a horizontal or downward position to ensure that the liquid phase contacts the packing, or a bellows seal might be used.

Corrosion is not a primary concern. Hardened trim surfaces (Stelliting) can produce slow wear on the seating surfaces. Carbon steel bodies are acceptable up to their pressure and temperature limitations. Adequate performance by 316 SST trim is seen to 600°F. For higher temperatures, 316/Stellite trim should be used for improved wear resistance. Self-lubricating packing is imperative since any packing lubricants are soluble in the heat-transfer fluid. TFE V-ring packing should be used up to 450°F; otherwise graphoil or live-loaded graphite packing should be considered.

References

1. Fisher Controls, Cat. 10, Sec. 3, Fisher Controls, Marshalltown, Iowa, 1990.
2. Bauman, H. D., *Control Valve Primer: A User's Guide*, ISA, Research Triangle Park, N.C., 1991.
3. Hutchison, J. W. (ed.), *ISA Handbook of Control Valves*, Chap. 6, Part 3, 1st ed., ISA, Research Triangle Park, N.C., 1971.
4. Fagerlund, A., and Considine, D. M., *Process/Industrial Instruments and Controls Handbook*, 4th ed., McGraw-Hill, New York, 1993.
5. PS Sheet 80.2:005(A), April 3, 1989, Fisher Controls, Marshalltown, Iowa, 1989.
6. Cory, J., and Riccioli, F., "What Are Fire-Safe Valves," *Chemical Engineering*, May 27, 1985.
7. Schafbach, Paul J., *Liquid Pressure & Velocity Through Control Valve Trim for Potentially Cavitating Service*, Fisher Controls, Marshalltown, Iowa, Feb. 1993.
8. Schafbach, Paul J., *Cavitation Control Requires Pressure Control*, Fisher Controls, Marshalltown, Iowa, Feb. 1993.
9. Riveland, Marc L., "Cavitation in Control Valves," in *Process Instruments & Controls Handbook*, reprints from Fisher Controls, Marshalltown, Iowa, 1984.
10. *The Silent Treatment, Bulletin 80:005*, Fisher Controls, Marshalltown, Iowa, 1898.

Chapter

7

Packings, Gaskets, and Seals

The three categories of valve parts—packings, gaskets, and seals—are commonly referred to as the "soft parts" or "soft goods," differentiating them from the metallic parts such as the valve body, bonnet, and valve trim. They will be covered here as a group because their primary function in all three cases is to provide some type of pressure seal, normally between two metallic parts. It also turns out that all three are made from the same types of materials: elastomers, TFE, graphite, or soft metals.

While the following discussion deals exclusively with gaskets in sliding-stem valves, the principles outlined apply equally as well to rotary designs. The reason for concentrating on the sliding-stem construction is that it has more gasketed joints, and the gasketing arrangements are generally more complicated than they are for the rotary designs. In contrast, the treatment of seals and packing will cover both rotary and sliding-stem constructions.

7.1 Gaskets

The one big difference between gaskets and the other two types of seals covered in this chapter is that a gasket is a *static seal*. In other words, it is a seal that is trapped between two nonmoving parts, which is a big advantage when it comes to performance because a static seal is much easier to maintain than a dynamic one.

7.1.1 Gasket basics

Gaskets come in many sizes and styles including standard flat-sheet, spiral-wound, metal-reinforced flat-sheet, ring-type, and metal O-ring, among others (see **Fig. 7.1** for examples). However, there are

Gasket Materials and Contact Facings[1] Gasket Factors (m) for Operating Conditions and Minimum Design Seating Stress (y)						
Gasket Material		**Gasket Factor** m	**Minimum Design Seating Stress** y	**Sketches and Notes**	**Use Facing Sketch**	**Use Column**
Self-Energizing Types O-Rings, metallic, elastomer, and other gasket types considered as self-sealing		0	0	. . .	. . .	. . .
Elastomers without fabric or a high percentage of asbestos fiber: Below 75 Shore Durometer 75 or higher Shore Durometer		 0.50 1.00	 0 200		(1a),(1b),(1c),(1d),(4),(5)	II
Asbestos with a suitable binder for the operating conditions	1/8″ thick 1/16″ thick 1/32″ thick	2.00 2.75 3.50	1600 3700 6500			
Elastomers with cotton fabric insertion		1.25	400	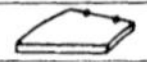		
Elastomers with asbestos fabric insertion, with or without wire reinforcement	3-ply 2-ply 1-ply	2.25 2.50 2.75	2200 2900 3700	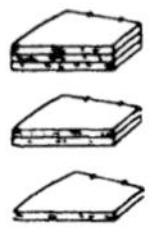		
Vegetable fiber		1.75	1100	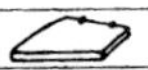		
Spiral wound metal, asbestos filled	Carbon Stainless or Monel	2.50 3.00	10,000 10,000			
Corrugated metal, asbestos inserted or Corrugated metal-jacketed, asbestos filled	Soft aluminum Soft copper or brass Iron or soft steel Monel or 4%—6% chrome Stainless Steels	2.50 2.75 3.00 3.25 3.50	2900 3700 4500 5500 6500		(1a),(1b)	
Corrugated metal	Soft aluminum Soft copper or brass Iron or soft steel Monel or 4%—6% chrome Stainless Steels	2.75 3.00 3.25 3.50 3.75	3700 4500 5500 6500 7600		(1a),(1b),(1c),(1d)	
Flat metal-jacketed, asbestos filled	Soft aluminum Soft copper or brass Iron or soft steel Monel 4%—6% chrome Stainless Steels	3.25 3.50 3.75 3.50 3.75 3.75	5500 6500 7600 8000 9000 9000	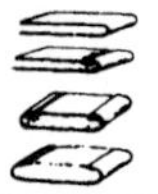	(1a),(1b),(1c),[2](1d),[2](2)[2]	
Grooved metal	Soft aluminum Soft copper or brass Iron or soft steel Monel or 4%—6% chrome Stainless Steels	3.25 3.50 3.75 3.75 4.25	5500 6500 7600 9000 10100		(1a),(1b),(1c),(1d),(2),(3)	
Solid flat metal	Soft aluminum Soft copper or brass Iron or soft steel Monel or 4%—6% chrome Stainless Steels	4.00 4.75 5.50 6.00 6.50	8800 13000 18000 21800 26000	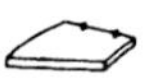	(1a),(1b),(1c),(1d),(2),(3)(4),(5)	I
Ring joint	Iron or soft steel Monel or 4%—6% chrome Stainless Steels	5.50 6.00 6.50	18000 21800 26000		(6)	

NOTES:
(1) This table gives a list of many commonly used gasket materials and contact facings with suggested design values of m and y that have generally proved satisfactory in actual service when using effective gasket seating width b The design values and other details given in this table are suggested only and are not mandatory.
(2) The surface of a gasket having a lap should not be against the nubbin.

Figure 7.1 Gasket configurations with M and Y factors. (*Courtesy of Fluid Sealing Association, Philadelphia, Penn.*)

some common traits among all of them that can help explain why they work the way they do and what might be the cause if they stop working. In any gasket design there are five elements that must be considered: surface preparation, gasket material, gasket shape, joint configuration, and load. All of these will be covered in this section, but they boil down to a relatively simple rule: If a seal (gasket) is to work properly, the surfaces to be sealed must be properly prepared and loaded so that they are in intimate contact with the seal and then sufficient load applied to maintain intimate contact after the pressure to be sealed against is applied.

Now, let's take a look at this rule and each of the five considerations in light of the types of gasket designs that we might encounter on a control valve. As far as surface preparation is concerned, the surface should be clean and relatively smooth, although not too smooth. A 125-rms finish is probably good enough, and many valve designs actually feature concentric serrations that are machined into the metal faces to assist in the seal. Experience has shown that gaskets, especially the softer ones, actually work better if the surface is a little rough. It probably has to do with raising the localized stress and increasing the deformation of the gasket so that intimate contact is better established. It also seems to keep the gasket from being blown out by the pressure it's sealing against. In terms of scratches or surface defects, be especially careful with those that cut across the gasket surface in the radial direction since they can result in leakage from the inside diameter of the gasket to the outside. Scratches or dents running circumferentially pose less of a problem because they don't result in an external leak path. If a surface has radial scratches that cut across the gasket face, it's best to completely remachine the face to eliminate them. Spot repair is not usually effective in this case. If concentric serrations were used in the original design, it's a good idea to duplicate them in the remachining process.

Once you're sure that the sealing surfaces are ready, you need to select a gasket material and configuration and determine the load necessary to "seat" it into the mating surfaces. This preload factor, as it is sometimes called, is a function of the type of gasket and the material and can be found in ASME Sec. VIII Division 1, App. 2, as the Y factor (**Fig. 7.1**). A typical Y factor for a gasket might be 5000 psi stress, and to find the load required to generate this stress, multiply this stress level by the effective contact area of the gasket. Once again, the area is defined in the ASME standard as the average contact diameter of the gasket multiplied by pi (3.14) times the effective contact width, which is not always the actual width of the gasket. It is adjusted for gasket configuration and effects like flange bending. The gasket then has to be properly placed and held between the two mating surfaces and this seating load applied.

There are two basic approaches to joint configuration that relate to how the gasket is held in place. The more common one, because it is the simplest and cheapest, is to simply place the gasket between two opposing flat surfaces and apply load to hold the gasket and to seal. This is cheaper but is somewhat limited in use due to the tendency for the internal pressure to blow the gasket out and because of the necessity for the gasket to carry the full bolt load. A better approach is for the gasket to be placed in a recessed area like a groove or a slot machined in one or both of the mating surfaces. This reduces the chances of the gasket being blown out and also allows a portion of the bolt load to be carried by the opposing surfaces rather than being completely absorbed by the gasket. Sharing the bolt load with the metal surfaces enables the joint designer to utilize a technique referred to as *preload,* which will be covered later but is very important in optimizing joint performance.

Before going on to the next step, there are two points regarding what has been covered so far. First, the gasket must be selected taking into account the type of pressure levels to be sealed against. The higher the pressure, the stronger the gasket needs to be to avoid being blown out. Relative gasket strength starts with nitrile rubber on the low end and increases in several steps all the way out to the solid metal lens-type ring for pressures greater than 10,000 psi. Containing the gasket in a recessed area will help somewhat with the problem of blowout, but pay strict attention to the pressure limitation of a particular material and configuration. Second, getting and keeping the gasket in the proper location before it is loaded can sometimes be very challenging, but it is critical for proper gasket performance. If chronic leakage is a problem, make sure that the gasket is where it's supposed to be and that it didn't slip out of position during assembly.

Once the gasket is positioned and properly preloaded, ensure that there is enough load present to keep the two mating surfaces from separating far enough with the application of pressure to permit the gasket to be blown out or extruded through the opening between the surfaces. Any relative movement of the surfaces also reduces the gasket load, which increases the chances of leakage. The load required to maintain the seal for a particular gasket material and configuration is defined by the M factor in **Fig. 7.1** and is a multiple of the pressure sealed against. So, getting back to the five factors mentioned earlier, if we properly prepare the surfaces, select our gasket material and configuration, and apply enough load to seat the gasket and hold the faces together, the gasket should seal properly.

7.1.2 Gaskets in control valves

If we examine this from a control valve standpoint, it is evident that there are two primary gasketed joints that prevent external leakage: the line flanges and the body-to-bonnet joint. Looking at the line flanges first, the standard approach for the chemical process industries is to use a full-face gasket in combination with a raised face flange. **Figure 7.2** shows the gasket in proper position on the flange face of the *body* and illustrates the dimensions mentioned earlier: contact diameter and width. For gasketed joints to work properly in a control valve, there are a number of items that need to be considered. They are covered in the following paragraphs.

Material. The gasket material needs to be selected taking into account chemical compatibility with the process fluid, service temperature, and the pressure to be sealed against. Asbestos used to be one of the most popular materials for this type of joint, but the concerns associated with asbestos exposure have nearly eliminated this from the marketplace. Typical gasket materials that are now used instead are listed below, along with some comments regarding their use:

Elastomers. This group is usually used for mild services and includes materials like natural rubber, nitrile rubber, neoprene, fluoroelastomer, EPDM, silicone, and polyurethane. They all conform very well to the mating surfaces under load but are somewhat limited in temperature and pressure use due to their low strength and

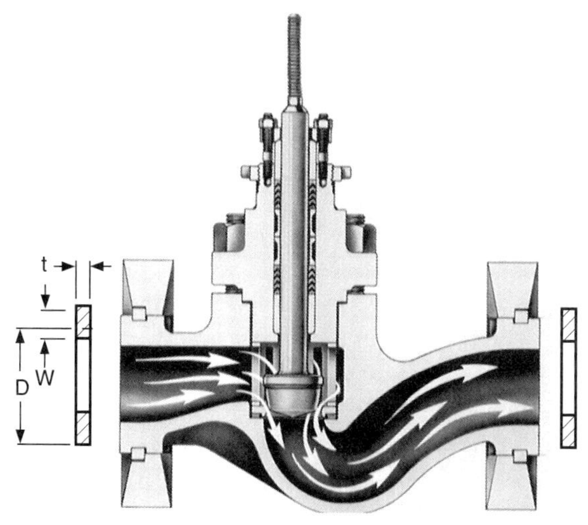

Figure 7.2 Standard line-flange gasketing. (*Courtesy of Fisher Controls International, Inc., Marshalltown, Iowa.*)

low melting temperature. Fabric reinforcement can help raise their inherent strength, which is very important in a full-face arrangement where they are not protected or confined in a recessed area. For nonrecessed designs, even with fabric reinforcement, they are generally so soft that it is difficult to build up sufficient bolting preload to withstand the pressure loading without damaging the gasket material. Therefore, their use is normally limited to 200 psi and less. Most are limited in temperature range from −30 to 200°F with the exceptions of silicone, which is good for −100 to 450°F and EPDM which is good for −50 to 300°F. In general they are reasonably good for acids and alkalis but do not stand up very well to solvents.

Aramid/elastomer blends. In this case, aramid fibers (or some other similar inorganic material) are mixed with an elastomer in a compressed sheet form that results in a much stiffer, stronger material that works well in applications up to 1400 psi even in nonrecessed designs, and exists in two blends that provide temperature service up to 300 and 600°F. A light coating of a gasket compound or lubrication can help this type of gasket to seat and seal better.

Fiber/PTFE blends. PTFE is used as the base material here, with the fibers (fiberglass, for example) added to improve pressure and temperature performance. These gaskets are good to 500°F and 1000 psi in nonrecessed designs. They also have very good chemical resistance since Teflon is the base material.

Graphite/graphite-reinforced. At the high end of the performance spectrum is graphite. It is chemically inert and good for temperatures up to 1100°F. It can either be used in a high-density formed sheet by itself or pressed around a wire mesh or metal foil for additional strength and ease in handling. It is good for pressures up to 1500 psi in nonrecessed designs and much higher if it is effectively trapped. This material also benefits from a light coating of gasket compound or lubricant.

Soft metals/flat sheet. Brass, copper, aluminum, stainless steel (annealed), nickel, Monel, and Hastelloy are all materials that might see use as gaskets. In general, the designer is looking for a soft metal that has good chemical compatibility and a wide temperature range. As a result, Monel is one of the more common materials used for metal gaskets in the chemical process industry.

Special metal gaskets. Flat-sheet metal gaskets work well in many applications but do not have much resilience. The springback or resilience properties can be improved by either corrugating the material or using a soft fiber filler inside a metal jacket. These options are shown in **Fig. 7.1.**

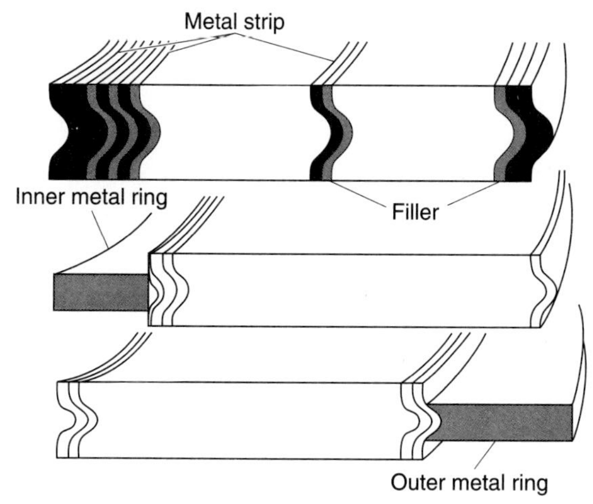

Figure 7.3 Spiral-wound gaskets.

Spiral wound. This is not really a flat-sheet gasket, although it can be used like one for the line-flange joint. **Figure 7.3** shows that the spiral-wound construction consists of alternating layers of metal and filler material wound from the I.D. to the O.D. in a spiral arrangement. The filler material is counted on to provide the seal, and the metal is there to provide a restoring spring force for resilience. The metal portion is normally V shaped to enhance its spring characteristics. Metal options include 316 SST as the standard and Inconel 600 for high-temperature applications. The fillers can be a fiber/elastomer blend for standard service and graphite for high-temperature (800°F and higher) or highly corrosive applications. The spiral-wound construction is not recommended for nonrecessed designs unless it is supplied with a metal backup ring, which is a key feature that can be supplied on either the inner or outer edge of the gasket and helps to control the compression on the gasket by absorbing any bolt load left over after the gasket has been crushed to its recommended thickness.

Preload. The concept of *preload* is very important in gasketing design and works to our advantage in two ways. Referring to **Fig. 7.4,** if we first examine a gasketing arrangement where there is no metal ring or recessed groove for the gasket, we can see that the bolt load is transferred through the flanges and sets up and is balanced by a compressive stress in the gasket. This compressive stress results in gasket compression, and the two flange faces move together ever so slightly as the bolting is torqued to the recommended level. Now, if

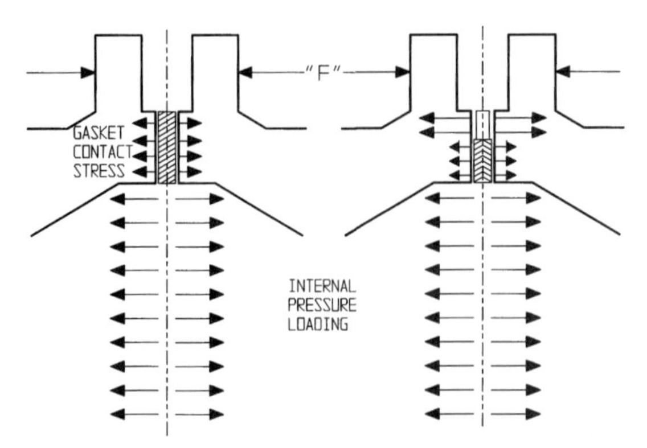

Figure 7.4 Bolting preload—gasket with and without compression limiter.

we begin to apply an internal pressure, it tends to relieve the contact stress in the gasket and increase the bolt load at the same time because the spring rate of the gasket permits the flange faces to move apart as the contact stress is reduced. In effect, the internal pressure load, which is the pressure multiplied by the area inside the gasket sealing diameter, is counteracted by a drop in the gasket contact stress and an increase in the bolt load. Neither of these effects is desirable because the drop in contact stress promotes leakage, and the increase in bolt load results in cycling stresses in the bolting which will lead to relaxation or even failure. Increasing initial bolt load will reduce the fluctuation as a percentage of overall bolt load, but too much initial load can damage the gasket since all the bolt load must be carried by the gasket.

If, on the other hand, we add the metal ring mentioned earlier, we end up with a much better situation. In this case, the initial bolt load can be very high since the load is partially absorbed by the ring or the flange face, and gasket compression is limited by the thickness of the ring or the depth of the groove. Now, if internal pressure is added, the first stress to drop is the ring or face contact stress because they have a very high spring rate compared to the bolting. (Their modulus is about the same as the bolting material and the cross-sectional area is on the same order of magnitude, but their spring length is much smaller.) Because the spring rate is very high, as its contact stress drops in response to buildup of internal pressure, the flange separation is essentially zero. This means that there is no appreciable change in gasket contact stress or bolt load with changes in internal pressure if the metal ring or gasket groove approach is used. The metal ring or groove also helps to contain the gasket. And, finally, if a ring is used, it can extend all the way out to the bolt circle and can be used to properly locate the gasket on the flange face by duplicating the bolt circle holes in the ring. As you've probably guessed by now, the use of a metal ring or groove in conjunction with a gasket is highly recommended. Of course, machining a gasket groove into one of the mating surfaces has the same advantages as the metal backup ring, but it does add some expense, so the metal ring is normally the best choice. Any controlled compression design, as just discussed, assumes that the gasket used has a relatively high resilience, so this approach is not usually recommended for flat-sheet metal gaskets.

Bolting. As far as *bolting* techniques are concerned, always employ a crisscross pattern to keep the flange faces straight as torque is increased, and liberal amounts of an acceptable lubricant should be applied to the threads and nut face to make sure that the torque is being properly converted into an axial bolt load. The lubricant will also help to facilitate disassembly. Torques should be tightly controlled if a flat-sheet gasket is being used without a metal ring or gasket groove. This is to ensure that the bolt load is high enough to prevent leakage but not so high as to damage the gasket. With a metal ring or groove, torque control is less critical and should lean toward the high end to ensure a tight joint and to reduce bolt load fluctuations. It's also a good practice to retorque line bolting after letting the joint set for a time to help counteract any relaxation. This is especially true of new bolting.

Line loads can add a significant amount of load to the line flange joint. They can result from thermal growth, fluid inertia, and pipeline weight. While good piping support design can help to minimize them, they'll never be completely eliminated and can cause leakage in what might otherwise be a tight joint. This is still another reason to use the metal ring or groove approach whenever possible since it is much less susceptible to leakage caused by this type of loading.

High pressure. In high-pressure applications, always use either a backup ring or a gasket groove to protect the gasket. At extremely high pressures, the gasket groove approach with a ring joint is usually used for line flange joints, and any one of several recessed designs might be used for the body-to-bonnet joint. Typical joint cross sections for high pressure are shown in **Fig. 7.5.** The most common joint for valve line flanges is the API octagonal, and the Delta and Bridgeman are sometimes used for bonnet joints.

The Bridgeman is sometimes called the pressure seal and is used in high-pressure designs to reduce the bolt load required to seal. As seen in **Fig. 7.6,** it utilizes a solid metal ring with an angle on it to seal against a hardfaced surface on the body. The ring is dropped down on top of the matching face on the bonnet, and then a slight preload is set up to force the seal out against the body. As pressure builds up inside the body, the upward thrust on the bonnet increases, forcing the ring

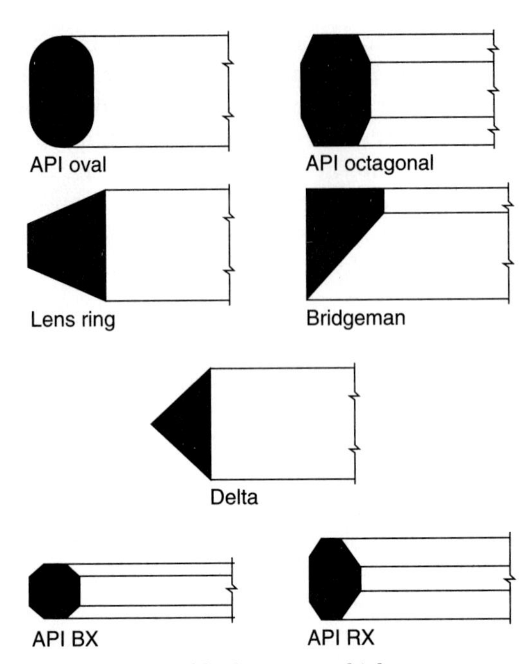

Figure 7.5 Typical high-pressure, high-temperature metal gaskets. (*Courtesy of Fluid Sealing Institute, Philadelphia, Penn.*)

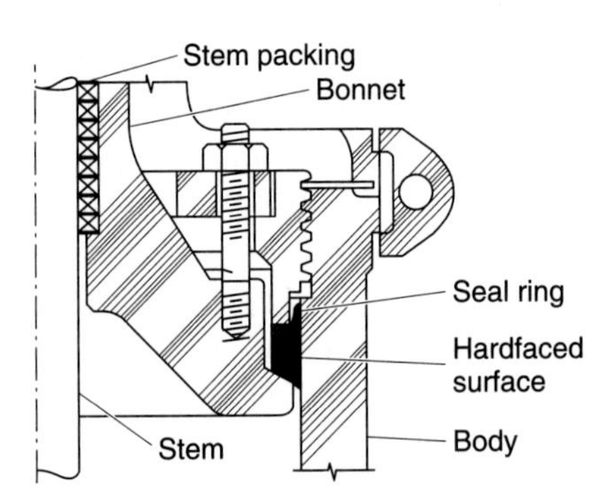

Figure 7.6 Pressure seal design.

to press even harder against the body, improving the seal. This pressure-assist helps to keep the seal leaktight while permitting the joint itself to be much less massive since no heavy bolting or flanges are required as they would be in a conventional design. Getting and keeping a tight seal in this type of design does require that the parts be machined to tight tolerances, and the surface finish on the seal and the hardfaced portion of the body have to be very good.

Line flange gasketing is fairly straightforward and can be accomplished successfully following the general guidelines covered to this point. The bonnet and internal gasketing on a control valve are a little more complicated and deserve some specific coverage.

First of all, on many control valves the bonnet-to-body bolting has to resist pressure as does a standard gasketed joint, but it also has to provide the load to seal the seat ring gasket along with resisting the actuator load transmitted to the body assembly. The valve shown in **Fig. 7.7** is an example of a typical construction. If we examine the gasketing arrangement in more detail (**Fig. 7.8**), we can see how all these parts work together to guard against leakage to the atmosphere and to prevent leakage across the valve.

The bonnet gasket that seals between the bonnet and body is typically a flat-sheet design contained in a recessed area in the body. While the gasket is contained, there is no metal-to-metal contact to protect the gasket from being crushed, so torque should be controlled. The outer half of the bonnet lip loads the gasket against the body to prevent external leakage, while the inside half provides the load for spiral-wound, cage, and seat-ring gaskets. These gaskets help guard against leakage around the cage. This is especially important for balanced

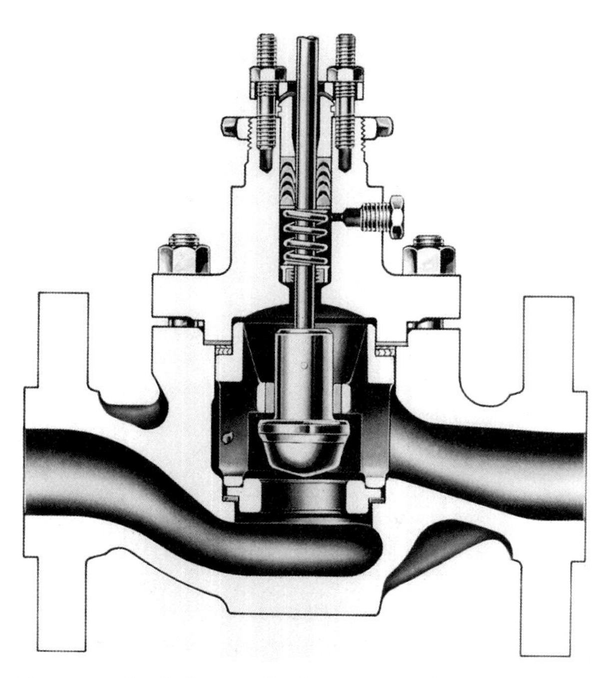

Figure 7.7 Typical control valve construction. (*Courtesy of Fisher Controls International, Inc., Marshalltown, Iowa.*)

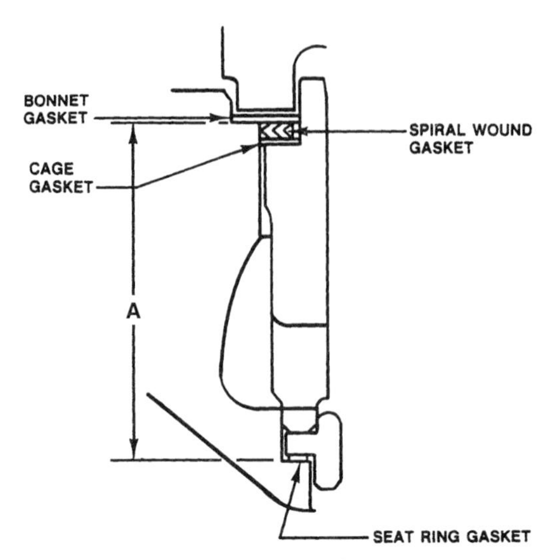

Figure 7.8 Close-up of gasket arrangement. (*Courtesy of Fisher Controls International, Inc., Marshalltown, Iowa.*)

designs where the cage can see the full shutoff pressure drop. The lower gasket keeps the fluid from leaking around the seat ring, which sees full pressure drop whether the valve is balanced or unbalanced.

The characteristics of the spiral-wound gasket are very important in ensuring that the three leak paths just identified are effectively sealed. In this type of assembly, the spiral-wound gasket actually serves as a spring as well as a seal. As described earlier, the bolt load is shared between the external and internal seals, and the extent to which it is shared is a function of the stiffness of the spiral-wound gasket, the stackup of the internal parts, and the face-to-face dimension *A* shown in **Fig. 7.8.** As we increase bolt load on the assembly, the first thing that happens is that the spiral-wound gasket begins to compress as the load on the inside of the bonnet lip increases. Eventually, if everything is sized properly, the load and deflection will be high enough that the outer lip will begin to load the outside edge of the bonnet gasket against the body. At the final bolt load levels, the outer edge of the bonnet gasket should be loaded enough to properly seat and seal against pressure, while whatever load that has built up in the spiral-wound gasket will also be seen in the cage and seat ring gaskets since they are in series.

If the spiral-wound gasket is too stiff, too much of the total bolt load will go into compressing it and the internal gaskets, and the bonnet joint will leak to atmosphere. If it is too flexible, not enough load will be built up in the internal gaskets, and they will leak, especially the seat ring gasket since it sees full pressure drop on a regular basis. In addition to the stiffness of the spiral-wound gasket, the stackup height of the internal parts and the face-to-face dimension *A* are also very important in determining how the bolt load is shared between internal and external gaskets. If *A* is too big or the stackup is too small, the internal gaskets won't seal properly. Conversely, if *A* is too small or the stackup is too big, the bonnet gasket won't be loaded properly and won't seal. All the elements have to be right to avoid leakage. Using *non-OEM gaskets* or *parts* or machining on the body or internal parts can all result in this fine balance being lost and need to be avoided if this type of gasketing arrangement is employed.

A simpler arrangement that is gaining favor in recent designs for the chemical process industry is shown in **Fig. 7.9.** In this case, a single gasket is used to seal the bonnet-to-body joint, and a second gasket seals under the seat ring. Given the machining tolerances on the body and internal parts, the lower gasket has to be relatively thick and flexible because the compression on this gasket can vary with the tolerances. Once again, the spiral-wound gasket is a good choice for this reason. If we make this valve unbalanced and characterize the

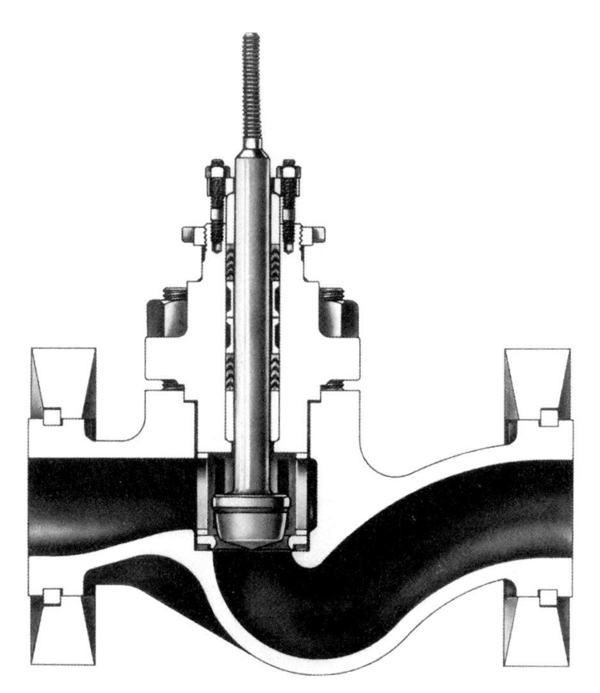

Figure 7.9 Alternate gasket arrangement. (*Courtesy of Fisher Controls International, Inc., Marshalltown, Iowa.*)

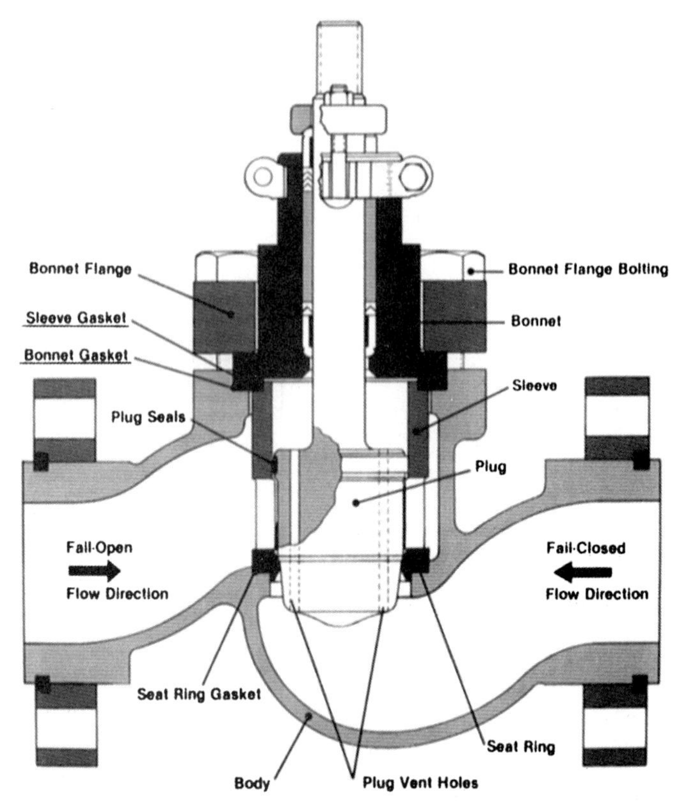

Figure 7.10 Balanced valve gasketing arrangement. (*Courtesy of Valtek International, Inc., Springville, Utah.*)

flow with the plug, we can eliminate the possibility of a pressure drop across the cage and don't need the cage gasket as a result. **Figure 7.10** shows the additional cage or sleeve gasket that needs to be added if a balanced design is used. This figure also illustrates that a separable flange arrangement can be used on the bonnet joint much like that on the line flanges.

These arrangements are all susceptible to problems caused by thermal or pressure gradients. The gradients, if severe enough, can result in internal gasket relaxation and leakage because the length of the cage changes and modifies the load on the gaskets. Therefore, for severe service applications, the use of some other method of seat ring retention and gasket loading is recommended. The use of screwed-in or bolted-in seat rings is one option since it doesn't rely on the cage to load the lower gasket. An example of this is shown in **Fig. 7.11** where

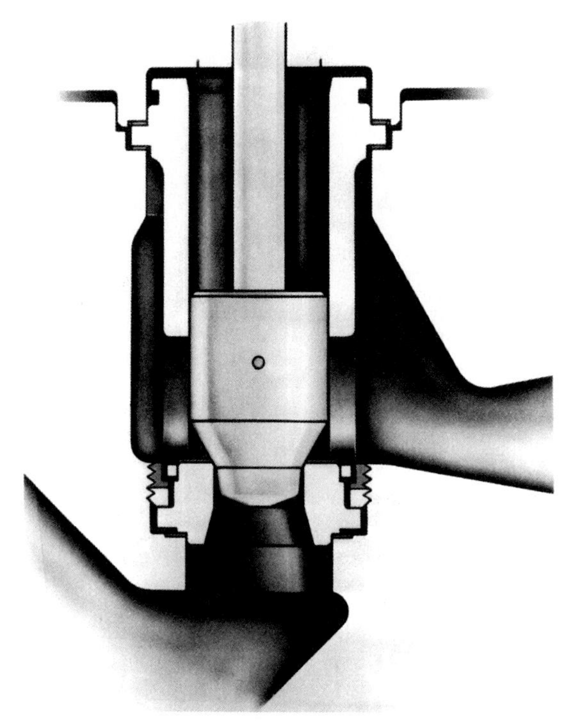

Figure 7.11 Severe service seat ring gasket design. (*Courtesy of Fisher Controls International, Inc., Marshalltown, Iowa.*)

the cage is held tightly between the body and bonnet at the top and allowed to grow axially without affecting the load on the seat ring gasket. In this case, a large threaded seat ring retainer is used to load the seat ring gasket, but a bolted arrangement could also be used. **Figure 7.12** shows a close-up of the area under the seat ring for two different seal designs.

The other gasketing approach that one might encounter in the chemical industry is the split-body arrangement where there is no bonnet or bonnet gasket (**Fig. G.4**). The seat ring is captured between the two halves of the body, and all that is required is a top and bottom gasket on either side of the seat ring to seal against the body half. For reasons explained in Chap. 2, this design has been losing favor because, while it does simplify the gasketing arrangement, it requires that the body come out of the line for even routine maintenance.

In summary, there are many different gasket materials, types, and gasketing configurations used in the chemical process industry.

Figure 7.12 Close up of seat ring gasketing for severe service. (*Courtesy of Fisher Controls International, Inc., Marshall-town, Iowa.*)

However, if we understand the basic rule that the effectiveness of the seal depends on surface preparation and load, and we follow the checklist outlined below, gasketed seals should not be a problem. This is particularly important now with the emphasis on reduced fugitive emissions in the chemical process industry since any gasketed joint represents an opportunity for leakage. The things to check are:

- Make sure that the surfaces to be sealed have been properly cleaned and that no surface damage is present that could result in an across-the-gasket leak path. Follow the manufacturer's recommendation on surface finish or the use of concentric serrations to improve the sealing capability of the gasket.
- Make sure that the gasket is stored and handled with care to avoid damage that might result in an eventual leak.
- Select an appropriate gasket for each application. Considerations include chemical compatibility, pressure, temperature, resilience, and, in some cases, stiffness or spring rate.
- Make sure that the gasket is properly located between the two mating surfaces.
- Utilize good bolting practices, including a crisscross torquing pattern to assure even loading, heavy lubrication of threads and nut faces, and torque control.
- Utilize recessed or contained gasket designs whenever possible due to the reduced risk of blowout, improved control of gasket compres-

sion, and the ability to utilize preload to reduce the load fluctuations in the bolting. Recessed designs work better with gaskets with high resilience.

- Stick with OEM gaskets and internal parts to make sure that the gasket characteristics and part dimensions are such that the gasketed joints can perform as designed.
- Be aware that any modification of the critical *A* dimension in the body or the dimensions of the internal parts can result in improper loading of the gaskets and leakage. Proceed with caution in remachining parts that could affect gasket loads, as described earlier in this chapter.

7.2 Seals

Seals, as discussed here, will differ from gaskets in that they are asked to seal between parts with *relative motion.* The two primary applications for control valves are the seal between the cage and the plug for a sliding-stem valve and the seal between the ball, disc, or plug on a rotary valve.

Looking at sliding-stem valves first, the plug is the only internal part that moves. As explained in Chap. 2, if the valve is a balanced design, a second leak path exists between the cage and the plug. This leak path needs to be sealed, which is accomplished through the use of a dynamic seal of some kind. A typical example is shown in **Fig. 7.10,** where the seal rides in the plug and seals against the cage. Occasionally the seal may be held in the cage rather than the plug, but there is really no functional difference. Close-ups of several seal designs are shown in **Figs. 7.13** and **7.14.**

In deciding which type of seal is best for a given application, consider fluid compatibility, seal friction, temperature limits, service life, required shutoff, and ease of installation and maintenance. The seals break down into two primary groups: low and high temperature, with the break occurring at about 450°F. The low-temperature group can then be split again with elastomeric materials being used up to 300°F and TFE-based seals used up to 450°F. At high temperatures, a piston ring approach is used rather than a seal, and the piston rings can be made from either graphite or any one of the softer metals. In general, the low-temperature seals provide much tighter shutoff than the piston rings because they can conform better to the I.D. of the cage. They also exhibit lower friction characteristics, which improves valve response. While elastomeric seals can be used, the TFE-based seals

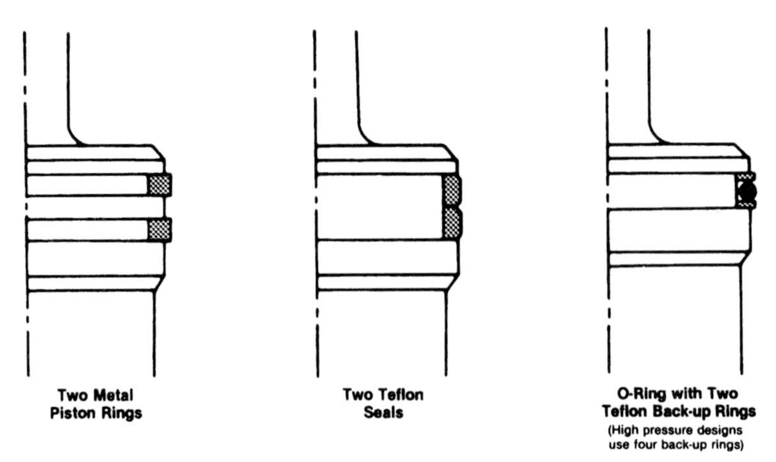

Figure 7.13 Balanced plug seals. (*Courtesy of Valtek International, Inc., Springville, Utah.*)

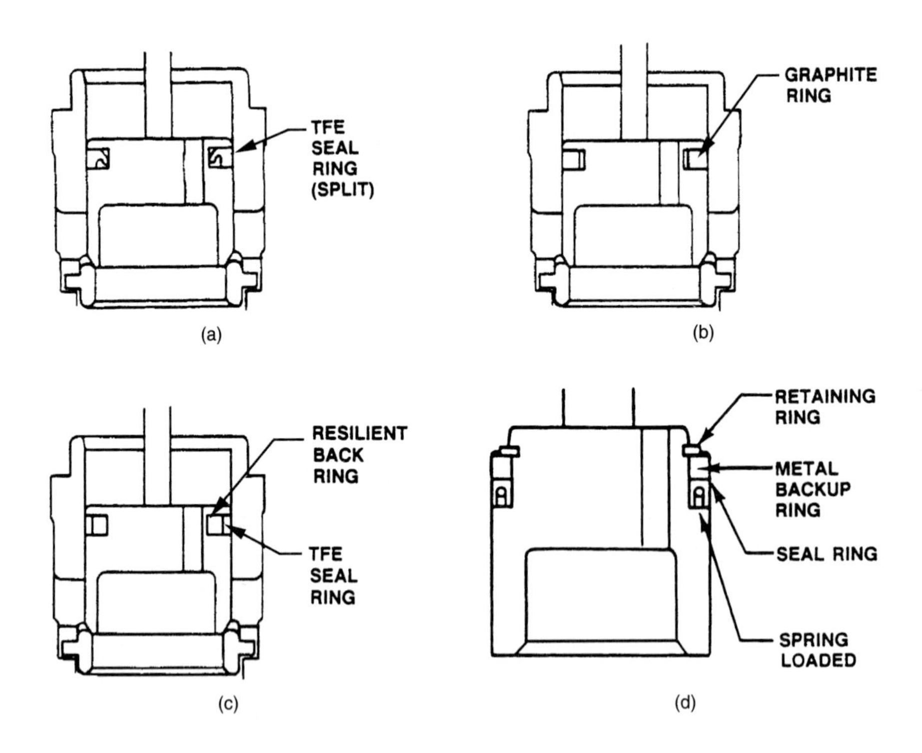

Figure 7.14 More balanced plug seals. (*Courtesy of Fisher Controls International, Inc., Marshalltown, Iowa.*)

provide broader chemical compatibility, temperature performance, and abrasion resistance. As a result, the TFE seals are recommended for nearly all low-temperature applications, and the balance of this low-temperature discussion will focus on their use.

In the low-temperature TFE seals there are a number of different styles that can be discussed from a leakage and installation standpoint. TFE-based materials have a tendency to *cold-flow* under load, so most seals of this type feature some type of fiber reinforcement to help with this problem. Even with the fibers, the seal will eventually take a set that reduces the ability to seal against the cage. As a result, the simple TFE constructions shown in **Figs. 7.13** and **7.14*a*** and ***b*** are generally used on class II and III shutoff valves where leakage is not a big concern. To improve the sealing performance, a more resilient material such as an elastomer or a metal spring can be used in conjunction with the TFE such as the seal in **Fig. 7.14***d*, which is formed in a U shape around a metal spring that helps to keep the seal pressed against the cage. The shape of the seal is specially designed to facilitate the buildup of pressure inside the seal, which helps to keep the lip of the seal tightly pressed against the cage. This design is the most effective of the TFE-based seals shown and can easily provide better than class V shutoff performance for the valve it's installed in. It is commonly referred to as a pressure-assist design and greatly improves the performance of the seal, but for it to work, be careful to insert the seal with the opening of the U shape pointing in the right direction.

An easy way to remember the proper technique for *seal orientation* is to always point the opening of the seal in the same direction as the flow through the seat ring (i.e., if the flow is up through the seat ring, the seal should be installed so that the opening is also pointed up). Because the valve is balanced on a flow-up application, the fluid goes up through the balancing holes in the plug and would attempt to flow back down between the plug and the cage and out through the outlet of the valve. If the seal is installed properly with the U-shaped opening pointed into the fluid, the pressure can build up inside the cavity and can help to obtain a tight seal. On flow-down applications, the seal has to be reversed. This might seem trivial, but it's amazing how many times excessive leakage can be traced to incorrect installation of this seal. Murphy's law being what it is, if something can be installed backward or upside down, it will be.

As for *installation,* there are three possible methods of getting the TFE-based seals into the cage grooves. The easiest technique is to cut the seal on a 30° angle so that it can be opened up and slid over the top of the plug into the groove. The cut will result in more leakage and could damage the plug or cage if fluid velocity past the cut is high enough to cause erosion. If multiple rings are used, it is always a good idea to stagger the cuts so that they are not lined up to the point where fluid velocity could become excessive.

Another procedure that is sometimes used is to heat the ring to about 300°F so that it grows in size and becomes flexible enough that it can be slid over the edge of the plug. As the ratio of diameter to seal thickness becomes larger, the ring is easier to slip over the plug, and in large sizes it may be possible to install the ring without heating it if it is pushed on very slowly so that TFE has a chance to cold-flow. Take care not to roll the seal over as it is pushed over the plug or it may be damaged. On seals with elastomeric backup rings, the technique is essentially the same since the backup ring can be very easily slid over the plug and into the groove.

Seals with integral metal spring elements can't be cut or the spring would lose its effectiveness. On larger sizes they can be slipped over the plug without too much trouble, but on the small sizes an arrangement like that shown in **Fig. 7.14*d*** is necessary. In this case there is no groove. The seal is dropped into place and then a metal backup ring and a retaining ring are added.

Service life depends to a great extent on the finish of the parts that the seal runs against, in most cases, the cage. The goal should be a 32 rms or better finish, and regular maintenance should include an inspection of the cage bore. Any roughness or pitting should be ground out if possible and the bore I.D. checked to make sure that wear has not resulted in excessive clearances between the plug and cage that would permit extrusion of the seal material. One aspect that improves service life is that the seal is not subjected to full pressure drop when the valve is open. Another concern is abrasive fluids that can cut into and damage the relatively soft seals. Higher fiber content can help delay the onset of damage in this case.

To remove seals that weren't cut during installation, simply pry them out or cut them since they should not be reused anyway. Since they are relatively inexpensive, it is recommended that they be replaced anytime the valve is disassembled, especially on valves where tight shutoff is critical.

The *high-temperature piston ring seals* used do not provide the same degree of shutoff because they are more rigid and do not conform as well to the cage I.D. The two primary material types in use today are metal and graphite. The metal rings can be any of the relatively soft metals that serve as good bearings, but make sure they can stand up to whatever chemicals are present (Ni-resist is one material that is frequently used). The graphite rings are very good from a corrosion standpoint and tend to be self-lubricating, so friction is lower for a given level of leakage. But they are also relatively brittle and tend to erode more quickly than their metal counterparts.

Additional rings can be added to help improve shutoff, and triple piston rings are generally good for better than class IV shutoff. However, they do add friction and there is not always room for more than one ring on the plug. As a result, the most common construction is a single ring, which gives no better than class II shutoff.

Installation can be tricky for these types of rings. A metal or graphite ring could be made in one piece and then a retaining ring used above it to hold it in place, but experience has shown that the clearance between the ring and the retainer allows the ring to vibrate until it breaks. As a result, other methods have to be used for the metal and graphite rings. Many metal designs are cut on one side so that they can be pried open and slid over the plug into the groove. If multiple rings are used, make sure to stagger the cuts to help minimize leakage. The graphite rings are usually supplied as a one-piece ring and need to be broken in two places just before installation. It is usually recommended that the ring be scored on the outside edge in two places about 180° apart. The line that is scored in the ring should run axially and should not be too deep. To make the first break, make a fist and then grasp the ring in one hand with the ring between the middle and ring fingers. The back of the ring should rest against the fleshy part of the palm below the thumb. The mark where we want to make the first break should be about 1 to 2 in away from the front of the fist. Now, strike the ring near the scored mark against a sharp metal edge of some kind while absorbing most of the impact on another surface with the bottom of the fist. The ring should break cleanly at or near the mark. It can then be pried apart until it breaks at the other mark. This does take some practice, and someone doing it for the first time may want to have an extra ring or two on hand.

Once the ring has been successfully broken into two distinct pieces, they can be inserted into the groove, making sure to properly match the ends of the rings together. Once again, if multiple rings are being used, be sure to stagger these breaks whenever possible.

Metal rings tend to last longer than graphite ones and are less likely to break due to vibration or shock, but they add more friction and don't seal as well. As with their TFE-based counterparts, service life for metal or graphite piston rings depends very heavily on the surface finish of the cage I.D.

To summarize, whenever the service temperature is less than 450°F, utilize TFE-based seals because they provide superior shutoff and lower friction and are easier to install than metal and/or graphite seal rings. If the temperature is above 450°F, or if there is a danger that the TFE seal may be eroded due to fluid characteristics, it may be necessary to use metal or graphite, but the shutoff will not be as good. TFE, because of its tendency to cold-flow, should be used with some sort of backup that will improve its resiliency.

7.3 Packing

Valve packing is the name given to whatever material is used to provide a dynamic seal between the primary pressure boundary of the control valve and the valve stem or shaft. It is one of the most troublesome and highest maintenance areas on a control valve for four reasons. First, it is a dynamic seal against a part that is moving continuously as it is being subjected to relatively high pressure drops, thermal transients, and significant side loads. It's obvious that this is tough service. Second, any failure is normally easy to detect since leaks go into the atmosphere where they can be detected visually or with some of the sophisticated "sniffing" equipment now available. Third, the recently enacted Fugitive Emissions restrictions (EPA Clean Air Act) have made leakage to the surrounding atmosphere a "hot button" with plant management, and packing leakage is the most common form of fugitive emission. And fourth, packing design and the factors affecting performance are not all that well understood by the typical end user. Each of these four areas will be addressed in more detail below.

7.3.1 Basic construction

A typical *packing box* construction is shown in **Fig. 7.15.** The packing box is the area of the valve where the packing is located and includes

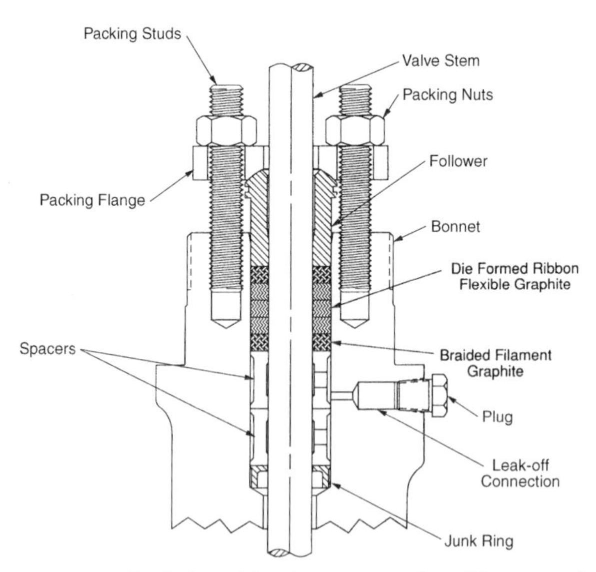

Figure 7.15 Typical packing box construction. (*Courtesy of Fisher Controls International, Inc., Marshalltown, Iowa.*)

the packing itself and all the parts that are used to contain and/or load the packing. In the construction shown, graphite packing material is held in a confined space surrounding the valve stem or shaft. The packing nuts are torqued to a specified level which stretches the packing studs and transmits a load to the packing flange. The packing flange contacts the packing follower which, in turn, applies a load to the packing itself. The packing is made up of two types of graphite material: die-formed ribbon, which forms the primary seal against the stem, and braided filament, which acts as an antiextrusion device to help contain the ribbon packing in the sealing area. Underneath the packing, *lantern rings* are employed as spacers to help fill a portion of the packing box cavity, and a *junk ring* completes the assembly by providing a flat base upon which the whole assembly rests. Junk rings are necessary because the easiest and simplest way to machine a packing box cavity is with a standard drill, which leaves a hole with a tapered bottom.

While this is a *typical construction,* there are a multitude of different arrangements that might be encountered in a plant. Some examples are shown in **Figs. 7.16** through **7.22.** While there are a large number of different arrangements, they can be broken down into three major groups: single, double, and leak-off. In a single construction, there is one set of packing that forms the seal against the stem. In a double construction, there are two separate groups of packing, with the lower group providing the primary seal against the process fluid. The upper packing set might be added for a couple of reasons. First of all, it may be there to contain any leakage through the primary set. It can simply serve as a backup, or it may permit the leakage to be contained and piped off through a *leak-off connection* to some type of secondary storage system. This prevents the leakage from

Figure 7.16 Typical packing arrangements. (*Courtesy of Fisher Controls International, Inc., Marshalltown, Iowa.*)

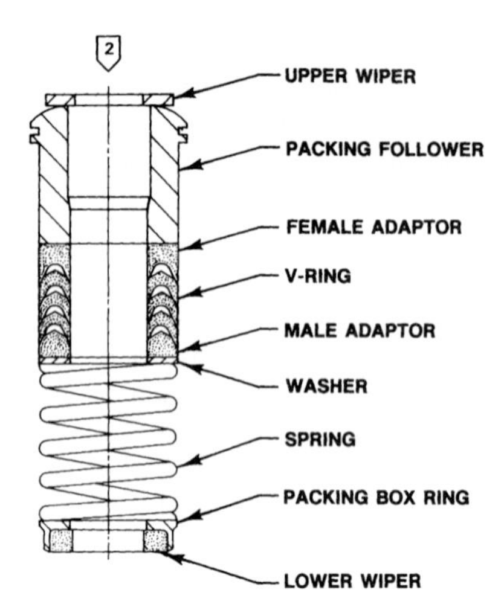

Figure 7.17 Typical single-packing arrangement for PTFE V-ring packing. (*Courtesy of Fisher Controls International, Inc., Marshalltown, Iowa.*)

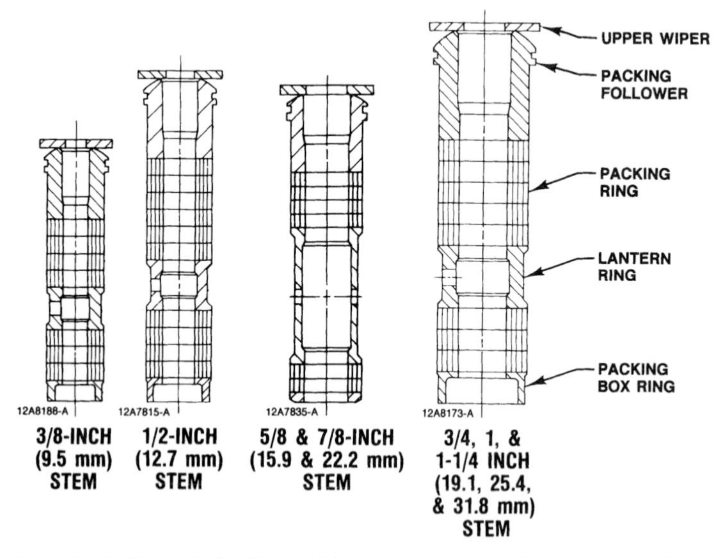

Figure 7.18 Typical double-packing arrangements for square packing. (*Courtesy of Fisher Controls International, Inc., Marshalltown, Iowa.*)

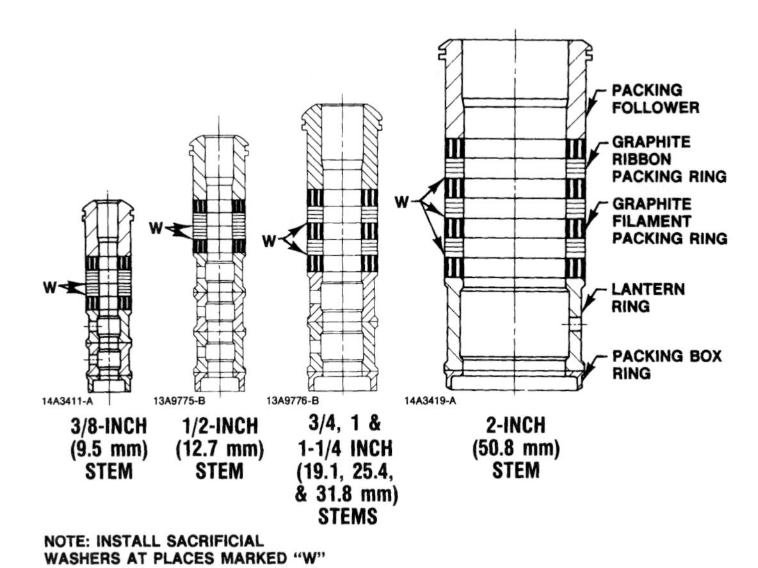

Figure 7.19 Typical single-packing arrangements for graphite ribbon/filament packing. (*Courtesy of Fisher Controls International, Inc., Marshalltown, Iowa.*)

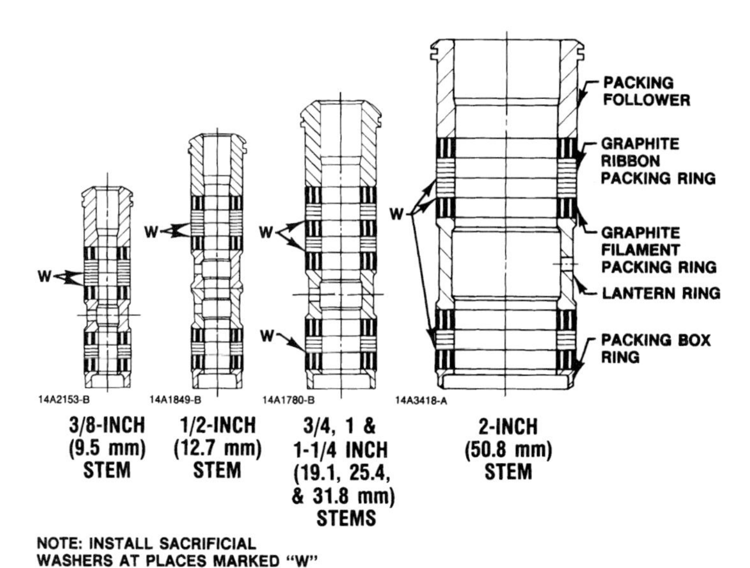

Figure 7.20 Typical double-packing arrangements for graphite ribbon/filament packing. (*Courtesy of Fisher Controls International, Inc., Marshalltown, Iowa.*)

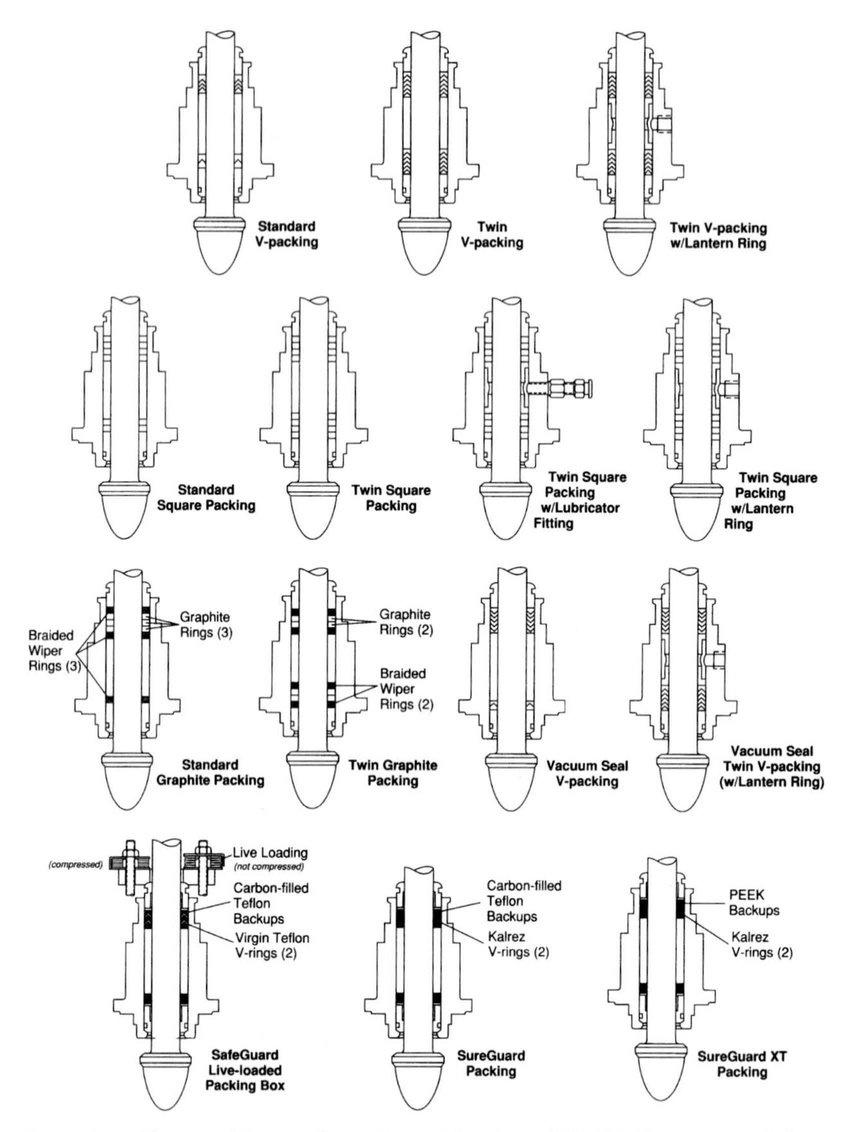

Figure 7.21 More packing configurations. (*Courtesy of Valtek International, Inc., Springville, Utah.*)

going directly into the atmosphere. In another case, the space between the two sets of packing can be used to inject lubrication that helps keep the packing friction low and also helps to reduce leakage since the lubrication is normally more viscous than the process fluid itself and, hence, easier to seal. An example of a *lube system* is shown

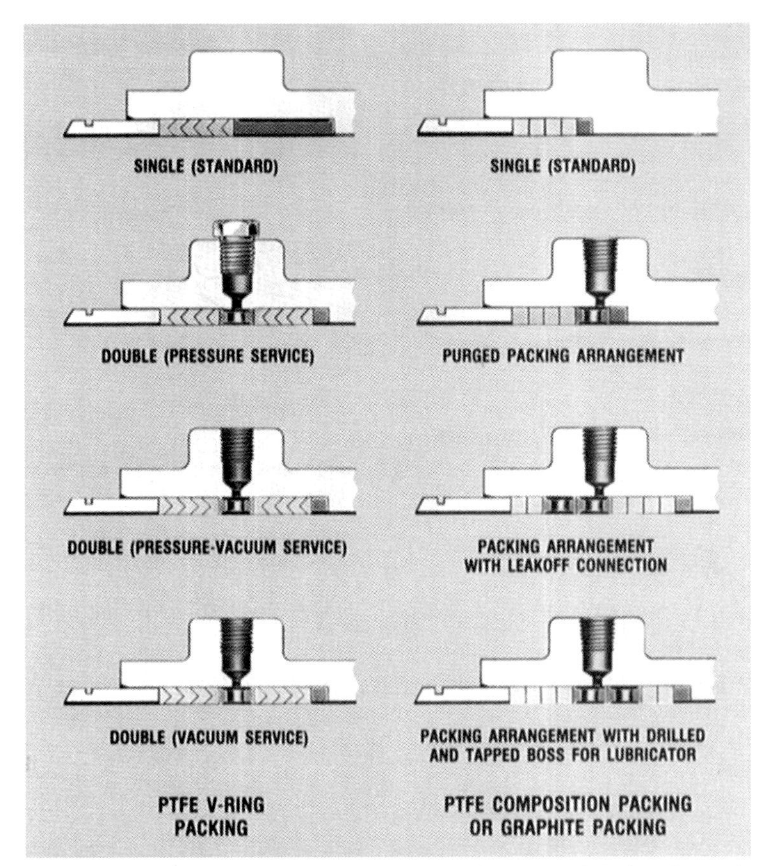

Figure 7.22 Rotary packing arrangements. (*Courtesy of Fisher Controls International, Inc., Marshalltown, Iowa.*)

in **Fig. 7.23.** This type of approach was common in the days when asbestos was used for valve packing, but it is not normally used with modern-day packing materials. Whenever possible, the lube arrangements should be removed and replaced with a plug since they serve no useful purpose and add to maintenance headaches since they represent another possible leak path. If a leak-off or lube system is used, a lantern ring needs to be used between the two packing sets. The lantern ring has cavities and through-holes that ensure that any leakage through the primary set can find its way to the leak-off connection or that any lubrication that is added through the leak-off system is equally distributed into the cavity between the two sets. If there is no leak-off or lube system, a simple spacer can be used to hold the two sets apart.

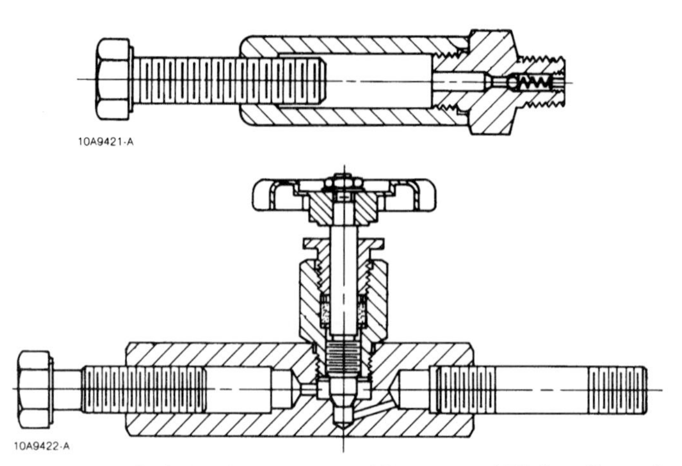

Figure 7.23 Lubrication systems. (*Courtesy of Fisher Controls International, Inc., Marshalltown, Iowa.*)

In addition to the three major groups, single, double, and leak-off, there are some other considerations that can affect the packing construction. TFE V-ring packing is pressure assisted due to the shape of the packing itself, but it usually includes a spring inside the packing box to help establish an initial packing load. The inside of the V shape should always run against the direction of leakage flow so that the flow tends to open the V and assist the seal. This means that the V should point down toward the fluid for standard applications but up toward the atmosphere for vacuum service.

V-ring pressure-assist packing is preferable because it provides the best seal with the lowest friction. The other option is *jam-style packing*. In contrast to the pressure-assist approach, jam-style packing requires that all the sealing load be developed in the packing through the load transmitted through the packing follower. There is no pressure assist. More massive designs are required in this case, and friction is higher because of the higher loads.

Like the gaskets covered earlier in this chapter, the sealing effectiveness of packing depends on surface preparation and sealing load. If the surface to be sealed against is properly prepared and the packing can be loaded to the point where it fills the microvoids in the material, a seal will be established. There are actually two areas that are sealed in a normal packing box arrangement—the inside diameter of the box and the valve stem or shaft. The inside of the box is a static seal since the box does not move in relation to the packing. For this reason, this seal is relatively easy to maintain. If the surface doesn't have any major flaws or scratches running parallel to the axis of the stem and the overall surface finish is 32 rms or better, this area should present no problems from a sealing standpoint.

The *stem seal* is another matter altogether. The valve stem is in constant motion, so it is difficult for the packing material to fill the voids in the stem material and stay there. For this reason, stem finish is critical, as is the ability of the packing to react to this constant motion while maintaining intimate contact with the stem. This will be covered in more detail in the next section.

Assuming that the proper sealing load can be reached, the next challenge in any packing system is in *maintaining the load* for an acceptable service life period. This is actually the most difficult part of any packing design. Nearly everything is acting to reduce load over time and cycling life. TFE, one of the most common packing materials, has a tendency to *cold-flow* under load, so it is always looking for an opportunity to escape the confinement of the packing box area. This makes it extremely important that *clearances* between the packing box, junk ring, lantern rings, and packing followers be kept to a minimum. Excessive clearances will permit most packing materials (but especially TFE) to escape past the metal parts, resulting in a loss of load and leakage. At the same time, if the clearances are made too small, or tend to close up during service due to differential thermal expansion, the metal parts can contact one another, resulting in scratches that promote leakage. The designer has to walk a very fine line in setting clearances to avoid these two problems. Recent designs have begun to use an intermediate antiextrusion ring between the metal parts and the primary sealing rings for this reason. The antiextrusion rings can be any one of several materials (carbon-graphite or fiber-reinforced TFE) that are stiffer than the primary sealing material and therefore less prone to extrusion but are softer than the metal packing box parts so that the chance of damage to the stem is greatly reduced.

In addition to extrusion, *wear* can also result in material loss that drops the effective sealing load on the packing. Keeping the sealing load to the minimum required to seal will help reduce wear, as will making the stem finish as fine as possible. The antiextrusion measures already discussed also help the wear situation since they tend to keep packing material that has worn away inside the packing box cavity.

The load that provides the seal between the stem and the packing actually acts perpendicular to the surface of the stem while the load generated in the packing studs is parallel to this surface. For V-ring packing, this is not important since the primary sealing force is developed by the internal pressure, not the packing studs. In jam-style packing, however, the studs do provide the sealing load, so this stud load has to be transformed into a load acting normal to the stem surface through the elastic-plastic deformation of the packing material.

This conversion differs with each packing material and will be covered in more detail in the following sections by material type, but it has to take place for the packing to function. In general, the stiffer the material is, the less efficient this transformation, so we need higher stud loads for the same sealing loads with stiffer materials. Also, note that, like gaskets, the sealing load required varies directly with the pressure to be sealed against. As the internal pressure increases, so does the packing stud load required to effect a seal.

One final point on packing performance and its effect on valve performance: The goal of obtaining a tight stem seal in the packing area flies in the face of securing optimum *control valve throttling performance*. The stem seal adds friction which has to be overcome by the actuator anytime the valve is called upon to change position. Improving stem seal performance nearly always means increasing friction, and increasing friction tends to slow down valve response time and can frequently result in limit cycling by the valve. Limit cycling is a phenomena where the valve cannot find the perfect position corresponding to the process setpoint and jumps back and forth across this position. The cause of this behavior can sometimes be traced to the difference between *static* and *dynamic friction* in the valve packing. The valve starts in a given static position that does not correspond to where the control system wants it. As the controller begins to change the input to the valve, the valve system reacts by changing the actuator pressure in the direction that would tend to move the valve to the desired position. To initiate movement, the actuator pressure has to change enough to overcome the static friction in the valve assembly, most of which is located in the packing box.

Unfortunately, once the movement has begun, the friction can drop drastically, and the valve tends to *overshoot* the ideal setpoint because of this drop in friction. After the overshoot occurs, the process starts all over again in the opposite direction. The higher the friction, and the larger the ratio between the static and dynamic friction, the greater is the tendency for this limit cycling to occur. Limit cycling results in unnecessary wear to all related components as well as high *variability* between the process variable and the setpoint, so it is something that should be avoided, if possible. Since effective packing seals tend to increase friction along with the static-to-dynamic ratio, they tend to hurt control performance when compared to a valve where the packing is leaking and friction is low. In the past, many valves were probably leaking through the packing, which ended up resulting in low friction and satisfactory control performance. Now, with the increased emphasis on reducing packing leakage, many loops are beginning to show signs of control problems that didn't exist when the packing was permitted to leak. While there are no easy solutions to this dilemma, the balance of this chapter will attempt to provide guidance on addressing this seeming contradiction between good control and tight packing seals.

7.3.2 Stem condition as it relates to packing performance

Over the past several years, as the emphasis on improving stem seals has increased, a number of firms have been marketing packing systems that are "guaranteed" to provide tight shutoff. These same firms have also made claims that over 50,000 valves have been retrofitted with their systems and have no leaks. First of all, their systems normally include some type of proprietary packing material or shape, sometimes coupled with a live-loading system that helps to maintain packing load over time and cycling service (see Sec. 7.3.5). While packing material and load are important in providing a good seal, they cannot guarantee good performance without considering a number of other elements, including stem condition. If the stem condition is not right, no system, no matter how sophisticated, will provide good performance. The stem condition and its effect on packing performance will be covered in detail in this section, with this in mind.

The second point that needs to be made regarding these claims is that, as noted above, packing seal performance cannot be examined in and of itself. It can have a very marked effect on overall valve performance. Any packing performance improvement program needs to recognize this, or a plant can end up with all its control valves sealing very well at the packing but no longer moving in response to changes in the control system output.

One of the most important elements of stem condition is *surface finish* since it affects sealing performance, friction, and packing wear. The better the finish, the lower the wear and friction and the better the seal. As a result, many manufacturers recommend a surface finish of 4 rms or better on the valve stem. What this means is that the stem must actually be ground or burnished to almost a mirror finish, and this finish must be maintained over the length of the stem that comes in contact with the packing as the valve strokes. To get this kind of finish, a very hard surface needs to be used for the valve stem, and manufacturers have found that strain-hardened SST, 17-4PH, and hard chrome plating provide the best combination of hardness and corrosion resistance for most applications. The high hardness also helps to resist wear in the area of the stem in contact with the packing. With the harder packing materials such as graphite, high packing stresses and constant movement can combine to result in stem wear of 0.002 to 0.003 an in in the area where the valve throttles most of the time. This wear reduces the packing stress and can result in leakage. It also requires the packing to react to the taper in the shaft as it passes from the worn portion to the area of the stem that has not changed dimensions. *Taper,* whether it results from wear or the original machining of the stem, makes optimum packing performance difficult to achieve since it requires a certain amount of resilience that materials like graphite don't have. Anytime graphite is used, try to ensure that stem taper is kept to a minimum, or packing performance will suffer. Proper machining techniques and hardened stem materials will help to accomplish this goal.

Another possible source of stem taper is *thermal gradients* that can exist along the stem, especially in high- or low-temperature applications. There's not much that can be done to prevent this since any insulation that would reduce the gradient will also result in raising or lowering the packing box temperature, which can hurt overall packing performance more than reducing the gradient will help.

A *bent stem* will affect the packing much like a tapered stem since it requires that the packing move laterally as the valve stem cycles. This, at best, results in increased packing wear, and at worst, the packing leaks because it can't react enough to the lateral movement to maintain a seal. Again, depending on the inherent resilience of the packing material being used, anything less than a perfectly straight stem can result in leakage. Graphite is particularly susceptible to this type of problem since it is much harder and less flexible than TFE-based packings. Like taper, a stem that is bent more than 0.003 in over the stroking length is cause for concern with graphite packing. TFE-based packings can stand 2 to 3 times as much lateral movement as graphite, but even with TFE, the stem should be as straight as possible to maximize service life.

The other major concern regarding the stem is how well it is *guided.* If, as the valve strokes, the stem is free to wander in the lateral direction with only the packing to resist it, the sealing ability of the packing will be compromised. This wandering can be caused by poor guiding in the valve or actuator, or it can be the result of misalignment between the valve and actuator stems. Guiding can be and has been improved through the use of carbon-graphite bearings above and below the sealing packing rings (**Fig. 7.24**). These bearing rings complement the guiding in the valve and actuator and also serve as antiextrusion devices since they are much harder than the sealing rings. It should be noted, at the same time, that they are softer than the metal packing box parts, so there is no risk of damage to these parts.

While the use of supplemental bearings in the packing area does reduce the amount of lateral movement in the packing and helps to reduce the risk of extrusion, it is no substitute for good guiding in the

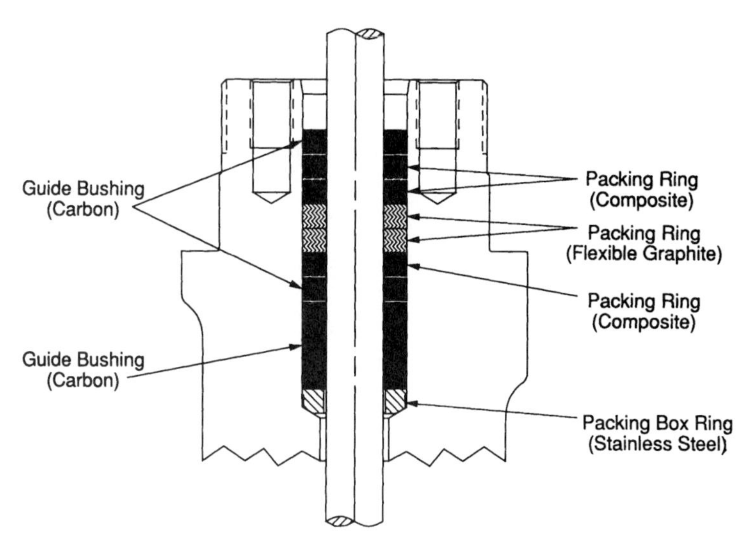

Figure 7.24 Supplemental packing box guiding. (*Courtesy of Fisher Controls International, Inc., Marshalltown, Iowa.*)

actuator and valve. *Poor stem guiding* or misalignment between the two stems will result in premature packing failure regardless of what type of guiding is used in the packing area. Anytime the two stems cannot be easily connected without bending or prying, the alignment should be corrected before the valve is reassembled or there is a risk of less than satisfactory packing performance. The "we'll make it fit!" approach is a good way to guarantee lots of overtime correcting packing problems that surface once the valve is brought back on line.

As a side note, one of the more frequent causes of stem misalignment is *spring side load* on the actuator stem caused by the ends of the actuator spring not being square with the axis of the spring. If the actuator stem cannot be easily centered, check the ends of the spring to be sure that they are parallel to one another and perpendicular to the spring axis. The spring ends should also be about 180° apart if side loads are to be minimized.

Reviewing this section, to get the most out of any packing system, the following guidelines should be adhered to:

- Keep stem taper below 0.002 in over the stroking length.
- Keep the stem straight to within 0.003 inch over the stroking length.
- Select a hard stem material compatible with the process fluid.
- New stems should have a finish of 4 rms or better.

- Attempt to maintain a stem finish, in service, of 8 rms or better.
- Try to avoid excessive side loads in the packing through good guiding design and proper assembly.
- Utilize supplemental guiding in the packing box area to help with side loads and extrusion.

7.3.3 TFE-based packing

All things considered, TFE-based materials should always be the first choice for packing. They seal better at much lower friction levels than anything else in the marketplace and are compatible with the majority of process fluids found in the chemical process industries. Their only real limitation is temperature. Most of them are limited to 450 to 500°F, but this range is being gradually expanded through the use of embedded fibers, more temperature-resistant backup materials, or newly developed materials like Kevlar (DuPont) or PEEK, which act much like TFE but have higher temperature limits.

In addition to the temperature limits, TFE also has a tendency to cold-flow, which can make it difficult to contain in the packing box. One of the side benefits of the attempts to improve temperature resistance through the use of embedded fibers is that the stability of the material has improved at the same time. Nevertheless, *containment* is still a major concern with TFE, which means that clearances have to be tightly controlled and that stiffer materials should be used as antiextrusion rings whenever possible. This problem is made worse by the fact that TFE has a thermal expansion rate that is 10 times higher than most metals used in control valve construction. If the valve is subjected to thermal cycling, the TFE will grow faster than the surrounding metal and tend to push into the clearances between the metal parts. Then, if the temperature drops, the packing load will drop as well since the TFE will also shrink faster than the metal surrounding it. Two methods of addressing this particular problem are to use a pressure-assist design as described in the following paragraphs or to use a technique called *live-loading* that will be covered in Sec. 7.3.5.

TFE packing comes in two forms, *V ring* and *rope*. The V-ring version is shown in **Fig. 7.25.** It is used more often than the rope and features a V shape that permits the fluid being sealed to press against the inside of the V, forcing it out in contact with the stem and the I.D. of the packing box (make sure that the inside of the V always points toward the fluid being sealed or the pressure assist will not work). This pressure assist makes the packing very easy to use since the initial load on the packing can be low and doesn't have to be controlled like it does with other forms of packing. As long as there is a

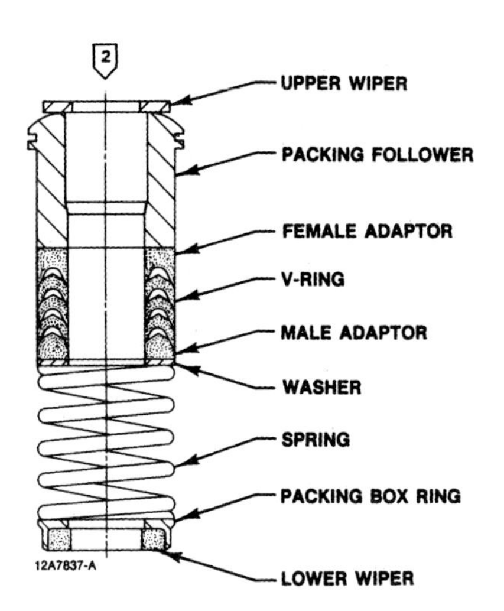

Figure 7.25 V-ring packing. (*Courtesy of Fisher Controls International, Inc., Marshalltown, Iowa.*)

nominal load that provides an initial seal, the internal pressure will provide whatever additional sealing load is required at full service pressure. The initial sealing load is usually provided through the use of an internal coil spring that is loaded by drawing the packing follower down against the top of the bonnet. If a chronic sealing problem exists with this type of packing, make sure that the initial preload is being properly applied.

V-ring packing is also less susceptible to variances in the stem and packing box diameters since the pressure assist will help it adjust to minor variations. Be aware, however, that deviations from design tolerances in either of these diameters could result in excessive *extrusion* if *clearances* have opened up as a result. As with any other type of packing, *stem finish* is critical to maximizing service life. If the stem finish is kept at 8 rms or better in service and the initial preload is properly applied, this type of packing should provide many thousands of cycles of satisfactory service.

The other form that TFE-based packings might take is a braided filament commonly called *rope*. Unlike the V-ring design, the rope is not pressure assisted and has to be fully loaded by the packing follower to whatever stress is required to fully seal at service pressure. In general, the loading stress should be about 1.5 times the pressure to be sealed against. This is because the TFE is not a liquid, and therefore a stress applied along the axis of the shaft is not directly converted into a contact stress normal to the shaft. (The stiffer the packing material, the higher this ratio has to be. The ratio is about 2.25 for graphite packing.) This also means the load has to be controlled so that adequate stress is reached to seal, but the load should not be so high as to result in unneeded friction or extrusion. Rope is normally used as a perceived cost-saving measure since it can be stocked on a reel and rings cut as required to fill a wide range of packing boxes. In contrast, V-ring packing has to be specially sized for a given packing box dimension. While the use of rope packing may appear to be a cost saving at first glance, it really isn't when sealing performance is considered along with initial cost.

Rope-packing *sealing performance* does not match that of V ring for a number of reasons. Because it is made up of braided filament, it tends to be permeable and does not seal as well against the mating metal parts. To get acceptable sealing performance relatively high loads have to be applied, which result in higher friction, and even at these higher loads, it still doesn't seal as well as the V-ring design. Proper adjustment is also more difficult, requiring that torque on the packing nuts be controlled to make sure that the appropriate stress levels are reached for a given internal pressure. If increased leakage along with the need for more controlled maintenance when using rope-style packing are considered, it actually makes more sense to use V-ring packing even though the initial cost can be slightly higher.

Friction for TFE-based packings is relatively low. This is good news since excessive friction can lead to performance problems or require the use of larger, more expensive actuators. The performance problems associated with high friction is covered in more detail in Chap. 8, but a few comments on packing friction for TFE-based packings need to be made here. First of all, minimizing absolute friction is very desirable, and V-ring TFE packing is the best way to do this. Since it is pressure assisted, it only develops as much friction as is required to maintain a proper seal. For example, tests have shown that V-ring packing on a ½-in shaft sealing to 1400 psi only develops about 50 lb of friction. A typical actuator used with this stem size might have an effective area of 100 in^2, meaning that the packing friction corresponds to only 0.5 psi pressure load on the actuator. Rope-style TFE packing for the same conditions exhibits about 150 lb of friction. While this is higher, it still is only about 1.5 psi on the actuator, and given that the normal pressure range for the actuator is 60 psi or more, it is a very small percentage of overall actuator capacity. From this analysis, it's easy to see why TFE is a popular choice. It provides a good positive seal while having very little effect on valve performance.

While absolute packing friction is low with either of the TFE-based packings, there is one other friction-related characteristic that can affect valve performance. Most materials exhibit a phenomena called *stiction* where the force required to overcome static friction is higher than the force required to maintain motion. TFE is no exception, and this can sometimes result in a stick-slip, jerky movement of the valve stem that makes good control difficult. While there is no way to eliminate this behavior completely, light lubrication of the stem during packing installation does seem to help. Experience has also shown that stiction drops off after an initial break-in period. It also appears that any increase in temperature along with normal vibration due to fluid flow will help to reduce these effects. What this means is that stiction is normally worst immediately after packing installation and gets progressively better the longer the valve stays in service. In any case, TFE-based packings, while not perfect, are still better than the other alternatives as far as stiction is concerned.

As noted above, *packing friction* is low for TFE packings. In fact, in the past, many valve vendors and end users ignored packing friction when it came to *actuator sizing*. Since V-ring packing friction is typically less than 1 percent of the overall actuator capacity, ignoring it didn't cause too many problems. This is no longer a recommended practice, however. Even with TFE V-ring packing, problems sometimes resulted if an actuator was marginally sized to begin with. And with the increased use of graphite and live-loading, packing friction can now be as high as 50 percent of the input range of the actuator. As a result, packing should always be included when sizing an actuator and determining the operating range. Typical packing friction levels are shown in **Fig. 7.26** for different packing styles, materials, stem sizes, and pressure ratings.

Two final comments on friction. Note that with the exception of the V-ring pressure-assist design, the estimated friction levels assume that the packing has been properly adjusted, taking into account the service conditions. In reality, this very seldom happens, with the packing either being too tight or too loose, and even if the torque recommendations for the packing nuts are followed, the conversion of torque into gland load is only accurate to within ±25 percent. When evaluating operational problems, never assume that the packing friction is at published levels until it has been checked and verified. There are numerous examples where excessive packing friction manifested itself in other areas of valve performance. Packing friction that is too high, for example, can result in leakage past the seat if the actuator doesn't have enough force to overcome the packing friction and apply proper seat load.

And finally, friction levels are always *estimates*. This is not by accident. Friction levels vary all over the map. They change with time in service, temperature, pressure, number of cycles, and other variables too numerous to mention. Experience has shown that actual friction levels can swing as much as 30 percent from nominal, and this needs to be taken into account if proper valve operation is to be achieved. The good news is that friction is usually highest right after installa-

Stem Diameter, Inches	ANSI/FCI Valve Class	Packing Arrangement			
		TFE		TFE/ Composition[1]	Graphite[2]
		Single	Double		
For Design CE Valve Body					
1/2	150 thru 600	---	75	70	135
3/4	150 thru 600	---	112.5	100	200
1	150 thru 600	---	150	105	215
1-1/4	150 thru 600	---	180	140	275
For All Other Valve Bodies					
5/16	All	20	30	---	---
3/8	125	38	56	63	---
	150			63	125
	250			95	---
	300			95	190
	600			125	250
	900			160	320
	1500			190	380
15/32[3] or 1/2	125	50	75	90	---
	150			90	180
	250			115	---
	300			115	230
	600			160	320
	900			205	410
	1500			250	500
	2500			295	590
5/8	125	63	95	109	---
	150			109	218
	250			145	---
	300			145	290
	600			200	400
23/32[4] or 3/4	125	75	112.5	175	---
	150			175	350
	250			220	---
	300			220	440
	600			330	660
	900			440	880
	1500			550	1100
	2500			660	1320
1	300	100	150	305	610
	600			425	850
	900			530	1060
	1500			650	1300
	2500			770	1540
1-1/4	300	120	180	400	800
	600			425	1100
	900			700	1400
	1500			850	1700
	2500			1020	2040
2	300	200	300	613	1225
	600			862.5	1725
	900			1125	2250
	1500			1375	2750
	2500			1623	3245

1. Values shown are frictional forces typically encountered when using standard packing flange bolt torquing procedures.
2. Values shown apply to laminated graphite and graphite laminate/filament packing.
3. Nearest U.S. equivalent to actual 12 mm diameter stem.
4. Nearest U.S. equivalent to actual 18 mm diameter stem.

Figure 7.26 Typical packing friction levels. (*Courtesy of Fisher Controls International, Inc., Marshalltown, Iowa.*)

tion and falls off after the valve has been in service, so if a valve works initially, it should only get better from a friction standpoint as time goes on.

7.3.4 Graphite packing

Graphite packing is generally used on high-temperature applications where TFE is not acceptable. It was developed as a replacement for asbestos, which was the standard for high-temperature use until concern over health problems related to long-term exposure to asbestos drove it from the marketplace. In many ways, asbestos was superior to graphite-based packing. It is more durable and less likely to be damaged due to storage and handling. It also exhibited lower friction readings for equivalent sealing performance. Nevertheless, graphite is now the standard.

In terms of chemical resistance, graphite stands up very well to everything but very strong oxidizers. It can be used up to 1100°F in a standard packing box arrangement, which covers most high-temperature control valve applications in the chemical process industry. It also has a coefficient of thermal expansion that is very close to the metals that surround it in the packing box, so thermal cycling does not pose any significant problems.

It comes in three primary forms: *graphite filament, graphite laminate,* and *ribbon grafoil.* The *filament* is a braided arrangement that is more resilient than the other two forms and is normally used as an antiextrusion ring to contain the sealing rings and to keep the stem clean (**Fig. 7.15**). Graphite packing materials tend to adhere to the stem and then be dragged through the packing area, causing leaks. The filament helps to wipe the stem clean, which keeps this from happening. Because the filament is braided, it contains voids that make it permeable, and it does not work very well as a sealing ring because of this.

Graphite laminate is made by pressing a number of sheets together, forming a ring with a square cross section. The rings have to be precisely cut and sized for a given packing box dimension. The "grain" of the ring runs perpendicular to the surface of the shaft, so fluid does not easily permeate through the ring. However, this also reduces the efficiency with which a gland load is transformed into a sealing load against the stem, so the ratio of loading stress to the pressure to be sealed against has to be higher (**Fig. 7.27**).

Ribbon grafoil, on the other hand, is made by wrapping a grafoil ribbon around a mandrel and then compressing it in a die to a predetermined density. The compression stress for both the laminate and the ribbon grafoil should be significantly higher than the service pressure to limit further consolidation in the packing box. Density should

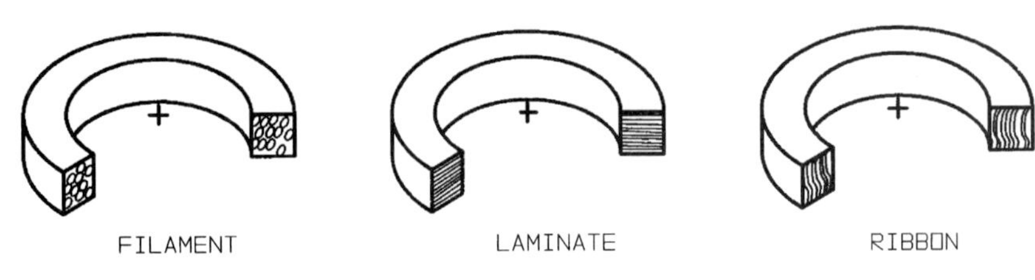

Figure 7.27 Graphite ring cross sections.

end up at around 100 lb/in^3. Because the rings are formed in this way, the grain structure tends to run parallel to the axis of the shaft, which can result in permeation through the ring at high pressures. On the other hand, the ribbon style is easier and cheaper to make and does a better job of converting gland load into sealing load on the stem. For this reason, ribbon grafoil has become the more popular form of graphite packing. In fact, some end users even stock the ribbon on reels and form the rings in the packing box itself. While this is cheaper than stocking rings, it is not recommended due to problems with forming high-quality, high-density rings when the packing box is used as the die.

The big problem with any graphite packing is achieving and maintaining enough packing load to properly seal without developing levels of friction that can hurt valve performance. While the same general restrictions also apply to TFE-based packings, there is much more latitude from a design and adjustment standpoint because the TFE seals better and the friction levels are *much* lower. **Figure 7.26** illustrates that the friction for graphite packing is 7 to 10 times as high as TFE for the same pressure conditions. This is why *whenever possible, you should use TFE rather than graphite.*

If, after all other avenues have been explored, you still must use graphite, the following discussion can help you do it successfully. But you must follow the recommendations very closely and recognize that, even when done properly, graphite packing will have a detrimental effect on valve performance.

First of all, as mentioned earlier, the supporting cast of metal parts must be designed properly and be in good condition. All the conditions set forth in Sec. 7.3.2 regarding stem condition become even more important when dealing with graphite. You should also minimize the effective height of the packing in contact with the stem. Additional rings only add friction without improving the seal. Designs vary somewhat, but experience has shown that a sealing height on the stem equal to one stem diameter is a good guideline for the primary seal. Add to this one antiextrusion ring at either end of the sealing rings for a complete assembly. Generally, double-packed valves add a second packing set consisting of a single sealing ring in combination with two antiextrusion rings. The preferred choices at this time are ribbon-grafoil for the sealing rings and braided filament for the antiextrusion rings.

Minimizing packing box depth, as just described, always helps to improve packing performance and can be done even on valves with deep boxes by adding metal or carbon spacers in the bottom of the box. Make sure that the spacers are sized so that they will not grab or scratch the stem. Carbon is generally preferable to metal since it can serve as a spacer and an additional guide at the same time.

At this point we have good guiding, a stem in good condition, antiextrusion rings to properly contain the packing and to ensure that it doesn't adhere to the shaft, and no more packing than is really needed to provide a good seal. The only thing left to ensure adequate sealing is to get and maintain just enough load to seal the packing against the shaft and packing box. This is not easy. Even ribbon grafoil is not very fluid, so it takes a contact stress between the sealing rings and the stem of 1.5 times the service pressure to get a good seal, even with a shaft burnished to a 4 rms finish. This contact stress is acting normal to the stem surface, so the stress along the axis of the shaft has to be even higher, again because the graphite is not a perfect fluid.

In addition, this radial stress is gradually dropping off as we go farther down into the stack. This is because the frictional force between the packing and the stem acts along the axis of the stem and resists the transfer of the gland load down into the stack. To get adequate sealing load into the bottom sealing ring, the gland load has to be increased even more (**Fig. 7.28**). Note that a light coating of lubrication on the packing will help reduce the rate of stress drop-off in the stack and that cycling the valve will also tend to equalize stress, but this uneven distribution never goes away completely.

What all this means is that the gland stress developed on the top of the first antiextrusion ring needs to be 2.25 to 2.5 times the service pressure to be sealed against to be sure of developing sufficient radial sealing load in the bottom sealing ring. This can sometimes result in very high gland loads. When the ±25 percent tolerance between torque applied to the packing nuts and actual gland load is added in, torque levels can go even higher. Already we can begin to see why graphite packing presents so many problems for the average end user.

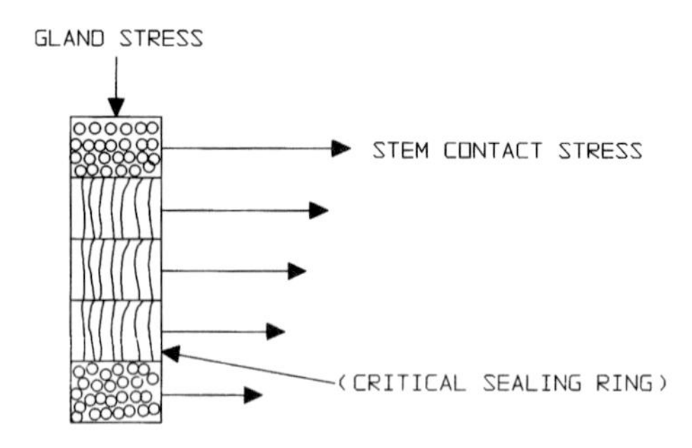

Figure 7.28 Stress reduction in graphite packing stack.

Now, assuming that the recommended loads to provide tight shutoff can be reached, how can they be maintained over an acceptable service life? This is really the key to good packing performance and is particularly difficult with graphite packing. When the packing nuts are torqued to the recommended level, the packing follower loads the packing and the packing studs stretch in response to the load developed. As long as the packing follower doesn't move in relation to the studs, the packing load will remain constant. Unfortunately, there are numerous things going on that tend to relax the load in the studs. As described above, the packing is precompressed in a die before insertion into the packing box, but over time and under heavy load it will compress even further, allowing the load to relax. This is called consolidation, and selecting high-density grafoil and retorquing the packing nuts from time to time are the only ways to reduce its effect.

Another problem is centered around the fact that with graphite packing the initial gland load is not equally distributed down the packing stack, as described above. However, as the valve cycles, the stress begins to equalize and the lower rings compress farther, also relaxing the load. Lubricating the packing and retorquing after an initial break-in period can help counter this phenomena.

Wear and *extrusion* are the other two things that can happen to graphite packing that will tend to reduce the volume in the box and permit the load to drop. Wear can best be controlled by making sure that the stem has a very good finish and that the loop is stable so that cycling is minimized. Extrusion can be minimized through the use of good tight clearances and antiextrusion rings to keep the packing in the box and off the stem. Extrusion is actually less of a problem with graphite than with TFE because it doesn't cold-flow as much, and the coefficient of thermal expansion is very close to the metal parts surrounding it, so thermal cycling does not relax the load as it does with TFE.

From a sealing standpoint, following the above recipe should provide good tight shutoff in a typical application for 40,000 full-stroke cycles or more. But to be successful with graphite packing, you must not let a leak start because it can cut a leak path through the packing, and this can be very hard to reseal.

Besides shutoff, the other major headache one has to deal with graphite packing is *friction*. High absolute friction can result in slow response, poor shutoff, high hysteresis, and a number of other valve problems. You have to be very careful to assure that the packing is configured and adjusted as outlined above so that the packing friction will be at or near the levels predicted. As with TFE, the predicted levels are ballpark estimates, and it's a very big ballpark, so be conservative in actuator selection to be sure that the valve performance will be consistent with the demands of the application. **Figure 7.26** shows some typical numbers. Fortunately, graphite packing friction tends to drop off as time passes and the valve cycles, so if the valve checks out properly at initial installation, performance should only get better. Note that lubrication tends to reduce the initial friction levels but that the ultimate values that result after break-in don't vary much whether lubrication is used or not.

While absolute friction does pose some problems with valve performance, the ratio of *static-to-dynamic* friction is always a concern regardless of how high the actual friction levels are. As explained for TFE packing, this ratio is sometimes called stiction and can result in a very jerky, stick-slip valve motion. If the stiction is severe, the valve can actually jump past the setpoint in closed-loop operation and set up a *limit cycle* where the valve continuously jumps back and forth, resulting in excessive wear and poor control. Graphite is usually worse than TFE in this respect, but the jerky motion can be controlled through the use of light lubrication in the packing and by running the valve through a number of break-in cycles. Once again, this behavior is more likely to occur when the packing is new and seems to dissipate with time and increases in temperature associated with putting the valve in service. It is also affected by the effective spring rate in the actuator and the gain in the positioner, if present, so changes to either of these properties may help to alleviate it.

The final point that needs to be made about graphite packing is that its use can sometimes result in *severe corrosion* and *pitting* of the valve stem. In effect, if there is any humidity present, the graphite facilitates the creation of a galvanic cell within the packing box that results in severe galvanic corrosion of the stem. The end result is pits and buildup on the stem that tear the packing and cause leakage once the stem is set in motion. This can occur even with stem materials that are normally very resistant to corrosion.

Fortunately, this pitting requires that the packing area be wet and normally only occurs under static conditions when the valve stem is not moving. If the valve in question is going to be out of service for several weeks, it is recommended that the graphite packing be removed so that this galvanic corrosion does not have a chance to take place. Check with your vendor on new valves to make sure that the packing box is dry if the valve will not be in service for some time after delivery. In response to this problem, some vendors are placing sacrificial zinc washers in the packing box between the rings to help guard against this corrosion before start-up. Anytime packing is replaced, the maintenance personnel should make sure that the zinc washers are properly inserted back into the packing box if they were there to begin with. Note that the zinc will normally dissolve during regular service, so it doesn't provide any real anticorrosion resistance except at start-up.

7.3.5 Live-loading and high-performance packing systems

The advent of the Clean Air Act has made packing leakage a very high-visibility problem within the chemical process industry. As a result, a number of "zero-leakage" designs have come onto the market from both the valve manufacturers and independent packing vendors. We put zero-leakage in quotes because too many things can happen in a control valve application that can cause leakage no matter what kind of packing system is being used. Most of the systems do improve performance, but do not go into a packing improvement program with false expectations of zero leakage across the board. The high-performance packing systems generally incorporate one or more of the following features in an attempt to improve performance. In most cases the features apply to both TFE-based and graphite packing assemblies.

The most common design improvement is to live-load the packing. All packings tend to lose volume over time due to things like packing consolidation, thermal cycling, wear, extrusion, and the equalizing of the load down through the packing stack. As the packing volume drops, it takes up less space, and the packing follower has to move down to compensate. As the follower drops, the load in the packing studs drops as well, and the sealing load in the packing falls off as a result. In a V-ring pressure-assist assembly this is less of a problem since the sealing load is developed by the internal pressure and not by the gland load. However, in any packing system where the primary sealing load is supplied by the gland and the studs, it's easy to see that this volume loss will eventually result in leakage. Since the studs are the primary spring in most systems, and because their spring rate is very high, it takes very little loss of packing volume to drop the load to the point where leakage can result. In fact, this is the cause for most packing leakage: insufficient load either due to volume loss or improper adjustment from the start.

To address this problem, the industry has come up with *live-loading,* which is a system where springs are added to the assembly somewhere between the packing studs and the packing itself. The springs are much more flexible than the packing studs and have more travel. As a result, as the packing volume decreases, the packing follower can move farther downward without losing any appreciable load. In many designs, the allowable volume loss can be 15 to 20 times greater than in a standard design, meaning that the load can be held constant for 2 to 3 years of normal service without any need for retorquing. Don't forget that constant load maximizes the chances for minimum leakage but cannot do the job by itself. If any of the other critical success factors already identified do not measure up, packing performance will still suffer. Another important point relates to controlling the maximum load on the packing. If the load goes too high, it will accelerate the wear and result in premature failure.

Live-loading also helps in the area of thermal cycling. As mentioned earlier, TFE packings grow much faster than the surrounding materials, so in a conventional packing box, they tend to extrude due to the greatly increased gland load that results when the packing is heated. Then, when the temperature drops, the stress drops and the packing leaks. Live-loading helps keep the load constant by permitting the gland to move ever so slightly in reaction to the changes in packing volume. This helps keep the packing from leaking.

The other benefit that comes with many live-loaded systems is that the adjustment of the *initial load* is made simpler through the use of an indicator that shows when the springs have been compressed to the proper point. You don't have to worry about controlling torque. Just take a conventional wrench and tighten the packing nuts until the indicator shows that everything is adjusted properly. Adjustment is easier to accomplish, and the packing load is more tightly controlled since the relationship between deflection and load in the springs carries a tighter tolerance than torque versus load. By routinely rechecking the indicator, you can even arrange to retorque the nuts from time to time to keep the load within acceptable levels so that a leak never has a chance to develop. One such design is illustrated in **Fig. 7.29.**

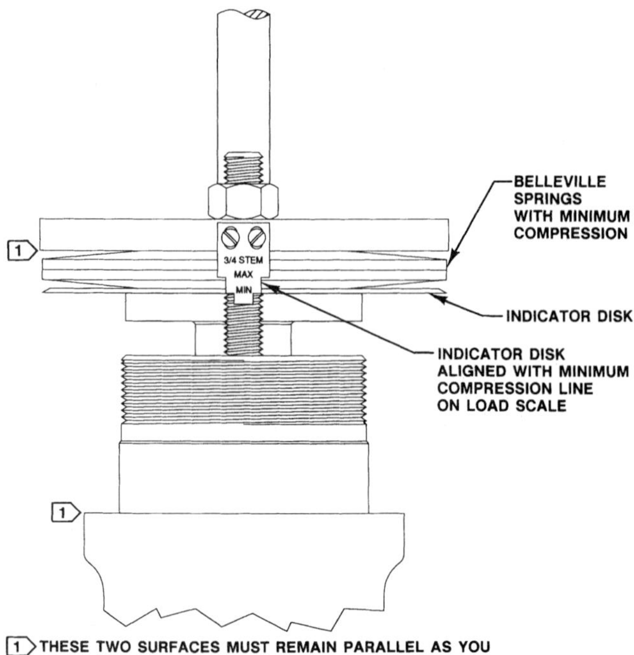

Figure 7.29 Live-loading with load indicator. (*Courtesy of Fisher Controls International, Inc., Marshalltown, Iowa.*)

The springs can be placed in any number of locations. Several different approaches are shown in **Figs. 7.30** and **7.31.** The simplest design is shown in **Fig. 7.30** and involves simply dropping Belleville springs over the packing studs between the flange and the nuts. While this is the simplest design, it does not produce the best performance. Because the space is limited between the stem and the packing studs on a typical control valve, small springs must be used, which cannot be designed to carry very much load. As a result, they need to be nested in parallel to increase the load-carrying capacity. This nesting results in friction between the springs that reduces their ability to respond to changes in load. The other problem with this approach is that because the springs are small, they don't have much deflection, so to increase the available movement, they have to be stacked in series (edge to edge). While stacking them in this way does increase deflection, its effect is limited due to space limitations between the flange and the stem connector. Overall, you end up with a spring system that doesn't permit much flange movement without

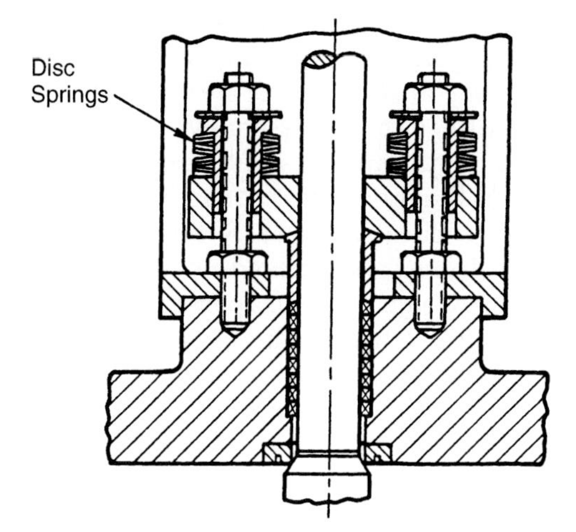

Figure 7.30 Live-loading with springs over studs.

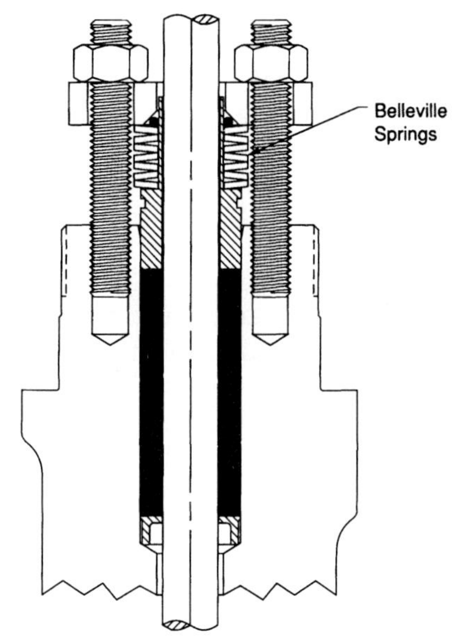

Figure 7.31 Live-loading with springs over stem. (*Courtesy of Fisher Controls International, Inc., Marshalltown, Iowa.*)

loss of load and does not respond well to changes in load due to friction between the springs. It's also relatively easy to make a mistake during assembly and end up with one or more of the springs installed upside down, which changes the stiffness of the assembly.

The design shown in **Fig. 7.31** is an improvement because it utilizes larger springs that go around the stem rather than the packing studs. The larger springs help to reduce the need to nest them, as with the previous design. They still must be stacked in series to get sufficient deflection. Note that this design also incorporates a spring-deflection indicator. The indicator is flush with the top of the flange when the springs are properly loaded. Note that disassembly for this design requires that the stem connection be broken.

The design shown in **Fig. 7.29** features springs that actually go around both packing studs. In this case the springs can be very large, with high loads and large deflection. As a result only two springs are required and provide very high levels of deflection. While improving the spring performance, this approach does carry the penalty of not being able to remove the actuator without unloading the packing. From a spring-material standpoint, most designs utilize either 17-7PH stainless steel or, if corrosion is a concern, Inconel 718.

In addition to live-loading, high-performance packings usually employ some type of improved antiextrusion measures. In TFE systems this might mean carbon-filled TFE or PEEK backup rings. (PEEK is a high-temperature plastic material.) For graphite, the use of carbon/composite rings help to keep the packing inside the packing box. In most cases, antiextrusion rings are used in combination with harder carbon-based guide rings to ensure that the packing does not have to absorb the side loads that might be present. The antiextrusion rings keep the sealing rings from extruding, and the guide rings help contain the antiextrusion rings and keep them from wearing out.

At least one high-performance system uses TFE washers in combination with graphite. The graphite is the primary seal, but the TFE helps the graphite conform to the stem and improves the seal, while reducing the inherent high friction of graphite-only designs. Because the graphite helps to confine the TFE, this type of system can be used up to 600°F, which is 150° higher than with TFE alone.

One other thing that helps improve packing performance with these types of systems is the fact that most employ double packing arrangements, so if something happens to cause a leak in one set, there is a backup seal. Note that in nearly all cases, the use of high-performance packing systems will result in higher friction when compared to their standard counterparts. This can be traced to the use of double packing arrangements and the fact that the packing loads used are usually higher to help guard against leakage.

7.3.6 Installation and maintenance

It would not be an exaggeration to say that many packing performance problems are actually associated with and introduced during routine maintenance. As a result, it is beneficial to provide some recommendations in this area in an attempt to reduce the chances of this happening. Initial installation normally occurs before delivery, so the only thing the end user should do at this stage is to check to make sure that the packing is properly adjusted before putting the valve in service. Once the valve is in service, make sure that it strokes smoothly. A valve with stick-slip action can limit cycle in closed-loop operation, greatly increasing the wear on the packing. Recommendations on how to reduce stick-slip action were reviewed above.

If the valve is not live-loaded, it is a good idea to retorque the packing after the valve has had a chance to stroke for several days or after the valve has warmed up. In critical service or on valves with a history of leakage, you may want to institute a program to have the packing checked and retorqued on a regular basis. Once a quarter should be enough.

If a valve has to be repacked, the actuator and bonnet should be removed, and this should be done out of the line in the protected environment of a valve shop. This will greatly reduce the chances of the stem or packing box being damaged or scratched and will help to keep dirt and/or debris from being introduced into the packing area during reassembly. It's also much easier to inspect and clean up the stem with this approach because the bonnet is not in the way.

Packing can be replaced in-line, but be very careful in extracting the old packing to be sure that the stem and packing box aren't damaged in the process. There is not much room, and the packing box is narrow and deep, so you are usually forced to use some type of hook or corkscrew arrangement to pull the old packing out. Even assuming that you are successful in getting the packing out, there is no easy way to tell what condition the stem is in. If it is damaged, simply repacking the valve won't solve any problems, and all that's been done by repacking in the line is that a lot of time has been wasted. In fact, if a valve is a chronic leaker and retorquing to normal levels does not seem to help, it's probably time to pull the valve or at least take it apart to find the root cause of the problem. Don't forget to check the packing parts that surround the stem for signs of unusual wear. They may need to be replaced if they are contacting the stem and causing side loads or scratches.

The other thing to look for in a valve that is leaking is *excessive side load*. Check to make sure that the valve and actuator stems are properly aligned. If they are not, figure out why and correct the situation. Don't try to force it. The packing won't last.

When all the peripheral problems have been corrected, and the new packing is ready to go back in, be especially careful when slipping it over the valve stem. The sharp edges and threads can damage the rings and cause leaks. Some packing is split along a 30° angle to facilitate installation, so you can open the ring before sliding it on the stem. (It also permits installation without removing the actuator, but this procedure is not recommended for reasons stated above.) While this does make installation easier, it also makes a leak more likely, so solid rings are preferred.

Graphite can be fairly delicate outside of the packing box, so be especially careful with it. As mentioned earlier, a very light coat of lubrication on the inner and outer edges of graphite packing can make it easier to install. For graphite, care should be taken to not trap air between the rings. This can be accomplished by only pushing a ring into the box to the point where it is flush with the top of the box. The next ring can then be used to push it further (**Fig. 13.19**). Once the packing is reinstalled, it can be torqued to the recommended load and the valve put back into service.

7.3.7 Rotary considerations

Much of what has already been stated above applies equally as well to rotary or sliding-stem designs, but there are a few characteristics that apply uniquely to rotary designs and bear further mention. First and foremost, the inherent action of a rotary valve stem means that the stem is not being dragged through the packing but is simply rotating within it. This is an improvement over the sliding-stem valve because a defect or debris on the stem moves from side to side within the packing but does not provide a leak path from one end of the packing to the other, as it would if the same defect were present on a sliding-stem valve.

There is also *less relative movement* between a rotary shaft and its packing for a full valve stroke. This can be explained by looking at what happens compared to a sliding-stem valve. Each valve has a 1-in stem and 1 in of packing in the box. The rotary valve moves 90° in one stroke, and the sliding-stem valve moves 2 in. As the rotary valve strokes, the contact area of the packing (3.14 in^2) sees a relative movement of 0.785 in (equivalent to 90° movement on a 1-in diameter). The sliding-stem valve contact area (also 3.14 in^2) sees a relative movement of 2 in, or almost 3 times that of the rotary. As a result, the packing tends to wear more quickly for equivalent service on a sliding-stem valve.

Taking these two factors into account, it is easy to see that rotary packing will generally provide superior performance when compared to its sliding-stem counterpart. If packing leakage is a primary concern with a particular application, rotary valves should always receive heavy consideration. Note that this is true for rotary valves with single packing boxes. Some rotary valve types have a second box on the end of the shaft away from the actuator which tends to reduce the rotary advantage. The final note on rotary valves is that they usually do not include a bonnet, which makes packing removal and maintenance more difficult.

7.3.8 Other stem seals

There are instances where no matter how good the packing design is and how much care is put into maintaining it, there is still a small risk of leakage. In handling toxic or lethal fluids and gases, even a small risk may be too much. In this case, alternate designs are available that eliminate the packing as the primary seal on the valve stem. One option and the one that is most commonly used for control valves is the *bellows seal.* A typical example is illustrated in **Fig. 7.32.** In this case, the bellows is used in combination with an extension bonnet that is needed because of the length of the bellows. The bellows in the figure is a formed design, and to get the travel and cycle life required, it must be relatively long. Most bellows are designed around a full stroke cycle life of 100,000 cycles and are made from stainless steel, Hastelloy, or Inconel. (There are bellows made from TFE, but they see application more from a corrosive standpoint than a leakage standpoint because fluids tend to permeate through the bellows.) An alternate to the formed bellows is the welded bellows shown in **Fig. 7.33.** It is much more compact than the formed bellows but can be prone to fatigue failure where the ends of the convolutions are welded together.

While the bellows eliminates the packing as the primary seal, it should always be *backed up with packing* in case the bellows fails. Failure can be brought on by excessive cycling (watch out for an unstable loop or pump fluctuations) or may be the result of the bellows being twisted or damaged in some other way during *assembly.* Most bellows designs feature an *antirotator element* to resist twisting, but damage can sometimes still occur, so care must be taken during assembly. If failure does occur, a *monitoring tap* like that shown in **Fig. 7.32** should permit bellows leakage to be detected before the leakage has a chance to pass through the packing.

The bellows approach does eliminate a dynamic seal and replaces it with a solid metal boundary, but it is very expensive to purchase and maintain and can still result in leakage if the bellows fails. For this reason, some of the new high-performance packing designs described above are now being used instead of a bellows seal, with good results.

The only other option in eliminating the packing is to use an elastomeric boot valve like that shown in **Fig. 2.6.** While this is a pack-

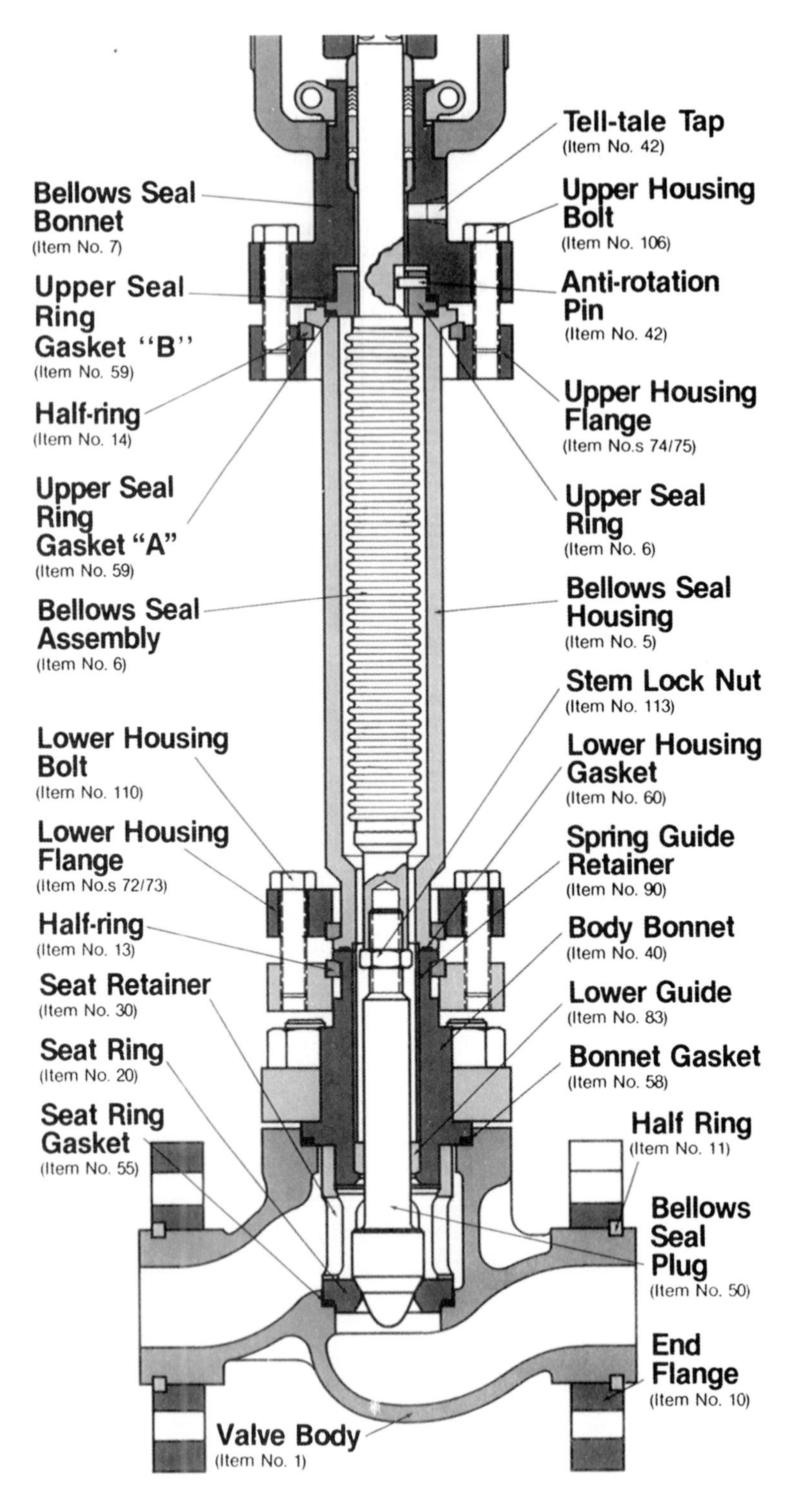

Figure 7.32 Formed bellows seal construction. (*Courtesy of Valtek International, Inc., Springville, Utah.*)

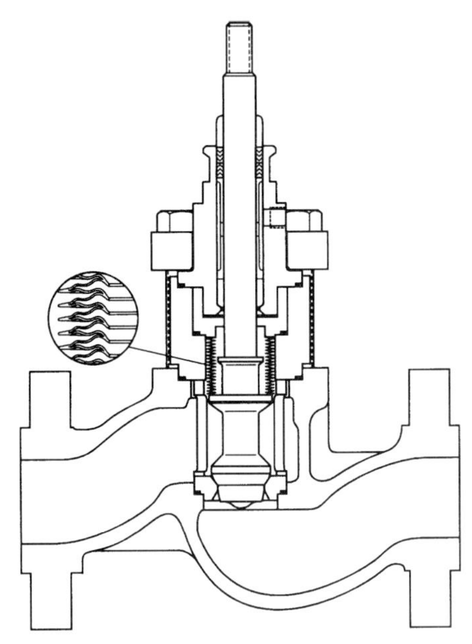

Figure 7.33 Welded bellows construction. *(Courtesy of Valtek International, Inc., Springville, Utah.)*

less valve, it's application is limited from a pressure and temperature standpoint, and it is still susceptible to fluids permeating through the membrane.

References

1. O'Keefe, William, "Gaskets & Static Seals," *Power Magazine,* Jan. 1985.
2. ASME Sec. VIII Division 1, App. 2, ASME, 1988.
3. *Fisher Controls ES208, Pressure Retaining Bolted Joint Design Standard,* Fisher Controls, Marshalltown, Iowa, Oct. 24, 1990.
4. *Technical Handbook Metallic Gaskets,* 2d ed., Fluid Sealing Assoc., Philadelphia, Penn., 1979.
5. *Non-Metallic Gasketing Handbook,* 3d ed., Fluid Sealing Assoc., Philadelphia, Penn., 1989.
6. Brestel, Ron, Hutchens, W., and Wood, C., *TM-38, Control Valve Packing Systems,* Fisher Controls, Marshalltown, Iowa, 1992.

Chapter

8

Valve and Actuator Sizing

Valve and actuator sizing will be addressed together since there is a growing trend in the industry for software programs to handle them this way and because much of the same information is required for both. There is actually a lot of math involved for both these operations, but it will not be covered in much detail since there are computer programs designed to carry out these calculations. Instead, the focus will be on a basic understanding of both techniques and on areas where you might run into problems.

8.1 Valve Sizing

For purposes of this discussion, a control valve can be modeled very simply as a variable flow restriction. As the primary flow control element is moved within the flowstream, the flow capacity changes in response to this movement and is a function of the actual cross-sectional flow area and the flow geometry of the valve. The challenge in selecting and sizing a valve for a given application is to determine the required range of capacity based on the service conditions and then select a valve that gives the best fit to the requirements.

You could test every valve with the fluid to be handled, at the service conditions specified, to determine if the flowing capacity meets the requirements, but this is not very practical and could get to be very expensive. Fortunately, the industry has found that flow through a restriction is predictable for a certain set of conditions and then corrections can be made for different fluids and different pressure drops. Essentially, a standardized test is run on a particular valve and a set of flow coefficients is developed for the different valve travels. Then, given the actual service conditions such as fluid type, inlet pressure, outlet pressure, temperature, and flow required, you can use a set of valve-sizing equations to calculate the required flow coefficient and compare it to that already determined for a given valve type.

There are some minor differences in how the valve vendors approach valve sizing, and an attempt was made to standardize the equations in ISA standard S75.01. While the standard exists, some valve manufacturers have maintained their own approach, but the bottom line is that, in most cases, the answers reached will be roughly the same, regardless of whose calculations are used. Since most valves are grossly oversized, these minor differences are of little consequence. The important thing to remember is that the basic equations are all relatively similar and work well. It's the special corrections that are applied to these base equations that really determine how accurate the fine tuning turns out to be. In particular, the corrections should not be too conservative since there is already an excessive amount of conservatism built into the whole sizing approach. The reasons for this will be covered later in this chapter.

As a point of reference, the inherent flow characteristics discussed in Chap. 2 are based on these published coefficients since the tests used to find them are done with a constant pressure drop. If we take each coefficient for a given travel divided by the coefficient for the maximum travel and plot them versus percentage of travel, we have the traditional characteristic curves shown in **Fig. 8.1.**

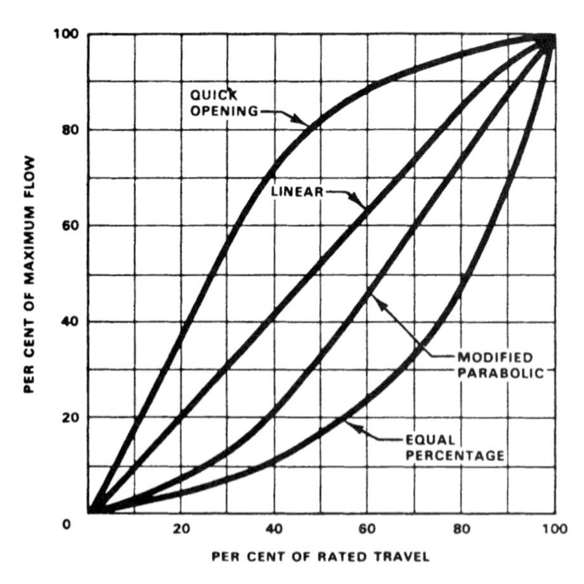

Figure 8.1 Flow characteristics. (*Courtesy of Fisher Controls International, Inc., Marshalltown, Iowa.*)

8.1.1 Notes on valve selection

Before covering specifics, a few more general comments need to be made. First of all, selecting the correct valve is not as simple as it might have sounded from the above description. Identifying the actual service conditions is difficult because of all the different system parameters that can affect flow and the interaction of the mechanical components that make up the fluid-handling system. As a result, guesses are made and in good engineering fashion, the approximations are conservative. There is also a problem in identifying all the different possible service conditions, including normal, upset, and emergency conditions. In many cases, the maximum capacity of a valve may be selected based on a theoretically possible emergency occurrence that may never happen in real life. If this capacity is much higher than that required during normal service, the overall performance of the valve will suffer since it will spend most of its time near the closed position. Throttling at low flows can increase damage to the trim due to high-velocity erosion and vibration. It can also result in poor dynamic performance since the hysteresis and deadband in the valve assembly is magnified as a percentage of input span. In other words, an oversized valve might be providing full required capacity with an input range of only 4 to 8 mA from the controller. This can also make loop tuning more difficult when compared to a valve where the usable input span is wider, say 4 to 16 mA.

The tendency to oversize is reenforced at valve selection time, as well. Since you will very seldom find a valve with exactly the capacity required, it's usually necessary to pick the next larger valve, and we get more capacity built-in. Whenever possible, take a good, hard look at what service conditions are really possible and try to keep the valve operating out at 75 to 80 percent of its travel under normal max. flow conditions.

At the other end of the spectrum, you need to check minimum flow required to make sure that the valve is not asked to operate for any extended periods at less than 5 percent of travel. Operation in this range can result in damage to the seat due to high-velocity flow. This is where valve rangeability is a real plus. A valve with high rangeability can meet the requirement for high capacity and still be able to throttle down to relatively low flows without going under 5 percent of travel. If the range of capacities required is too broad, two valves will have to be used in parallel: one covering the low flows and a larger one to cover the high-capacity requirement.

Note that the published flow coefficients are developed in an ISA standardized test using water at 68°F that requires relatively long, straight lengths of pipe on either side of the valve being tested. It also specifies controls on the testing instrumentation that keep the cumulative tolerance on the results at about ± 5 percent. Manufacturing tolerances from one valve to the next add to this error, so actual installed capacity factors can vary from those published by as much as 10 percent and still be within accepted industry practice.

Along with the possible variances just described, piping layout can have a very significant effect on installed capacity since it very seldom looks like the nice straight piping used in the ISA test. The other point to be made about differences between installed and rated capacity is that the methods used to measure flow in service have their own error tolerances that can further contribute to the differences. These points are mentioned here to make you sensitive to the fact that sometimes installed capacity does not seem to agree very well with what was predicted for the valve. This does not necessarily mean that anyone has made a mistake. It may simply be the accumulation of errors just described. This is one of the reasons why the target for normal maximum valve capacity is 80 and not 100 percent.

8.1.2 Capacity calculations for liquids

Liquid sizing forms the basis for all sizing of fluids handled by control valves, so it will be covered first. Without worrying about how it's derived, the standard equation for liquid (noncompressible) flow through a control valve (or any restriction) is as follows:

$$Q = C_v \sqrt{\frac{P_1 - P_2}{G}}$$

where Q is the flow capacity, in gallons/minute.

P_1 is the inlet pressure, in psig (measured 1 diameter upstream of valve).

P_2 is the outlet pressure, in psig (measured 6 diameters downstream of valve).

G is the specific gravity of the fluid; specific gravity is defined as the dimensionless ratio of the density of the fluid in question divided by the density of water at about 60°F (62.4 lb/ft^3).

C_v is a sizing coefficient that is determined experimentally using the ISA test referred to in Sec. 8.1.1. It is defined as the amount of water in gallons that will pass through a valve in 1 min with the valve in a given travel position and with a pressure drop of 1 psid.

In reviewing this equation, it becomes apparent that the flow through a valve varies directly with the C_v and the square root of the pressure drop and inversely to the square root of the specific gravity of the fluid. This is great. Once a set of C_v's is established for a valve at its various travels, the flow through the valve for any travel and any service condition can then be predicted if we know the service pressure drop and the specific gravity of the fluid. Or, looking at it as we did in the previous section, if the service conditions, the fluid, and the required flow are known for a given application, we can select a valve to meet the requirements based on the published C_v.

This works for the majority of the situations, but, like always, there is some fine print. The limitations in applying this basic equation to all liquid flowing conditions come about because liquids can change phase as they pass through a valve. This phase change was covered in great detail in Sec. 6.1. Very briefly, as a liquid accelerates as it passes through the reduced cross-sectional flow area of a valve, the pressure of the liquid has to drop due to the physics of the situation. As the pressure drops, there is a tendency for the liquid to change to the vapor phase, as in boiling, as the average pressure approaches the vapor pressure of the liquid at the given temperature.

This change in phase creates bubbles in the flowstream that take up more space than the equivalent liquid, and because they take up more space, the ability of the valve to pass fluid flow is reduced when compared to when bubbles are not forming. In other words, the capacity of the valve no longer follows the nice, clean equation given above. All is not lost, however. Once again, tests can be run to determine when and if these bubbles will form, depending on valve travel, geometry, and service conditions, and then corrections can be made to the equation to permit its continued use.

Before talking about how these tests are run, several new terms have to be defined. Fluids flow through a valve in response to a pressure drop, $P_1 - P_2$. The critical factor in determining the degree to which flow chokes in a valve is how close the average pressure comes to the vapor pressure as the fluid passes through the valve. Modeling a valve as a restriction, with a certain inlet pressure, we can see that the fluid accelerates through the valve and then exits the valve at a reduced velocity. Theoretically, since the exit velocity is essentially the same as the velocity at the inlet, the outlet pressure should be the same. We know, however, that this does not happen. Some of the energy is absorbed and transformed as the fluid passes through the valve; therefore, the outlet pressure is always less than the inlet pressure. Different valves behave in different ways when it comes to this energy absorption, and, as a result, for the same inlet pressure and same vena contracta velocity, they will have different outlet pressures. The degree to which the outlet pressure approaches the inlet pressure is called the recovery coefficient and is a measure of the efficiency of the flow path in a given valve. As you might imagine, rotary valves with their line-of-sight flow paths are high-recovery valves when compared to globe-style valves where the flow path is relatively tortuous.

To summarize, a high-recovery valve will have a high outlet pressure for a given inlet pressure. A low-recovery valve will have a low outlet pressure. Looking at this another way, if we fix the inlet and outlet pressures for a given application, the high-recovery valve will have a much lower minimum pressure as the liquid flows through the valve when compared to the low-recovery valve. This is illustrated in **Fig. 8.2.** It follows that the high-recovery valve will be more prone to bubble formation and choking because its minimum pressure will be closer to the vapor pressure for the fluid, all other factors being equal. With this in mind, we can characterize the choking potential for a valve by looking at its recovery characteristics.

This is done by testing the valve at a number of different points over its travel range. For each point, a flow test is run where the pressure drop is gradually increased, keeping the inlet pressure constant. The results of this test can be plotted as flow versus $P_1 - P_2$. Typical results are shown in **Fig. 8.3.** As can be seen here, the valve capacity begins to deviate from the straight line as bubble formation begins and then fully chokes shortly thereafter with no further capacity increases. A valve-recovery coefficient is defined based on these results; it relates the pressure drop at which choked flow occurs to the difference between the inlet pressure and the vapor pressure for a given inlet temperature. The recovery coefficient is sometimes

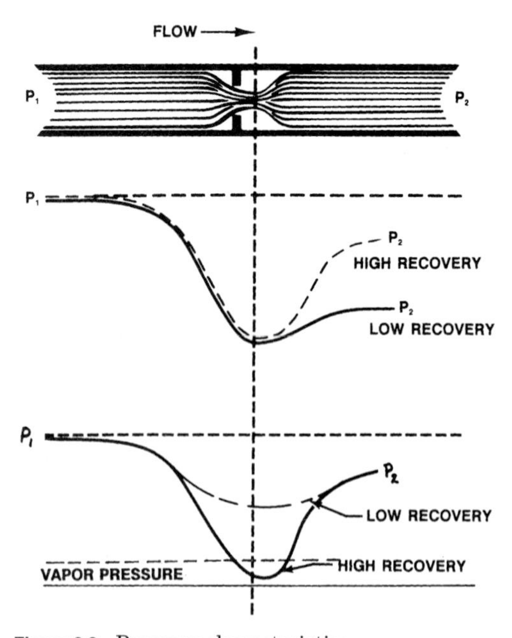

Figure 8.2 Recovery characteristics.

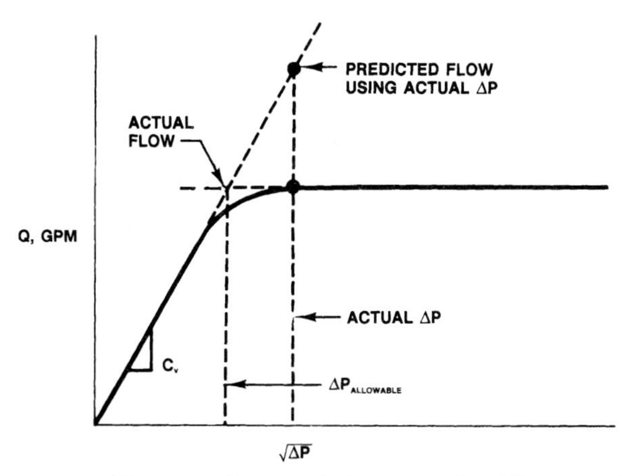

Figure 8.3 Recovery characteristic test results. (*Courtesy of Fisher Controls International, Inc., Marshalltown, Iowa.*)

referred to as K_m and is defined by the following relationship and is based on the test results shown in **Fig. 8.3:**

$$K_m = \frac{(P_1 - P_2)_{\text{choked}}}{P_1 - r_c P_v}$$

where $(P_1 - P_2)_{\text{choked}}$ is the pressure drop at which the flow chokes in the above test. It is also sometimes referred to as the allowable pressure drop.

r_c is a term called the critical pressure ratio that varies as defined by **Fig. 8.4.**

P_v is the vapor pressure for the fluid at the inlet temperature.

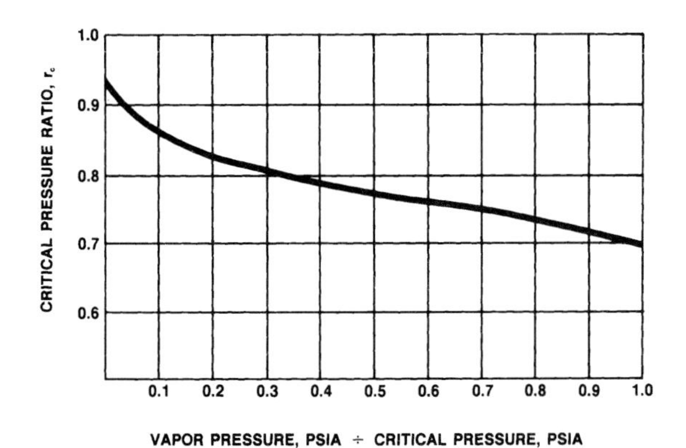

Figure 8.4 Critical pressure ratio. (*Courtesy of Fisher Controls International, Inc., Marshalltown, Iowa.*)

In practice, K_m is determined experimentally for all valves and all travels and is published by the valve vendors. Then, if you know K_m, P_1, P_v, and r_c based on the fluid type, you can calculate $(P_1 - P_2)_{allowable}$ based on the above equation. If the actual pressure drop is less than this calculated allowable pressure drop, you can use the actual pressure drop in the standard sizing equation because no choking occurs. However, if the actual pressure drop is greater than the calculated allowable pressure drop, the latter must be used in determining valve capacity because the valve will choke.

It should be noted that the name *allowable pressure drop* is not meant to imply that the valve cannot be subjected to a higher pressure drop given the service conditions. It only means that this is the highest pressure drop that can be used in determining flow capacity. The ISA standard uses the term F_L for the recovery coefficient and F_F for the critical ratio term. The relationship between these terms and the terms used earlier is shown in the following equations:

$$K_m = F_L^2$$

$$r_c = F_F$$

As discussed in Chap. 6, bubble formation, as just covered, is called flashing and can damage trim and cause problems with capacity. When the bubbles form and then collapse due to pressure recovery, this is called cavitation and is very hard on trim. See Chap. 6 for more information on both these phenomena.

As a final comment, it is interesting to note that while a rotary valve has higher inherent capacity than its sliding-stem counterparts, it also has a lower K_m, so in cases where bubble formation is possible, the real capacities of the two different valve types are nearly the same.

8.1.3 Gas flow

In many ways, gas flow is similar to liquid flow through a control valve. The major difference is that gases are compressible, and this makes the form of the equation much more complicated. Nevertheless, the various equations utilized all work well, particularly given the discussion regarding accuracy of flow predictions covered in Sec. 8.1.

One approach uses a term called C_g that is analogous to the C_v term used for liquids. It is a compressible fluid critical flow coefficient that is also established experimentally for each valve at each travel and can be plugged into the following equation to determine valve flowing capacity if the pressure drop, inlet density, and inlet pressure are known. Of course the same equation can be solved for the required C_g if the required flow capacity is known:

$$Q = 1.06\,(D_1 \times P_1) \times C_g \sin\left[\left(\frac{3417}{C_1}\right) \times \frac{P_1 - P_2}{P_1}\right]$$

where Q = flow, in lb/h
D_1 = inlet density, in lb/ft^3
P_1 = inlet pressure, in psia
C_1 = C_g / C_v and is usually published for a particular valve
P_2 = the outlet pressure, in psia

C_1 normally varies between 15 and 40 and is a numerical indicator of the valve's recovery characteristics. The higher the number, the lower the recovery. It is analogous in use to the K_m factor for liquids.

Like liquids, gases also can choke under certain circumstances but for a different reason. As the pressure drop is increased across a valve, the fluid accelerates, reaching a maximum velocity where the flow cross section is the smallest (the vena contracta). The gas can travel no faster than sonic velocity, and if this velocity is approached as the pressure drop increases, the capacity increase versus pressure drop will drop off. This is called critical flow, and the effect is much like choked flow for liquids in that we reach a point where no matter how much additional pressure drop we apply, the capacity will not increase. This is handled in the above equation by limiting the sine factor to no more than 90°, even though its actual value may be higher. Note, once again, that this is just one approach to gas sizing and is not meant to imply that it is any better or worse than the other methods that might be encountered in the marketplace.

8.1.4 Two-phase flow

Liquid and gas flows have been covered separately, but what happens if the fluid being handled is a mixture of the two phases before it enters the valve? In this case, the required flow coefficient C_{vr} is a combination of the required C_v for the liquid flow and the required C_g for the gas flow. The coefficients are each calculated separately using the individual gas and liquid flows and the equations given in the two previous sections and then combined using the following equation:

$$C_{vr} = \left(\frac{C_v + C_g}{C_1}\right)(1 + F_m)$$

where F_m is the correction factor based on the gas volume ratio V_r. The relationship between F_m and V_r is shown in **Fig. 8.5.**

$$V_r = Q_g\left(\frac{284 Q_1 P_1}{T_1 + Q_g}\right)$$

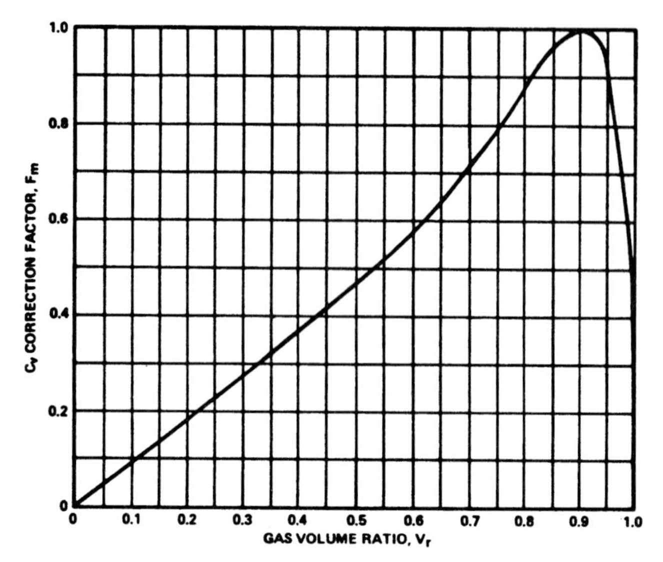

Figure 8.5 Correction factor. (*Courtesy of Fisher Controls International, Inc., Marshalltown, Iowa.*)

where Q_g = gas flow, in scfh
Q_1 = liquid flow, in gpm
P_1 = the inlet pressure, in psia
T_1 = the inlet temperature, in degrees Rankine

Note that when the flow coefficients are calculated for the liquid and gas phase, you need to check actual pressure drop against allowable pressure drop and critical gas flow, respectively. If the actual pressure drop is higher than either of these, the limiting pressure drop should be used in each case. **Figure 8.6** shows an easy way to check for critical gas flow using C_1 for the valve.

8.1.5 Highly viscous flows

Everything covered so far in this chapter has been related to fully developed turbulent flow. If the Reynolds number is above about 3500, this is appropriate, but below this, a correction needs to be made to account for the additional friction or flow resistance associated with flow of a viscous fluid. This is accounted for through the use of a correction factor called the Reynolds number factor, which is plotted versus the Reynolds number in **Fig. 8.7.** This figure shows three curves for predicting flow rate, valve size, and pressure drop. Once the Reynolds number has been determined for a given valve, fluid,

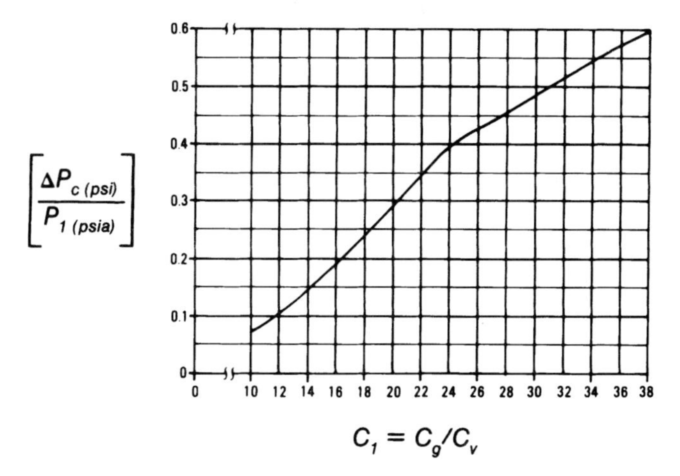

Figure 8.6 Pressure drop ratio for critical gas flow. (*Courtesy of Fisher Controls International, Inc., Marshalltown, Iowa.*)

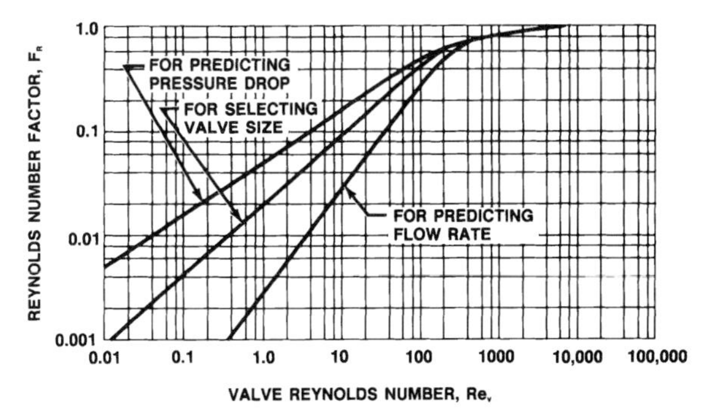

Figure 8.7 Reynolds number factor. (*Courtesy of Fisher Controls International, Inc., Marshalltown, Iowa.*)

and service conditions, you can look at the graph and find the appropriate factor on the vertical axis.

The most common approach is to use the middle line to get a factor that can be used in the following equation to determine the C_v really needed:

$$C_{vr} = \frac{C_v}{F_R}$$

where C_{vr} = the C_v really needed to pass the required viscous flow
C_v = the original C_v calculated based on the standard equations and turbulent flow

If the flow rate curve were used, the factor found would have to be multiplied by the calculated turbulent flow to find the actual flow under viscous conditions. If the pressure drop curve were used, the actual pressure drop would be found by dividing the calculated pressure drop for turbulent conditions by the correction factor.

8.1.6 Piping effects

The piping arrangements for a control valve will nearly always be different from the perfectly straight sections used in the standard sizing test, so you would expect that the installed flow characteristic might differ from the ideal as a result. It turns out that the most frequent piping effects to be considered are those that result from using reducers on the ends of the valve. Once again, the easy way to address the change in flow capacity due to pipe reducers is to apply a piping geometry correction factor F_p. **Figure 8.8** is offered as a quick guide to estimating F_p using the valve C_v divided by the pipe end diameter squared (see reference 7). The calculated flow can then be multiplied by F_p to arrive at the actual flow with reducers in place. These factors obviously have a greater effect when dealing with high-capacity valves such as balls and butterflies. When using very high-efficiency valves, other piping effects can come into play. The valve vendor should be consulted in these cases.

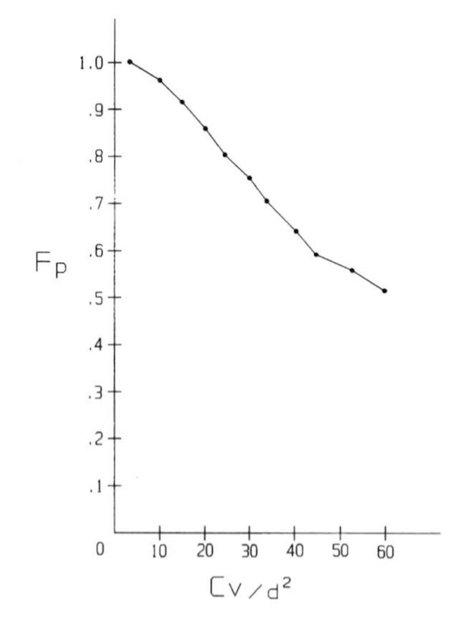

Figure 8.8 Piping correction factors.

8.1.7 Extremely low flows

For valves designed to control extremely low flows, flow prediction can be fairly difficult because not only do we have to worry about transitions between laminar and turbulent flow which can change the actual flow, but we also have to be concerned about normal manufacturing and travel tolerances changing capacity. There *has* been some work done in applying a correction factor to the calculated flow coefficient such as the approach discussed in Sec. 8.1.5. A modified Reynolds number is calculated based on the type of valve used and then the factor is selected based on this Reynolds number. Reference 7 carries a good discussion of this technique, but minor changes due to the manufacturing tolerances could still throw these calculations off. Whenever possible, try to test under actual conditions to verify flow. One of the primary problems with determining in-service capacity with valves in these flow ranges is the difficulty in establishing turbulent C_v measurements under test conditions.

8.2 Actuator Sizing

Once again, as for valve sizing, much of the drudgery associated with actuator sizing has disappeared with the arrival of customized software programs that can size an actuator for a valve and a given set of application conditions in about 10 s. Therefore, the details of these calculations will not be covered as much as the general principles involved will be. One concept that *will* be covered in detail is the term *benchset* since it continues to cause many problems in process plants all over the world.

8.2.1 Basic principles

An actuator should perform five basic functions:

Shut the valve and prevent leakage

Stroke the valve fully open

Respond quickly to load or setpoint changes

Position the valve accurately and with good repeatability

Operate in a stable fashion

While this is a relatively short list, it's amazing how many actuators in the field are having problems meeting one or more of these. Most of the problems can be traced to a lack of understanding of the fundamentals behind good actuator sizing practices. The next section will review these fundamentals for a common type of valve and actuator and will give some useful tips on how to avoid problems and, more importantly, how to get the maximum performance out of equipment.

8.2.2 Spring and diaphragm sizing example

The example that will be covered here is for a spring and diaphragm actuator on a sliding-stem, globe-style valve. This is one of the more common models found in service today and is also one of the more complicated actuators to size because it includes the concept of bench-set. The basics of actuator sizing will be covered for this model and then deviations from this approach will be reviewed for the other types of actuators in use today.

The valve assembly in question is shown in schematic form in **Fig. 8.9.** As mentioned, it is a spring and diaphragm model mounted on a balanced, globe-style, sliding-stem valve. Since the valve is balanced and it is best to have the unbalance force on the plug acting to open the valve, it is set up to flow down through the seat ring. The valve has a conventional packing box seal and the whole assembly is what is called *push down to close, air closes, spring-open.* These terms mean that the valve plug moves down into the body to shut the valve and that the force required to close the valve is supplied by air pressure loading the top of the diaphragm. Valve designs may be encountered that are reverse acting, where the plug is below the seat and moves down to open, but these are rare. In the same way, you can also have an actuator where the air pressure enters under the

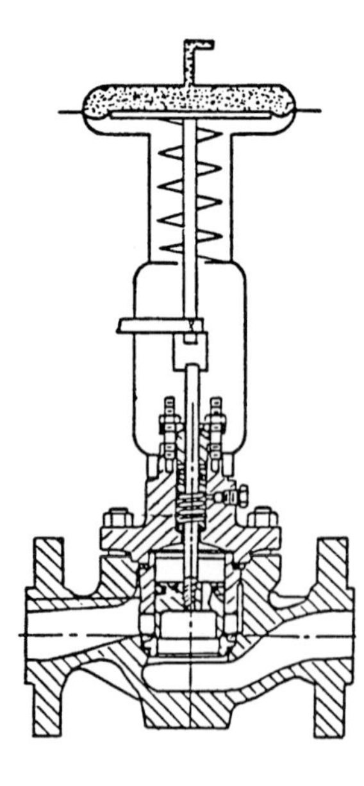

Figure 8.9 Typical spring and diaphragm actuator.

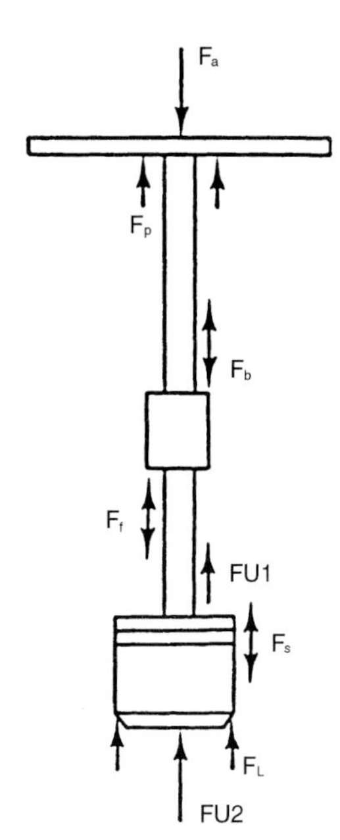

Figure 8.10 Simplified free-body diagram.

diaphragm and forces it up to open the valve. In any case, the principles covered here apply to these different models as well, with the understanding that the direction on some of the forces may change.

If we examine the valve in **Fig. 8.9** in more detail, we can identify the moving parts within the assembly along with the forces that act on these moving parts. This is illustrated in **Fig. 8.10,** which shows the diaphragm plate, the actuator stem, the stem connector, the valve stem, and the plug as moving parts. This same sketch, a free-body diagram, also shows the forces that act on these parts. Each of these forces is then defined in **Table 8.1.** In its simplest form, actuator sizing is nothing more than determining the values for each one of these forces, depending on valve construction and service conditions and then checking to make sure that the forces balance in such a way that the operational characteristics defined in the previous section are achieved.

A general force balance on the parts shown in **Fig. 8.10** yields the following equation:

$$F_a = F_p \pm (F_b) \pm (F_f) + F_{u1} \pm (F_s) + F_L + F_{u2}$$

Combining the three friction terms into one stem friction value, the equation can be simplified into:

$$F_a = F_p + F_{u1} + F_L + F_{u2} \pm (F_t)$$

TABLE 8.1 Definition of forces

Force symbol	Description
F_a	The product of the air pressure (psig) in the actuator and the effective area of the diaphragm. The effective area can change with valve travel.
F_p	The spring load acting against F_a and expressed as $F_p = F_i + T \times K$.
F_i	Initial spring preload with the valve wide open. This is a function of the valve benchset or, in other words, the initial spring adjustment (lb).
T	Valve travel with zero as the wide-open position (in).
K	Spring constant (lb/in).
F_b	The actuator bearing friction force, which always acts in the direction opposite to stem movement (lb).
F_f	Packing friction, especially important with graphite or asbestos-type packing where improper packing adjustment can result in relatively large variation in packing friction (lb).
F_{u1}	The stem unbalance caused by the internal pressure (psig) acting on the unbalance area (in^2) of the stem (lb). (Always act up, except for vacuum applications.)
F_s	The friction force between the seal and the I.D. of the cage. This changes as a function of the type and diameter of the seal (lb).
F_L	The seat load between the plug and seat ring. This is critical in determining the shutoff capabilities of the valve in question. There is an approach that defines a load requirement in lb/linear inch of seat circumference and that depends upon the leak class required as defined in ANSI/FCI 70.2 1976 (R1982). Note that this only exists when the plug is on the seat.
F_{u2}	The pressure unbalance on the plug. It can be in either direction depending on whether the valve is flow-up of flow-down. In a balanced design, it is very small and acts in the direction opposite to flow through the seat ring. It is the product of the unbalance area times the pressure drop across the plug.

where F_t is the total stem friction and carries a $\pm$ sign because it can act in either direction since it always opposes the direction of motion. What this equation implies is that for equilibrium to be established (no movement), the diaphragm force has to be balanced by the spring force, the stem unbalance, the plug unbalance, the seat load, and the friction. For movement in the closed or open direction, the diaphragm force has to be greater or less than these same forces, respectively. Using this equation, we can now begin to determine the actuator size and adjustments needed to meet our previously stated goals.

Let's begin by verifying what load is required to properly shut the valve. We can do this by first examining the forces inside the valve.

As the valve is stroked into the closed position, movement is resisted by the friction forces, the stem unbalance, and the plug unbalance, all acting upward. As noted in **Table 8.1,** the seat load acts upward but only when the plug is in contact with the seat. The spring force also opposes the diaphragm force but acts outside the valve body. We can see that it will be equal to the initial wind-in on the spring plus the spring rate times the travel of the valve. (This initial wind-in of the spring is called the benchset and will be covered in more detail later.) So, for the valve in the closed position, the equation becomes:

$$F_a > F_i + K_t + F_{u1} + F_{u2} + F_L + F_t$$

F_a has to be slightly more than this total for the valve to move in the closing direction.

The other major objective for the actuator is for the spring to be able to then turn around and stroke the valve fully open. In this case, with the valve fully open, the friction forces reverse with the motion reversal and now act downward, the seat load disappears completely since the plug is no longer in contact with the seat, the plug unbalance also disappears since the plug is out of the flowstream, and the stem unbalance helps to open the valve. The equation now becomes:

$$F_a < F_i + F_{u1} - F_t$$

F_a has to be slightly less than this total for the valve to go open.

If we examine this equation, it begins to yield some interesting results. We can re-arrange it to put the initial wind-in term on one side and see that it must be slightly more than the diaphragm force plus the friction minus the stem unbalance. Depending on the range of available air pressure, we can then calculate the initial wind-in on the spring necessary to open the valve. This initial wind-in is the starting point for the actuator selection process since it tells us how much spring load we need to get the valve open. At this point, an actuator size is tentatively selected and some numbers run to see if the actuator size and available springs can make the above two equations work.

The easiest way to illustrate this technique is with an example using real numbers. Let's use the following valve and service conditions and see how it works:

4-in balanced globe-style valve

Unbalance area of plug = 0.5 in^2

600 # rating

4 ⅜-in diameter, single-port, metal seat

Elastomeric piston ring seal

Class IV shutoff

½-in stem, single graphite packing

Stem area = 0.196 in^2

2-in travel

P_1 = 300 psig

P_2 = 0 psig

Flow down

Given this information, we can determine the following forces: F_t = 320 lb.

F_t is made up of seal friction, actuator bearing friction, and packing friction. Normally we can safely ignore bearing friction because it is so low in comparison to the other forces present. Since we're using an elastomeric piston seal, this friction is also negligible. (Note that with a graphite or metal seal, you might have to include a friction term for the seal.) Given these assumptions, the only significant friction is in the packing. Most vendors can estimate friction based on the pressure to seal against (300 psig), the packing type (single graphite), and the stem diameter (½ in). Based on the values given, the total stem friction was pulled from a vendor table giving an estimated packing friction of 320 lb (**Fig. 7.26**):

$$F_{u1} = P_2 \times \text{stem area} = 0 \qquad \text{(valve closed)}$$

$$F_{u1} = P_1 \times \text{stem area} = 300 \times 0.196 = 58 \text{ lb} \qquad \text{(valve open)}$$

F_{u1} is the stem unbalance and with the valve closed and flowing down, the pressure around the stem is equal to P_2, which in this case is 0. Note that P_2 is not always 0, so this term may not be negligible with the valve closed under a different set of service conditions. With the valve open, the assumption is made that the pressure around the stem is equal to P_1, so we multiply this pressure by the area of the stem. Note that the area used for the stem is always defined by the sealing diameter in the packing since this is the area that defines the pressure drop across the stem between internal and ambient pressure.

$$F_{u2} = (P_1 - P_2) \times \text{unbalance area} = 0 \qquad \text{(valve open)}$$

$$F_{u2} = (300 - 0) \times 0.5 = 150 \text{ lb} \qquad \text{(valve closed)}$$

F_{u2} is the plug unbalance, and normally it's preferable to have this force tending to open the valve to avoid problems with negative gradients. Even though this valve is balanced, there is still a small unbalance force due to the difference in sealing and seating diameters. This is covered in more detail in Sec. 2.2, which also explains why the force acts upward even though the flow is down through the seat ring. In the open position, it is assumed that the unbalance across the plug is 0 because the pressure drop is very low and the plug is more or less out of the flowstream. In the closed position, the plug sees the full shutoff pressure drop.

$$F_L = 3.14 \times \text{port diameter} \times \text{required seat load/in}$$

$$F_L = 3.14 \times 4.375 \times 40 = 550 \text{ lb} \qquad \text{(valve closed)}$$

$$F_L = 0 \qquad \text{(valve open)}$$

F_L is the seat load required to shut off the valve after the actuator has overcome all the other forces present and has placed the plug in contact with the seat ring. Most vendors have developed guidelines that determine the amount of contact load required, per inch of contact length, to shut off a valve to a particular shutoff classification. Guidelines that are defined in terms of contact length enable the force to be calculated as a function of the size of the port. The contact length is the circumference of the port for this configuration, and one set of vendor guidelines states that the load per inch required is 40 lb for a class IV shutoff.

At this point, we know all the forces that need to be overcome inside the valve for the valve to be able to stroke fully closed and shut off and to get fully open. Now, an actuator size needs to be selected and tried to see if it can supply the necessary loads just calculated. The actuator size is defined by the effective area of the diaphragm, and if we pick an actuator with 100 in^2 of area, it will simplify the example calculations and is typical of a size that might be used for this type of valve, in this service.

Let's first look at the spring force required to get the valve open. The governing equation derived above is:

$$F_a < F_i + F_{u1} - F_t, \text{ rearranged} \qquad \text{or} \qquad F_i > F_a - F_{u1} + F_t$$

where F_i is the initial spring force. Substituting the numbers just developed:

$$F_i > F_a - 58 + 320 = F_a + 262$$

Now, we have a choice to make regarding the diaphragm force in the open position. Some sizing programs assume that the diaphragm pressure will go to zero at the low end, so we could assume that the F_a term is zero and the spring force required would be 262 lb. A more conservative approach would be to assume that the diaphragm pressure drops to only 3 psi as it would if the actuator were connected to an I/P or controller. We'll take the more conservative approach here, and if we're wrong the only thing that happens is that we have a little more force than we calculated to get the valve open. With this 3-psi load on the diaphragm, the equation becomes:

$$F_i > 300 + 262 = 562 \text{ lb}$$

This tells how much initial spring load is required to get the valve fully open, and a convention in the valve industry is to convert this load into an equivalent diaphragm pressure by dividing by the diaphragm area (100 in^2). The pressure found in this way is 5.6 psi, which we'll round up to 6 psi. The reason it's done like this will be clear in the next step.

If we need 600 lb of spring load to be sure to get the valve open, we select a spring for our actuator, install it, and begin to tighten the adjuster to develop the necessary spring load. Let's try a spring with a spring rate of 300 lb/in. In practice, we adjust the initial setting or wind-in on the spring by hooking up a regulator to the diaphragm inlet port and then gradually tightening the spring until the actuator just begins to move off its upper stop with 6 psi on the diaphragm. This assures us that we'll have the load we need to get the valve open if the pressure drops to at least 3 psi. Now let's look at what happens at the other end of the stroke.

To stroke the valve closed and shut it off, we need to overcome the internal loads at the closed position as well as the spring load. The spring load will be the initial wind-in just calculated plus the force necessary to compress the spring through 2 in of travel. Given the initial wind-in, the spring rate, and the travel already identified above, the spring load at the closed position is found to be 1200 lb, or 12 psig of air pressure on the diaphragm. To this we need to add the unbalance forces, the friction, and the seat load. This is expressed in equation form as follows;

$$\text{Total force at seat} = 1200 + 320 + 0 + 550 + 150 = 2220 \text{ lb}$$

$$\text{or 23 psig on the diaphragm}$$

Based on this, we have established that the actuator and valve combination will operate properly if the air pressure can be varied from 3 psig, valve open, to 23 psig, valve closed. This is the operating air-to-diaphragm range for this valve, and normally we would add about 5 psi to the high end and set the airset at this pressure to help account for any minor pressure losses in the air lines and accessories.

Our actuator sizing is now complete except for the additional checks covered in Sec. 8.2.4. An actuator has been selected that will position the valve appropriately given the service conditions stated.

Before going into more detail about the benchset and stroking ranges in the next section, there are a few caveats that need to be applied to what was just covered. The first, actuator sizing, is only as accurate as the *service conditions* defined for the valve. In many cases trouble with valve operation can be traced to the fact that the stated service conditions are not reflective of what is really happening at the valve. If an actuator is having problems getting the job done, the first thing to check is actual service conditions against those that were assumed when the valve was originally supplied.

The second thing to watch out for is *friction* in the bearings and seals. Note that we assumed above that there was no significant friction in the actuator or valve bearings, and the seal friction was negligible. In reality, friction can exist in these areas, especially after the valve has been in service. Friction should be checked from time to time, and if it is not zero, it needs to be corrected or taken into account in the sizing calculations.

And finally, watch out for *packing friction*. It is usually estimated in the above equations based on test results that assume that the packing is properly adjusted. As was mentioned in Chap. 7, packing friction can vary widely from specified levels, and this can mean that the sizing calculations will not be valid. To help ensure that a plant will not have constant, recurring problems with actuator performance, measures should be taken to ensure that packing is installed and adjusted properly and that packing friction is relatively close to the estimated values in the sizing equations. Now let's take a closer look at benchset, one of the most confusing terms in use today for control valves.

8.2.3 Benchset and stroking ranges

It is very important to note, first of all, that the *air-to-diaphragm pressure range* of 3 to 23 psi calculated in the previous section has nothing to do with the benchset for the actuator. The actuator benchset is defined as the pressure range required to begin moving the actuator from its top travel stop all the way through the rated travel for the valve. It is defined and meant to be checked with no frictional or valve forces present. This is intuitively obvious from the fact that it is always given as a single pressure range. If it were checked with friction present, there would have to be two ranges given, since the stroking range would change depending on the stroking direction because of the hysteresis shift associated with friction. In fact, the name *benchset* comes from the practice of setting the spring adjustment on a workbench before installing the actuator on the valve. Using this definition, it is clear that the benchset for this valve would be 6 to 12 psig since the initial spring adjustment was calculated to be 6 psi and it would take an additional 6 psi to compress the spring 2 in, which is the rated valve travel, based on a spring rate of 300 lb/in.

There are two additional terms that apply to any pneumatic valve with a spring: *stroking range* and *operational stroking range*. Stroking range is the pressure required to stroke the valve assembly from the open to the closed position and back again with packing friction present but without flow or pressure in the valve. The operational stroking range adds pressure and flow effects. We begin to see why there is so much confusion about setting up and verifying an actuator with a spring. There are at least four stroking ranges defined depending on the conditions under which the stroking is done. Confusing one of these ranges with another can lead to mistaken conclusions regarding the operation of the assembly.

An easy way to sort this out is to look at all four ranges in graphical form. If we plot diaphragm pressure versus travel for each case, the situation begins to make more sense. In **Fig. 8.11,** we start first with the *benchset*. If the actuator spring is tightened as specified above, and the air pressure is slowly increased from zero, the actuator should begin to stroke at about 6 psi. If the spring rate is right, the stroke should continue with increasing pressure until reaching 2 in at about 12 psi. Note that the range will never be exactly 6 psi due to minor variations in actuator effective area versus travel and to manufacturing tolerances on the spring. The measured benchset range

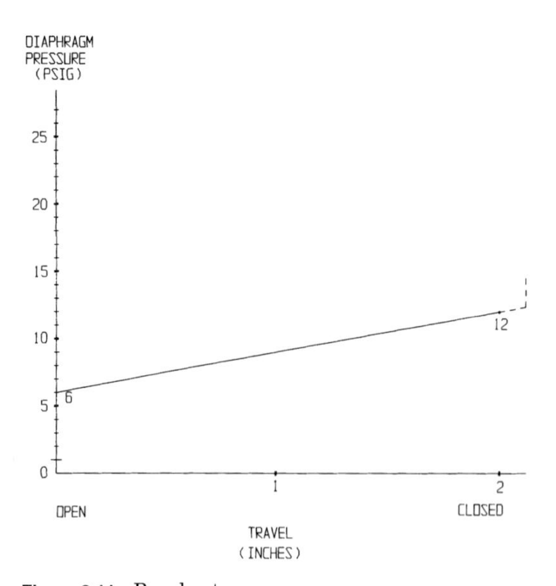

Figure 8.11 Benchset.

should be equal to the specified benchset range ±10 percent to account for these errors. Set the spring so that measured benchset is accurate to within ±0.2 psi at the end where the valve seats. The reason for using the seating end is that this is where the actuator load is more critical. For the valve in question here, this would result in an acceptable range for the benchset limits of 5.2 to 6.8 at the low end, and 11.8 to 12.2 at the high end. Note that the dotted line in **Fig. 8.11** reflects that fact that most actuators have about 10 percent more stroke than the valve is rated for to ensure that the actuator will not be the limiting factor for the valve stroke.

Figure 8.12 shows the *stroking range* where we have added friction to the equation by mounting the actuator on the valve and checking the stoke with no flow or pressure in the valve. This is the verification that is carried out most frequently by the end user. Given a friction level estimated to be 320 lb, or 3.2 psi, on the diaphragm, the valve and actuator will now begin to stroke closed at 9.2 psi instead of 6 psi and will reach the rated travel and the seat at 15.2 psi because the spring rate still defines the range. Note that once the plug is in contact with the seat, no additional travel occurs even as the pressure increases but that 7.8 psi (780 lb) is left over at the end of the stroke to help overcome unbalances and apply seat load, assuming a supply pressure of 23 psi. Now if the pressure is reduced, it is interesting to note that no movement occurs until the level drops all the way to 8.8 psi because of the friction reversal with the direction change. The

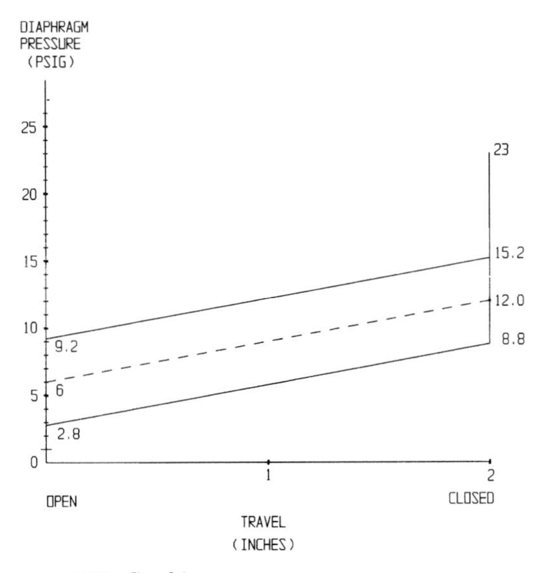

Figure 8.12 Stroking range.

valve then returns to the full open position at about 2.8 psi. In essence, the friction has been superimposed on either side of the original benchset curve. As long as the measured friction is no more than the assumed friction level as shown here, the actuator sizing is still valid. However, if the friction is higher than the assumed levels, corrections will have to be made.

To better illustrate what can happen if friction levels are too high, let's look at **Fig. 8.13.** In this case, the benchset is still 6 to 12 psi but the friction is 600 lb instead of 320. (This is not all that unusual with jam-style packings, if adjusting torque is not tightly controlled.) The stroking range now runs from 12 to 18 psi in the closing direction and there is only 5 psi left to overcome unbalances and provide seat load, so you can see that one consequence of excessive friction is that seat load is diminished. On the return stroke, the valve doesn't get fully open until the pressure drops completely to zero, which means that friction that is too high can also begin to affect the ability of the valve to fully stroke.

Looking at either of these figures, it is clear that the benchset has nothing to do with stroking range and if you mistakenly confuse the two, you can end up thinking that an adjustment is required when it isn't. In some cases we've seen a plant mistakenly check the stroking range for a valve and compare it to benchset. Since the valve didn't move until 9.2 psi, the spring was loosened until movement began at 6 psi, the value of the benchset. The spring load was reduced so far

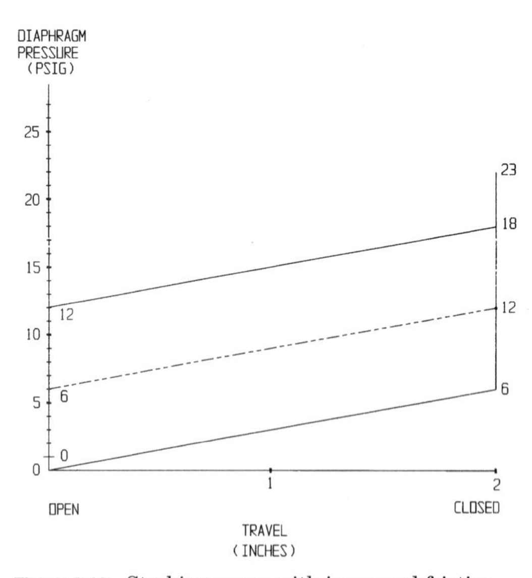

Figure 8.13 Stroking range with increased friction.

that the valve no longer reached full open on the return stroke and valuable capacity was wasted. This is a common problem at most plants and has come about because if the friction in a valve is low as it would be with TFE V-ring packing, the error associated with this practice is also small. In other words, the difference between benchset and stroking range is the friction in the valve, and if the friction is small, the difference between the two readings is also negligible. Unfortunately, once this practice is established, it tends to be applied indiscriminately regardless of actual friction levels, and the problem just described starts to crop up.

The last stroking range to be covered here is the *operating stroking range* and is where the effects of flow and pressure are added in. This type of check is done on-line (with pressure and flow), so it is not as common as the stroking range check but needs to be discussed, just the same. It is illustrated in **Fig. 8.14.** As the graph shows, there is a shift from the stroking range curve due to the stem unbalance at the open end of the stroke, and the shift gets progressively larger as the plug approaches the seat since the plug unbalance dominates near the seat. The circled numbers show the approximate stroking range under operating conditions. This curve also illustrates that the air-to-diaphragm range of 3 to 23 is accurate since the valve gets fully open at 3 psi and closes at 17 psi with 6 psi left over to apply seat load (600 lb versus 550 lb required).

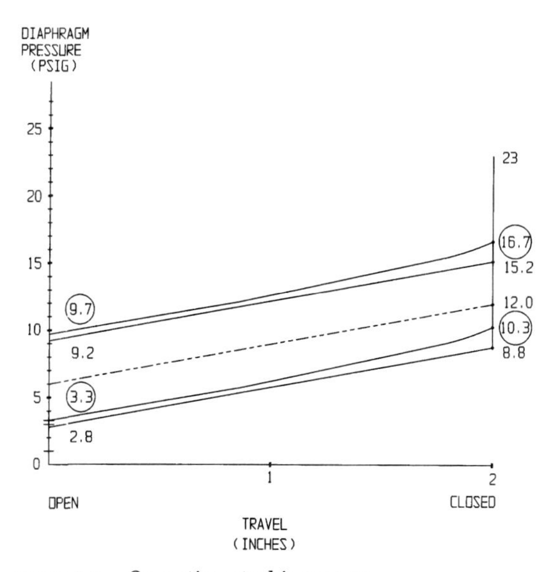

Figure 8.14 Operating stroking range.

One other practice that is sometimes employed at a typical plant is to look at the upper benchset reading and add 5 psi to determine the setting for the *airset*. In this case, this rule of thumb would yield a setting of 17 psi when the valve actually needs 23 psi to work properly. A better practice is to add 5 psi to the air-to-diaphragm range. Unfortunately, this range is not always given for a valve by the vendor, so it can be difficult to find. Vendors need to acknowledge the importance of this bit of data and start putting it on the valve name tag.

One final comment on spring and diaphragm actuators relates to the fact that many valves today work on a standard input pressure range such as 3 to 15 or 6 to 30 psi because they are directly connected to an I/P or a pneumatic controller. In some cases, this can cause a problem because, while the benchset may be 3 to 15 or 6 to 30, the required air-to-diaphragm range can be considerably different, as we've just seen. Each of these cases needs to be reviewed to be sure that the valve will work properly with the standard range input. If the valve in question is equipped with low friction TFE V-ring packing and only throttles and doesn't have to shut off, running with an input range that approximates the benchset if okay. However, if there is any appreciable friction and/or if the valve needs to shut off, either the benchset will have to be changed or the input range will have to be increased for this arrangement to work. In an attempt to increase the input range, some I/Ps now come with a *snap-action* feature, such that when the signal input drops below 4 mA, the output pressure immediately drops to 0 psi. This essentially provides another 3 psi of actuator load that can be used to either load the seat, if the valve is spring-to-close, or to fully open the valve, if it is spring-to-open. While this feature doesn't always solve the problem completely, it is a step in the right direction.

8.2.4 Additional checks on actuator sizing

The primary goal of any actuator sizing exercise is to make sure that the valve can be fully stroked in both directions and properly shut off, as we just covered in the previous section. However, to do a complete job there are some additional items that need to be verified to be sure that the assembly will perform as expected.

First of all, a check needs to be made to be sure that the valve stem will carry whatever the *maximum load* was calculated to be. Stems, if they fail, usually fail by buckling so the load, the stem diameter, the stem material, and the unsupported length of the stem all need to be considered when performing this check. Most valve vendors have standardized calculations that cover this.

The *safe working load* for the spring also needs to be verified to be sure that the spring won't be overstressed and break. This is done by taking the lower value of the benchset times the actuator area and adding it to the travel times the spring rate. This load is then compared to the maximum safe load specified by the spring designer. The pressure limits of the actuator casing, diaphragm, and accessories should also be checked against the maximum air-to-diaphragm pressure to make sure that their limits are not exceeded. In some cases, the pressure limits can be as low as 50 psi, so this is something to watch out for.

In actuators where the spring can be adjusted, the range of possible adjustment needs to be consistent with the benchset specified. In some cases, the calculated benchset cannot be reached because the *spring adjuster* can't be tightened enough.

One of the specific goals mentioned at the start of Sec. 8.2.1 was *stable operation*. The actuator needs to be stiff enough to resist negative gradients if they are present, and even if they are not a problem with a given valve, the actuator still needs to have a spring constant high enough to avoid problems with input range or load sensitivity. Negative gradients are explained in more detail in Chap. 2 but, briefly, are the result of fluid forces acting on the plug that tend to push or pull it in a direction that results in instability. Basically, if the rate of change in the fluid force associated with a negative gradient is greater than the spring rate in the actuator, the valve will be unstable and difficult to control in the applicable travel range. In most cases, the negative gradients are determined experimentally as a function of pressure drop for a given valve, and then these coefficients are used to calculate a gradient for an application based on the actual pressure drop. If the spring rate chosen is smaller than the gradient, a stiffer spring should be selected. Note that if a positioner is used with a pneumatic actuator, the *effective spring rate* of the assembly is a combination of the actuator spring and the stiffness of the positioner, so in many cases, adding a positioner will increase the stiffness to the point where the negative gradient is no longer a problem.

As far as *normal stability* is concerned, a good rule of thumb to follow is that for a stem or post-guided valve, no more than half the benchset range should be used to overcome the unbalance forces in the valve. What this means in the example above is that the 6-psi range multiplied by the area of the diaphragm area must be greater than 2 times the vector sum of the plug and stem unbalances. In equation form:

$$\tfrac{1}{2}(6 \times 100) > (150 + 58)$$

$$(6 \times 100) > 2(208)$$

$$600 > 416$$

Since 600 is greater than 416, the spring is stiff enough in relation to the forces inside the valve, and as a result, the valve will not move excessively due to random variations in these unbalance forces. (Note that for a cage-guided valve, the factor ½ is replaced with ⅔ in the equation, reflecting the inherent stability of the cage-guided design.) This is what is called load sensitivity, and you want to keep the sensitivity low like this to ensure stable operation. The other problem that can sometimes require the selection of a stiffer spring is when the benchset range is calculated to be something very small, based on static considerations: 12 to 15 psi, for example. With a range this small, the valve ends up being too sensitive to minor changes in input since it would fully stroke with an input change of only 3 psi. A minimum benchset range of 6 psi is a standard recommendation.

8.2.5 Piston considerations

Pneumatic pistons differ from spring and diaphragm actuators in several ways. First of all, because the pressure seal is provided by an O-ring, the piston can be loaded on both sides, eliminating the need for a spring. If no spring is needed, there is no benchset adjustment required, and the only thing that needs to be checked is that the actuator has enough force available to overcome the forces identified in the valve body for both stroking directions. The equations above change only in the elimination of the spring-load term. Note, however, that the friction term may need to include the friction from the piston O-ring if it is judged to be significant. The vendor can help in making this decision.

Pistons can also be single sided (pressure on only one side) with a spring providing the restoring force. In this case, the sizing is done just as for a spring and diaphragm actuator. Note that the effective area of the piston does not change with travel, however, and that the maximum pressure for pistons is typically higher than for the diaphragm designs.

Some piston operators have springs but without any provisions for adjusting spring tension, so there is no need to adjust benchset, but you still need to know what the spring tension is so that it can be properly factored into the equations. Other pistons are double acting but still have what is called a bias spring that helps to push the actuator in one direction or the other. The spring is there to provide a semblance of fail-action, pushing the valve in one direction or the other on loss of air. In most cases, a bias spring won't stroke the valve completely by itself, so assumptions regarding fail-action need to be checked out based on the actual spring load.

Stiffness still needs to be considered as above, but double-acting piston operators with positioners are comparatively stiff, so this does not normally pose a problem. The other checks covered in Sec. 8.2.4 need to be performed, as well. For other differences between piston and diaphragm actuators see **Table 4.1.**

8.2.6 Rotary considerations

For *rotary valves,* actuator sizing principles and equations follow the same logic but are expressed in torques rather than loads, and some of the operating torques that develop inside the valve are of a different nature than for a sliding-stem valve.

Seat load, for instance, deserves some discussion. On many butterfly and ball valves there is no contact between the disc or ball and the seat. Travel is limited in the actuator, and there is no term covering seat load or torque in the sizing equations. Other types of rotary valves have the ball or plug riding in continuous contact with the seat during the complete stroke, so this term appears as a frictional load did in the sliding-stem equations. The nature and magnitude of this torque depend on the type of seal and the pressure drop across the valve since most of these seals have a pressure assist. Of course, the highest pressure drop is with the valve in the closed position, so this is where the highest seat torque would be found, but there is some seal friction throughout the stroke.

The third type of rotary valve is much more like the sliding-stem version from a sizing standpoint because the ball, disc, or plug is forced into the seat at one end of the stroke to provide shutoff, so this torque is only seen with the valve near the seat. Once again, the torque for this type of seat depends primarily on the pressure drop and the type of seal used.

In addition to seal torque, there is also a friction component due to the bearings on the shaft that is a function of the shaft radius, the pressure drop across the flow element (disc, ball, or plug), the effective flow area of the flow element, and the material of the bearings. For a full ball valve floating between two seals, the torques are calculated in the same way, but the radius of the ball is used instead since it serves as the bearing surface. While the bearing friction will be higher for this type of design, there is no separate seal component of friction to be added since the bearing and the seal are one and the same. The equation form is:

$$T_b = (P_1 - P_2) \times A_e \times R \times K_f$$

where T_b = the bearing torque, in in-lb

P_1 = the upstream pressure for a given flow element position, in psig

P_2 = the downstream pressure, in psig

A_e = the effective area of the flow element for a given position, in in^2
R = the shaft radius in inches (or ball radius for a full ball valve floating between seals)
K_f = a friction coefficient for the bearing material

This bearing friction does not have a corresponding component in a sliding-stem valve since the bearings in that type of design see very little direct load and, as a result, are ignored from a frictional load standpoint. As you can see from the above equation, bearing loads and bearing friction are real concerns on rotary valves and are many times the limiting factor in the use of this type of valve. *Packing friction* needs to be taken into account, as well, and can be calculated based on the type of packing, the pressure to be sealed against, and the shaft radius.

The last type of load that is of concern is the *dynamic torque* on the flow element. This nearly always tends to close the valve and is a relatively complicated function of the pressure drop required to produce choked or critical flow and the geometry of the valve and the flow element. Factors are derived based on actual flow tests and then published for each different valve size and type. For most valve types, the dynamic torque is highest with the valve near the open position.

What all this leads up to is that there are two places where torque needs to be checked, full-closed position (breakout torque) and full-open position. Once both values are found, they can be compared to the available torque from the actuator at the two endpoints to make sure that the assembly will work. Available actuator torque is found just as it was done for the sliding-stem calculation except that the net force from the actuator varies with travel not only because of the effect of the spring but also because the effective moment arm changes. See **Figs. 4.12** through **4.15** for graphical indications of how torque varies with travel for several different types of rotary actuators. Also note that there may be additional actuator bearing friction torque that needs to be included for rotary actuators that wasn't significant for the sliding-stem valve.

Rotary sizing calculations can be simplified by combining the above terms into two very short and uncomplicated equations for breakout torque required to move the valve out of the seat and dynamic or running torque present with the valve in the open position:

$$T_b = A \times (P_1 - P_2) + B \qquad \text{and} \qquad T_d = C(P_1 - P_2)_{\text{eff}} + B$$

where T_b = the breakout torque, in in-lb
A = a factor that takes into account seal and bearing friction with pressure drop across the flow element

B = a factor that takes into account friction that is present that does not depend on pressure drop—packing friction, for example

T_d = the dynamic or running torque on the flow element near the open position

C = a factor that defines the dynamic or running torque for a given size and geometry as a function of pressure drop

$(P_1 - P_2)_{eff}$ = the smaller of either the actual pressure drop at the full-open position or the calculated pressure drop necessary to produce choked or critical flow

The factors *A, B,* and *C* are all available from the vendors and are based on valve type, size, geometry, bearing material, seal type and material, and shaft size. Once the required torques are calculated, they need to be compared to the allowable torques specified for a given shaft size and material to make sure that the shaft can safely transmit the required torques. The other additional checks outlined in the previous section also need to be carried out for a rotary valve. Sometimes operating torques need to be checked at intermediate rotations because of the possibility of maximum torques occurring there. Check with the valve vendor to see at what positions the calculations should be run.

8.2.7 Dynamic response and accuracy

Getting back to one of the recurring themes of this book, the performance of any process control system is only as good as the performance of the control valve. Three of the five basic requirements listed in Sec. 8.2.1 have already been addressed: shutoff, stroking the valve fully open, and stability. These are primarily static concerns. The other two requirements, dynamic response and positioning accuracy, deal more with the ability of the valve to adequately control the flow in response to the input from the control system. These two factors have more bottom-line impact than the other three, but do not usually receive the attention they deserve from an actuator sizing standpoint.

First of all, the ideal control system would respond instantaneously and accurately to changes in setpoint or load to keep the process variable from deviating from setpoint. We know that this doesn't happen in real life due to lags in the system and poor valve positioning performance and because corrections dictated by the controller occur in response to deviations, so the deviations have to exist for the control system to react. However, as controllers become more sophisticated, the theoretically possible deviations become very small, and the real deviations depend almost entirely on the response of the control valve assembly.

What this means is that any improvement in *valve response* shows up almost immediately in process performance. Conversely, less than optimum valve response means that the process performance is suffering unnecessarily. Given this fact, actuator sizing should be carried out with an eye toward permitting the valve to respond quickly and accurately to the appropriate signal.

Addressing *speed of response* first, many end users fail to realize that pneumatic accessories operate on a pressure drop principle where the flow through the device is a function of the pressure drop across the pneumatic relay up to the critical flow rate through the relay. If actuator sizing is marginal (based on the idea of saving up-front money on a smaller actuator), such that the pressure available through the airset is relatively close to the maximum pressure required to stroke the valve, near this pressure the pressure drop across the accessory will be small. If the pressure drop is small, the air flow will be, too, and the speed of response of the valve will suffer.

There have been numerous instances with valves where the response was sluggish at one end of the stroke because the supply pressure was only 3 or 4 psi higher than the pressure required to position the valve. This can be corrected by changing the spring, the spring adjustment, or the actuator size to reduce the required stroking pressure to the point where the difference is larger (7 or 8 psi, for example). The easiest remedy in most instances is to simply increase the supply pressure up to the maximum allowed for the actuator and accessories. This maximizes response speed over the complete travel range and also makes the system stiffer since the air spring rate in the actuator increases with increasing air pressure.

Also, while we're on the subject of *stroking speed,* make sure that the accessories selected for use with a particular valve are compatible with the flow requirements of the actuator. If the flow to the actuator is being starved because something like a solenoid valve has an orifice that is too small, big money can be lost over the service life of the valve because its stroking speed is being adversely affected. Most vendors can do a very credible job of looking at how a given accessory is affecting stroking speed based on the rated flow coefficient for the accessory. Don't forget to look at these flow coefficients when deciding on the accessories to use. The widespread use of solenoid valves for safety interlocks makes the chemical industry especially susceptible to this type of problem.

Positioning accuracy is a little harder to get a handle on but is no less critical to successful process control. Hysteresis and deadband readings for a given valve assembly will provide a qualitative look at how well it will perform in putting a valve into the theoretically correct position. It could be argued that repeatability is actually more important than overall accuracy since the control system should correct for accuracy problems if the system is repeatable. Positioning resolution is also important. As used in this book, this is defined as the smallest possible change in valve position that can occur in response to a change in input signal. The better the resolution, the closer the valve can get to the "perfect" valve position.

All these properties change with accessory selection and valve force characteristics, which is why this is being covered in a chapter on actuator sizing. If the actuator sizing is marginal, or if high friction or flow loads are present, positioning performance will suffer. To avoid problems with positioning accuracy, take every measure possible to reduce friction and to make sure that the actuator has enough thrust to counter the forces present in the valve. This may sometimes mean spending more money on a larger actuator up-front, but the payback over the life of the valve will more than justify it in most applications. Some general guidelines for acceptable accuracy and speed of response numbers will be suggested in the next chapter.

References

1. *Catalog 10,* Fisher Controls, Sec. 2, Marshalltown, Iowa, Aug. 1974.
2. Buresh, Jim, and Schuder, Charles, *TM 15, Development of a Universal Gas Sizing Equation for Control Valves,* Fisher Controls, Marshalltown, Iowa, 1974.
3. Riveland, Marc, *TM 30; Fundamentals of Valve Sizing for Liquids,* Fisher Controls, Marshalltown, Iowa, 1985.
4. *Control Valve Handbook,* 2d ed., Sec. 3, Fisher Controls, Marshalltown, Iowa, 1977.
5. *ISA—Control Valve Capacity Test Procedure,* ANSI/ISA-S75.02, Research Triangle Park, N.C., 1981.
6. Stiles, G. F., "Development of a Valve Sizing Relationship for Flashing & Cavitation Flow," *Proceedings of the First Annual Final Control Elements Symposium,* Wilmington, Del., May 14–15, 1970.
7. Bauman, H. D., *Control Valve Primer: A User's Guide,* ISA, Research Triangle Park, N.C., 1991.
8. Fitzgerald, Bill, "Automated Troubleshooting of Pneumatically Operated Control Valves," *Proceedings of the ASME Pump & Valve Symposium,* Kansas City, Mo., 1987.
9. Fitzgerald, Bill, "Benchset and Seatload Considerations," *Air Operated Valve Users Group Meeting,* Toledo, Ohio, June 1991.
10. Luthe, Fred, *Diaphragm Actuator Sizing,* Fisher Controls, Marshalltown, Iowa, April 10, 1990.
11. Barb, Gayle, "Pneumatic Operated Valves Evaluation Review," *Advances in Instrumentation,* vol. 47, ISA, Research Triangle Park, N.C., 1992.

Chapter

9

Acceptable Valve Performance

Throughout this book, the point has been made that end users who accept anything less than optimum performance from an installed base of control valves are not getting the best performance out of the control system, and money is being wasted. If we accept this as being accurate, some kind of benchmark has to be established so that the user will know when the valves are providing optimum performance. While every valve and every application cannot be covered here, the following is offered as a general blueprint in helping to establish reasonable levels of expectation for control valve performance. Note that predicting useful service life for a control valve is difficult because of the very wide range of possible services. Nevertheless, even general guidelines that include some broad assumptions are better than no guidance at all. We must start somewhere in deciding how well a control valve is performing relative to how well it could perform.

9.1 Pressure Integrity

At a basic level, the control valve needs to do a satisfactory job of containing the process fluid, with very little risk of catastrophic failure of the pressure boundary. Fortunately, the standards that have developed over the years governing design of the pressure boundary parts have worked very well in reducing this risk to a reasonable level. Proof tests, assembly line tests, and line hydros (hydrostatic tests) provide an excellent method of identifying any potential problems with the pressure boundary before they become a safety hazard. As a result, any valve in use today designed with acceptable materials in conjunction with recognized standards should not have a problem with containing the fluid for the specified service conditions.

Erosion and corrosion present the biggest risk of compromising the pressure boundary but can be managed as described in Chap. 6 through the proper selection of materials and trims. Any valve that is experiencing regular penetrations of the pressure boundary due to erosion or corrosion should have an application review carried out and the trim, materials, and/or body style changed as a result.

9.2 Leaks: External and Internal

For *external leaks,* what are covered here are leaks past *gasketed joints* or through the packing, differentiating them from the pressure integrity problems just covered where the primary pressure boundary has been compromised. As stated in Chap. 7, if good gasketing practices are followed and the surfaces to be sealed are properly prepared and loaded, a standard gasketed joint on a control valve should not leak. It is understood that in this context, *not leaking* means that the average joint will meet the EPA requirements defined in Chap. 16 for at least a 3-year period, which corresponds to the standard period between maintenance interventions for an average control valve. Many valves can be expected to last much longer than this, but individual performance will depend on factors that cannot be adequately determined here.

The *packing* is another matter, altogether. As stated in Chap. 7, the packing tends to be one of the biggest maintenance headaches for anyone dealing with control valves, so acceptable life is difficult to define. However, if the guidelines in Chap. 7 are followed to the letter, standard TFE, V-ring packing, TFE jam-style packing, and standard graphite packing should all provide leak-free service for a period of about 2 years on an average control valve in standard service. Leak-free in this case means less than about 5 times the EPA allowable limits for hazardous chemicals, or 2500 ppm (see Chap. 16). While this sounds like an excessive amount, it is still relatively low by conventional standards.

If the EPA limit of 500 ppm needs to be met, one of the high-performance packing designs should be used, including live-loading and improved guiding and antiextrusion measures. Again, assuming standard service for a control valve and that the recommendations from Chap. 7 are followed, 18 months of service with leakage under the EPA limits should be achievable for either graphite or TFE.

Note that the terms *average valve* and *standard service* are used in the previous two paragraphs. If the valve is in extremely high-cycling service or if the loop is unstable, these guidelines will probably not be met. If the valve is unstable and that is what is causing premature packing failure, attack the root cause by finding out the source of the instability and eliminating it.

For *internal leaks,* acceptable service is even more difficult to predict. For valves in throttling applications that are not counted on to shutoff at better than class II or III levels, normal service should be 3 or 4 years because the valves are not often closed. For class IV shutoff valves, the leakage should remain below this level for 2 to 3 years assuming that the fluid does not include entrained solids. If solids are present, they can get trapped between the plug and the seat and result in accelerated seat damage and leakage. Valves that are called upon to provide tight shutoff (class V or VI) should be routinely checked through some type of nondestructive leak-detection means to make sure that leaks are not developing. Once the leaks start, they tend to erode the seating surface fairly rapidly because these valves are closed most of the time and are frequently subjected to high pressure drops. Eighteen months of acceptable shutoff for this type of valve should be considered to be appropriate.

9.3 Dynamic Response

Dynamic response is the aspect of valve performance that actually has the most impact on the bottom line of an average chemical plant, but it receives the least amount of attention. It includes things such as *frequency response, stroking speed,* and *positioning accuracy* for a control valve assembly. You need look no farther than the standard valve specification sheet to see that how a valve needs to respond to the control system input is not a high-priority item with end users or vendors. **Figure 9.1** shows a control valve data sheet developed by ISA (ISA-S20.50-1983) that was meant to indicate the type of information needed to correctly select a valve for a given application and to document the type of valve selected.

It's clear from reviewing this sheet that there is a heavy emphasis on proper sizing of the valve for the flowing conditions and on the proper selection of the actuator, taking into account the forces that must be overcome in the valve body. Nowhere is there any mention made of the type of loop that the valve will be used in. Is it a fast loop? Is it a slow loop? What is the time constant of the process? How much error between the process variable and the setpoint is acceptable in both absolute terms and integrated over time? What are the consequences if these limits are exceeded?

The answers to any and all of these questions are important in determining what kind of *dynamic response* the valve should have. Part of the reason that the emphasis on these characteristics has

©ISA S20.50, Rev. 1 CONTROL VALVE DATA SHEET Second Printing

ISA®

PROJECT ____ DATA SHEET ____ of ____
UNIT ____ SPEC ____
P.O. ____ TAG ____
ITEM ____ DWG ____
CONTRACT ____ SERVICE ____
*MFR. SERIAL

1 Fluid ____ Crit Press PC

			Units	Max Flow	Norm Flow	Min Flow	Shut-Off	
2	SERVICE CONDITIONS	Flow Rate					—	
3		Inlet Pressure						
4		Outlet Pressure						
5		Inlet Temperature						
6		Spec Wt/Spec Grav/Mol WT					—	
7		Viscosity/Spec Heats Ratio					—	
8		Vapor Pressure P_v					—	
9		*Required C_v					—	
10		*Travel	%				0	
11		Allowable/*Predicted SPL	dBA	/	/	/	—	
12								

LINE
13 Pipe Line Size | In ____
14 & Schedule | Out ____
15 Pipr Line Insulation

VALVE BODY/BONNET
16 *Type ____
17 *Size ____ ANSI class ____
18 Max Press/Temp ____
19 *Mfr & Model ____
20 *Bony/Bonnet Matl ____
21 *Liner Material/ID ____
22 End | In ____
23 Connection | Out ____
24 Flg Face/Matl ____
25 End Ext/Matl ____
26 *Flow Direction ____
27 *Type of Bonnet ____
28 Lub & Iso Valve ____ Lube ____
29 *Packing Material ____
30 *Packing Type ____
31

TRIM
32 *Type ____
33 *Size ____ Rated Travel ____
34 *Characteristic ____
35 *Balanced/Unbalanced ____
36 *Rated C_V ____ F_L ____ X_T ____
37 *Plug/Ball/Disc Material ____
38 *Seat Material ____
39 *Cage/Guide Material ____
40 *Stem Material ____
41 ____
42

SPECIALS/ACCESSORIES
43 NEC Class ____ Group ____ Div ____
44 ____
45 ____
46 ____
47 ____
48 ____
49 ____
50 ____
51 ____
52

ACTUATOR
53 *Type ____
54 *Mfr & Model ____
55 *Size ____ Eff Area ____
56 On/Off ____ Modulating ____
57 Spring Action Open/Close ____
58 *Max Allowable Pressure ____
59 *Min Required Pressure ____
60 Available Air Supply Pressure: ____
61 Max ____ Min ____
62 *Bench Range ____ / ____
63 Actuator Orientation ____
64 Handwheel Type ____
65 Air Failure Valve ____ Set at ____
66

POSITIONER
67 Input Signal
68 *Type ____
69 *Mfr & Model ____
70 *On Incr Signal Output Incr/Decor ____
71 Gauges ____
72 *Cam Characteristic ____
73

SWITCHES
74 Type ____ Quantity ____
75 *Mft & Model ____
76 Contacts/Rating ____
77 Actuation Points ____
78

AIR SET
79 *Mfr & Model ____
80 *Set Pressure ____
81 Filter ____ Gauge ____
82

TESTS
83 *Hydro Pressure ____
84 ANSI/FCI Leakage Class ____
85 ____
86

Rev	Date	Revision	Orig	App

*Information supplied by manufacturer unless already specified.

ISA FORM S20.50, Rev. 1

Figure 9.1 Control valve data sheet—ISA S20.50-1983. (*Courtesy of Instrument Society of America (ISA), Research Triangle Park, N.C.*)

dropped off over the last 15 or 20 years is the heavy emphasis that has been placed on the control room side of the process control equation. The new control systems that have been developed over this same period are very powerful and flashy and make for a much better trade show display than traditional "pig-iron" valves. As control systems became more sophisticated, the market migrated away from being concerned about valve performance and went after whatever was the latest bell or whistle in DCS. Control valves came to be looked upon as a commodity, where all valves were more or less alike and could be purchased on price, without bothering with a comparison on engineering terms.

As with many trends, the underlying assumptions associated with this one were flawed. Control systems *were* becoming more powerful and more efficient. Overall, process control improved as a result, but what many end users (and vendors) are now coming to realize is that the control system can only accomplish so much. You reach a point where further improvements are limited by the performance of the control valve. If the valve doesn't respond accurately and rapidly to changes dictated by the control system, much of this new control system sophistication is being wasted.

With this as background, the industry needs to start to take a closer look at *valve performance,* with an eye toward raising expectations. The ideal valve should respond instantaneously and accurately to any change in controller output. It would also be immune to changes in the loads on the valve plug and stem so that there would be no chance of position change for the flow control element (plug, ball, or disc) without a change in the input from the controller. In other words, it would be extremely responsive but very stable.

In the real world, this is very hard to achieve, but there are some things that can be done and/or asked for at specification time that will move us in the right direction. As far as any of these performance measures are concerned, it is important to remember that performance should be evaluated in relation to the requirements of the process. As long as the valve is relatively fast, accurate, and stable when compared to the process, the performance should be good enough.

Looking at *accuracy* first, there are a number of characteristics that need to be evaluated. First of all, the performance of the control valve assembly needs to be linear, and the hysteresis plus deadband should be low. Linearity and hysteresis plus deadband are found by plotting the input to the valve versus the valve position for a number of different input points, as shown in **Fig. 5.6.** This curve can then be transformed into what is called a deviation plot by plotting the error as a percent of output span, as shown in **Fig. 5.8.** Deviation, defined as the opposite of accuracy, is plotted as the error between the actual readings and the ideal, straight line in this figure. The total deviation, or error, can be separated into two components with two different causes, linearity and *hysteresis plus deadband.*

The *hysteresis plus deadband* is a measure of the combination of the play and friction in a control valve assembly and is an indication of the range of possible valve positions for a single input value. It is usually defined for an assembly as the widest spread on this curve between the upstroke and the downstroke. Hysteresis plus deadband is a problem because it means that for a given input signal, the valve position can vary within these specified limits, resulting in a range of possible flows. It is a general indication of how well the process can be controlled because the process variable will vary with flow through the valve. If the maximum hysteresis plus deadband is 1 percent, for example, in the worst case, the travel can vary by as much as 1 percent of its total span. To find the accuracy limits, in this case, for the process variable, multiply this hysteresis plus deadband for the valve times the following ratio, the process-travel gain:

$$\frac{\text{\% change in process variable}}{\text{\% change in valve travel}}$$

In other words, let's say that we have a valve with a linear installed characteristic and that the process variable is flow. The above ratio would then be 1, since the flow would be a linear function of travel. In this case, we could expect to do no better than to control the flow within a bandwidth of 1 percent given the performance of the valve. Note that the process-travel gain will change with changes in valve characteristic and with changes in process variable and can be less or greater than 1.

What this means from an end-user standpoint is that the desired control range for the process variable needs to be determined as a percentage of span for a given application, along with the process-travel gain. The acceptable control range then needs to be divided by the process-travel gain to determine the acceptable hysteresis-plus-deadband performance levels for the valve. When this type of review is performed, many valves will be found that exhibit unacceptably high error levels, while on others the performance will be better than it has to be. Rather than just accepting average control valve performance for all applications, this review will show end users where they really need high-performance valves and where they can get by with cheaper, low-performance models.

Linearity deals with the error associated with the average curve between the two extremes in **Fig. 5.8** and is important in that nonlinearity in the valve adds to the overall nonlinearity in the loop. It has been determined that an overall loop gain of 1 is desirable because this means that the full span of the transmitter will correspond to the full travel of the valve. This helps to keep the loop from being too sensitive to minor changes in valve position. Note that this is desired but very seldom happens because most valves are oversized, and therefore their actual usable travel range is compressed and biased toward the closed position.

The reason for wanting to keep the linearity error low in the valve assembly is that changes in total linearity for the loop make it hard to come up with a single set of tuning settings that work well over the operating range for the loop. In most cases, the valve linearity error is a small percentage of the overall total for the loop, so it is not all that critical from a performance standpoint. A 2 percent limit on a valve with positioner is typical and should be acceptable. A 5 percent limit should be good enough for a valve with no positioner.

One other accuracy term that has not received as much emphasis as the two terms just covered is *resolution.* Recent thinking is that it may be even more important than linearity and hysteresis plus deadband in determining how well a valve can control. Resolution, as used in this book, can be split into two different terms, *input resolution* and *positioning resolution.* Input resolution is a measure of how well a valve responds to small changes in input and is related to hysteresis plus deadband in the assembly. This concept has been discussed in detail by Entech Control Engineering Inc. (see reference 2) and is measured by inputting a series of oscillating step changes, around an average value, into a loop while it is on-line.

The steps start out small, usually 0.5 percent, and gradually increase in magnitude, while changes in the process variable are monitored. An example is shown in **Fig. 9.2.** The advantage of checking valve response in this way is that because the test is run on-line, the results include any hysteresis or deadband that exists all the way down to the plug or disc inside the valve. The tests described earlier for hysteresis plus deadband normally just look at an external indication of plug or ball position, with no loads on the internal parts of the valve, so any additional error introduced by hysteresis or deadband between the external indication and the plug is ignored. This is especially important on rotary valves because the linkage between the lever, shaft, and ball or disc can introduce a large amount of error, if not designed correctly, and the flow loads can greatly increase the wind-up in the shaft.

The *resolution* as measured in this test is the input level where the process variable begins to respond in a linear fashion to the changes in input. The loop in **Fig. 9.2** would be judged to have a resolution of

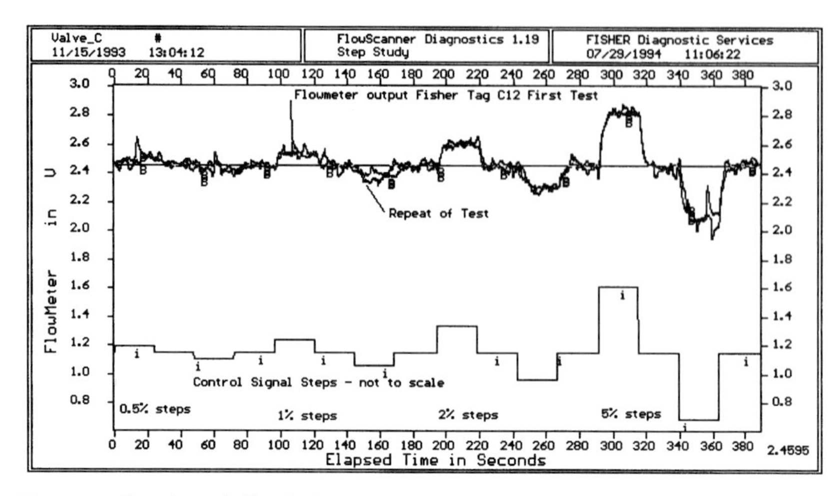

Figure 9.2 Input resolution test.

about 1 percent because even with the noise on the process variable, it does appear to be tracking well at this level. The input resolution should be approximately the same as the desired tolerances for the process variable, assuming that the input and output ranges are equivalent for the loop (100 percent change in input causes a 100 percent change in process variable).

Positioning resolution is an indication of the smallest step that a valve can make in response to a change in input. As we learned in Chap. 8, all valves exhibit friction, and one aspect of friction that is hard to get around is the fact that the static friction is higher than the dynamic friction. This characteristic, along with the gain in the valve accessories and actuator, ends up causing the valve to jump from one position to the next when the input is changed very slightly. The size of this jump as a percentage of total travel again gives an idea of how closely the valve will be able to control. If the smallest change that a valve can make is 5 percent of travel, on average, it will jump past the ideal setpoint by half this amount anytime a correction in position is called for. This will tend to set up a limit cycle that will force the valve and loop into continuous oscillation around the setpoint which, in addition to creating error, also wears out equipment. This behavior will never be eliminated completely because friction will always be present and static friction will always be higher than dynamic. But it can and should be controlled down to where it gets lost in the noise of the process variable.

It can be effectively controlled by minimizing friction in relation to the available actuator force and by proper design of the actuator and accessories. To get good performance, the positioning resolution of the valve assembly as a percentage of total travel should be about half of the overall accuracy desired for the loop. This can be determined for the valve alone, but it is more accurate to check it at the loop level with the valve in service, for the same reasons mentioned for the input resolution test.

Accuracy has been covered; now let's turn to *speed of response.* The desired behavior here would be a valve that responds very rapidly with no oscillation and no overshoot. Once again, this is hard to achieve and is not always needed. In very slow loops such as level control, the speed of response of the valve is not an issue because it's lost in the time constant of the process. For fast loops, valve response does become an issue, but it is very hard to generalize on how fast the valve should be in relation to the process. One approach suggested by Entech is shown in **Table 9.1.** The values shown apply to step changes of 2 to 10 percent, and for step response of the general shape shown in **Fig. 9.3.** The overshoot is limited to 20 percent of the size of the step in this case but may have to be reviewed depending on the consequences of overshoot for a given application.

Load sensitivity, as used here, is a function of the stiffness of the actuator in relation to the size of the unbalance forces acting on the internal parts. The lower the sensitivity, the higher the stiffness and the less the valve position is going to change in reaction to random variations in these internal loads. The general rule of thumb described in Chap. 8 states that as long as the internal loads are no more than half of the load represented by the benchset times the area of the actuator, the stiffness is sufficient. If we assume that the ran-

TABLE 9.1 Speed of Response*

Valve size, in	Td, s	T63, s	T98, s	BW, Hz
0–2	0.1	0.3	0.7	1.6
>2–6	0.2	0.6	1.4	0.8
>8–12	0.4	1.2	2.8	0.4
>14–20	0.6	1.8	4.2	0.27
>22+	0.8	2.4	5.6	0.2

*The above specification allows for large valves to be progressively slower than small ones. The Td, T63, T98, and BW specifications are a compatible set of numbers for a second-order critically damped system with deadtime, representing positioner hesitation equal to one time constant. All the responses have the shape illustrated in **Fig. 9.3.** Responses for larger valves are identical, except that time is rescaled appropriately.

SOURCE: Entech Control, Inc., Toronto, Canada.

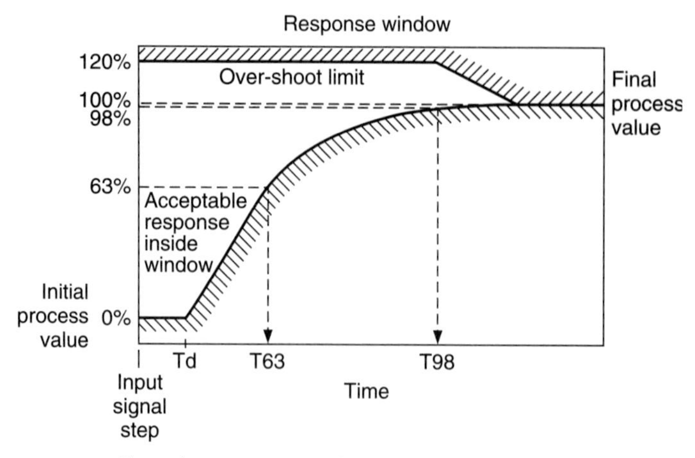

Figure 9.3 Step change response limits. (*Courtesy of Entech Control, Inc., Toronto, Ontario, Canada.*)

dom variations in these loads is about 5 percent due to turbulence, this would translate into a load equal to about 2.5 percent of the benchset, based on the above rule of thumb. This would mean that, in the worst case, the travel might fluctuate 2.5 percent due to flow loads, ignoring the damping affect of friction and the air spring rate of the actuator. Including these effects reduces the actual movement to something less than 1 percent, which should be acceptable in most cases. Piston operators with positioners have much higher stiffnesses, so load sensitivity is not usually a problem.

9.4 Life-Cycle Costs

The "perfect" valve that was referenced from a performance standpoint in the preceding sections could only be improved upon if it could achieve these high levels of performance at a very low initial price, with no maintenance required over the life of the plant. Vendors' claims notwithstanding, this valve doesn't exist yet. *Initial costs* are usually the focal point in the purchasing exercise and are generally felt to reflect value in terms of performance. In truth, the costs associated with maintaining a valve over a typical 20-year life are 10 to 20 times higher than the initial cost, and the real costs that result from poor efficiency or process shutdowns can dwarf even these relatively high maintenance costs.

As a result, the smart way to buy a valve is to *consider performance* as noted in the previous sections, uptime (or reliability), and maintenance costs, along with the initial costs for the valve in question. Only then can the real bottom-line impact on the plant be properly taken into account. A detailed discussion on how this evaluation should be carried out is covered in Chap. 15.

References

1. *ISA-S20.50 Specification Forms for Process Measurement and Control Instruments, Primary Elements, and Control Valves,* ISA, Research Triangle Park, N.C., 1983.
2. *Control Valve Dynamic Specification,* version 2.0, Entech, Inc., Toronto, Canada, July 1993.
3. Lloyd, Sheldon, and Anderson, Gary, *Industrial Process Control,* Fisher Controls, Marshalltown, Iowa, 1971.

Chapter

10

Valve Selection Summary

This chapter will attempt to pull everything together in terms of how to go about choosing the proper valve for a given application. In contrast with previous practices in this area, the selection criteria will include consideration of valve performance as outlined in the preceding chapter. All of the various elements will culminate in a recommended new form for a valve specification sheet that will ensure that, together, the valve vendor and end user will have considered all pertinent issues in coming up with their selection. This new specification sheet will also serve the role of a valve data sheet, encompassing the information required to effectively maintain the valve over its service life and to evaluate its on-going performance.

Before getting into the details on this, it needs to be pointed out that valve selection is a very complicated process and attempting to summarize it in one chapter is a little ambitious. All that is really meant to be done here is to portray the thought process that should be followed if the exercise is to yield the desired results.

10.1 Application Inputs

This section briefly describes the information that needs to be known about the application before an intelligent decision can be made regarding valve selection, referencing **Fig 10.1,** the control valve data input sheet. It covers a lot of information that is not usually included in the initial information transferred from end user to vendor. This form can save both parties time by making sure that all the necessary requirements for the valve are understood up-front rather than being picked up piecemeal in the course of the project. The form in the figure *is* much more detailed than previous documents designed to serve

CONTROL VALVE INPUT DATA SHEET

A. General
1. Customer ____
2. Plant Site ____
3. Unit ____
4. P.O. Number ____
5. Contract Number ____
6. Specification # ____

7. Drawing # ____
8. Item # ____
9. Tag # ____
10. Serial # ____
11. Quantity ____
12. Application ____

					2nd Flow Path		
					☐ 3 Way	☐ Mixing	☐ Diverging
B. Flow Requirements	Units	Normal	Min.	Max	Normal	Minimum	Maximum
1. Flow Rate							
2. Inlet Pressure							
3. Outlet Pressure							
4. Inlet Temperature							
5. Fluid Spec. WT/Density							
6. Viscosity							
7. Vapor Pressure							

8. Fluid Type ____
9. Critical Pressure ____ psig
10. Critical Temp. ____ °F

11. ☐ Two Phase?
Liquid Flow Rate ____ GPM
Gas Flow Rate ____ SCFH
12. ☐ Entrained Solids?
Size ____
Type/%by Weight ____

C. Valve Rating
1. Maximum Temperature ____ °F
2. Maximum Pressure ____ psia
3. Minimum Temperature ____ °F
4. Body Material ____
5. Other Standards ____

D. Pipeline — Inlet / Outlet
1. Size ____ ____
2. Schedule ____ ____
3. End Connection ____ ____
4. Special Facing ____ ____
5. Face-to-Face ____ in.
6. ☐ Insulation?
Type ____
Material ____
Thickness ____
7. Pipeline ☐ Vert. ☐ Hor.
8. Valve ☐ Vert. ☐ Horiz.

E. Materials
Body/Bonnet ____
Stem or Shaft ____
Seals/Gaskets ____
Packing ____
Plug/Ball/Disc ____
Other ____

F. Leakage
Shut-Off Class ____
Shut-Off Δ P ____
Leakage Limits ____
☐ EPA Valve? ____

(a)

Figure 10.1 Valve input data sheet.

this purpose, but the additional information included will save time in the long run.

The form should be filled out as follows:

Section A—General. The end user should supply the listed information that is required to properly identify the valve and the job.

Section B—Flow requirements. Valve selection nearly always starts with a definition of the flow that is required through the valve. To

G. Actuator Selection

1. Type: ☐ Piston
   ☐ Spring & Diaph.
   ☐ Other ____________
2. ☐ Spring? ________
3. Fail Action: ☐ None
   ☐ Open ☐ Closed
   ☐ Last Position (Time? ____ Sec.)
4. Supply Pressure ☐ Maximum ____ psig
   ☐ Minimum ____ psig
   ☐ Volume ____ scfh
5. Actuator ☐ Hor. ☐ Vert.
6. Clearance Above Pipeline :________ in.
7. Push down to ☐ Open ☐ Close

H. Accessories

1. NEC Class______ Div ______ Grp. ______
2. ☐ Positioner Req'd
   ☐ Direct ☐ Reverse ☐ Lin. ☐ E.P. ☐ Q.O.
   ☐ Gages ☐ Quick Disconnects ☐ High Perf.
3. ☐ I/P ☐ E/P
   Input ____________
   ☐ Rack ☐ Valve Mounted: Distance to Valve ____ ft.
4. ☐ Pneumatic Controller
   Process Variable ____________
   Output to Valve ____________
   ☐ Rack ☐ Valve Mounted
5. Limit Switches
   Rating (Voltage,etc.) ____________
   ☐ Open ☐ Closed
   Throws ____________
   Contacts (Poles) ____________
6. ☐ Volume Booster
7. ☐ Airset; Capacity ____________ scfh
   ☐ Rack ☐ Valve-Mounted ☐ Gage
8. ☐ Position Sensor, Type ____________
   ☐ Local ☐ Remote
   Output Type & Level ____________
9. ☐ Solenoid Valve; Qty. ____________
   Type ________ Input ________

   Flow Cap. ____________ scfh or CV)

I. Performance

1. ☐ Throttling ☐ On-Off
2. ☐ Loop Characteristics
   Process Variable ____________
   Loop Type ____________
   Process Variable Limits ____________
   Desired Control Accuracy +/- ____________ %
   Loop Time Constant ____________ sec.
   Loop Importance (1-5 Scale) ____________
3. Valve Linearity ____________ %
4. Valve Accuracy ____________ %
5. Positioning Resolution ____________ %
6. Input Resolution ____________ %
7. Freq. Resp. ____________
8. Stroking Speed:

	10% Step	100% Step
T63		
T99		

9. Overshoot Limit ____________ %

J. Maintenance

1. Valve Clearance
   Top ____________ ft.
   Side-to-Side ____________ ft.
2. Service Schedule ____________ months
3. ☐ Bypass ?
4. Importance: (1-5 Scale)____________

K. Special Tests/Handling

☐ Hydro ☐ Leakage ☐ Noise
☐ Material Cert's. ☐ Performance
☐ Packing/Shipping ☐ Oxygen Cleaning
☐ Inspections
☐ Other ____________

L. Preferences (List any special preferences/limits)

____
____
____

M. Paperwork

☐ IM'S ☐ Dim. Dwgs.(Wts.& CG's) ☐ Material Certs.
☐ Parts List ☐ Spares List ☐ Other________
☐ Special Test Reports ________

N. Miscellaneous

____
____
____

(b)

Figure 10.1 *(Continued)*

adequately size the valve, the listed information needs to be known and is usually specified for several different operating conditions, including normal, maximum, and minimum flows. This information is then plugged into the appropriate formulas in Chap. 8 to determine the required valve flow coefficient for each of the service conditions specified.

Section C—Valve rating. The *pressure* and *temperature rating* of the valve needs to be determined according to ANSI B16.34, in most cases. The listed information is necessary to accomplish this: Armed with the name of the material, the pressure, and the temperature, we then enter the tables in B16.34 to find the pressure rating that meets or exceeds the service conditions for a given material.

Section D—Pipeline. The *pipeline* can have a marked influence on valve selection since the body size is usually selected to match the piping end connections. Reducers or expanders are sometimes used, but they add expense and complexity that most end users would like to avoid. The *end-connection* details should include a description of the gasketing arrangement so that the valve body will duplicate the piping approach. The insulation is a consideration in determining noise levels for an application.

Section E—Materials. In many cases, the end user will have preferences regarding the choice of material for certain key valve parts. This should be limited as much as possible to permit the vendor to make what he or she feels are the best choices in this area.

Section F—Leakage. This section covers both through-leakage past the seat and leakage to atmosphere through the packing and gaskets. The shutoff classification and pressure drop help determine how much seat load is required and what type of trim should be selected (see Chaps. 2 and 8). If EPA limits are applicable to a given valve, it may mean that special high-performance packing systems are required (see Chap. 7).

Section G—Actuator selection. This section covers the information required to select the type of actuator. In some cases, the fail action is to hold the valve in a given position, and there is normally a time period specified that indicates how long the valve must remain in this position. The minimum supply pressure helps in picking an actuator size since it tells how much force a given pneumatic actuator can develop. The maximum range helps in determining whether an airset is required to limit the pressure to the valve and what the rating and size needs to be. The volume of the supply pressure will help determine the stroking speed for the valve. The orientation bullet tells us if there are any special requirements for something other than vertical orientation. Clearance dimensions are needed to be sure that the actuator selected will fit in the available space.

Section H—Accessories. This section relates to any special requirements on or for the accessories. See Chap. 5 for a more detailed discussion. Once again, the message here is that, whenever possible, leave the vendor as much freedom as possible in the choice of accessories and where they should be mounted. Admittedly, some of the information listed in this section will not be readily available, but the questions should be asked and the information supplied, if it can be found. Every one of these items can have an effect on the final valve configuration, even though on the surface it might appear unimportant.

Section I—Performance. This is an area that is not usually covered when the inputs for a control valve are assembled, but it is a very important aspect of valve selection and, as stated throughout this book, will have a major bottom-line impact on process control performance. See Chap. 9 for a more complete discussion of some of these performance measures. In some cases, the loop characteristics will help to determine the required performance for the valve. In others, the valve performance may be specified directly based on plant standards or previous experience.

Section J—Maintenance considerations. Much more money will be spent in maintaining a valve over a 20-year service life than is usually spent in the initial purchase. And yet, *maintenance issues* are not normally identified until the plant is up and running. Taking a look at the listed items at specification time can sometimes have an impact on valve selection that will save significant time and money over the service life of the valve. For instance, if access is limited to the valve or if it is in a critical service, it may make sense to pay for a more reliable design up-front to save money over the long run.

Section K—Special tests and handling. Operational or QA tests may sometimes be required for a given valve, but these should be limited in number and scope, because they add significant cost to the price of the control valve.

Section L—Preferences and limits. Some customers have developed in-house standards and policies dealing with preferred valve configuration and limits on valve performance. These should be clearly identified at specification time. They can include but are not limited to the items listed. It is a good idea to limit the number of these in-house standards, because they can occasionally keep you or the vendor from selecting the best valve for the job. On the other hand, some standardization is good because it can reduce the number of different valve designs in a given plant or unit, simplifying inventory and maintenance.

Section M—Paperwork. The listed documents may sometimes be required with the supply of a control valve. Like special tests, the need for these documents should be periodically reviewed to make sure that they serve a useful purpose and are not just adding cost.

Section N—Miscellaneous. This is a catch-all for any other requirements or guidelines that don't seem to fit in any of the above categories.

10.2 Valve Selection Considerations

Assuming the information called for in **Fig. 10.1** has been pulled together, the process of picking a valve to satisfy the application can begin. As mentioned earlier, valve selection is a very complex process and every contingency cannot be covered in the format selected here. What will be done in the next few paragraphs is to spell out the thought process associated with valve selection. This can then be applied in a more general sense to other cases.

10.2.1 Valve type, size, rating, and end connections

This is usually the first step in the selection process. The required flow capacity is calculated based on the information in Sec. B of **Fig. 10.1.** Before picking the size, we need to decide which type of valve best fits the application because the different valve types have different capacities. The basic valve types include sliding stem, rotary ball, rotary-eccentric plug, rotary-eccentric disc, and rotary butterfly, in descending order of cost for equivalent capacity. Cost versus C_v is mentioned here because, naturally, we'd like to pick the cheapest valve that best meets the requirements. The thought process regarding type would be as follows:

1. How much flow is required in relation to the line size? Higher flows can be supplied by the ball, eccentric disc, and butterfly valve designs.
2. What is the maximum pressure drop? The rotary valves are limited in terms of the pressure drops that they can handle. Higher pressure drops would point toward the sliding-stem group.
3. What kind of fluid is being handled? Is there a potential for cavitation? If so, the sliding-stem models might be preferable because of the high-recovery characteristics of the rotary valves.
4. Is the fluid erosive? The eccentric plug valves with their streamlined flow paths and massive trim parts could be the best choice in this case.
5. What is the maximum temperature and is the valve subjected to thermal transients? Very high temperature may point toward the

sliding-stem family because more models are available in the higher pressure and temperature ratings.

6. Is there an end-user preference or standard indicating one valve type over another for the application in question?
7. What kind of control is desired? For instance, butterfly valves probably don't provide the same degree of tight control as the other types of valves listed.
8. What is the pipeline size and what are the end connections like? The valve size and end connections should be compatible with the pipeline, including any special facings on the flanges.

Once these questions have been answered, the valve type can be selected. Knowing the required capacity and the type, along with pipeline size, we can select a valve body size, appropriate trim, and travel to provide the flow and to match up with the piping.

10.2.2 Body and bonnet materials

These are selected based on the type of fluid to be handled due to erosion and corrosion concerns, along with the maximum pressure and temperature conditions. Other considerations include the piping material and end-user preference.

10.2.3 Trim type and materials

The trim is selected based on things such as customer preference, inherent characteristic desired, presence of entrained solids in the fluid, erosion and corrosion concerns, thermal transients, maximum pressure drop, actuator force available, and shutoff required. Lined trim might be an option considered for highly erosive or corrosive applications. Typical decisions here include balanced or unbalanced trim and flow up or flow down through the seat ring.

10.2.4 Soft parts

The gaskets, seals, and packing are selected as a function of such things as service temperature, allowable leakage, fluid compatibility, and end-user preference.

10.2.5 Actuator type and size

Once the valve assembly details have been set, the service conditions can be used to determine the actuator force required using the approach outlined in Chap. 8. The final choice of type and size is based on this force, along with end-user preference, available air supply, failure mode, inclusion of a spring, and the availability of a positioner. Options like a manual operators and/or travel stops depend almost entirely on how the end user plans to use the valve. The normal preferred orientation is vertical to reduce the risk of side loads affecting service life.

10.2.6 Accessories

The need for a positioner is usually a function of things such as required actuator force, hysteresis-plus-deadband limits, end-user preference, stroking time requirements, and loop time constants. Except for very fast loops, where the time constant of the positioner is more than half that for the loop, a positioner should nearly always be used. It will improve overall performance of the valve assembly in every case but these fast loops.

The choice of direct or reverse acting and whether the positioner is characterized is usually made based on control logic and loop characteristics. In some cases, high-performance positioners are needed if fast, accurate response is a requirement of the loop.

If the input to the valve is an electrical signal, a transducer will have to be used to convert this signal into a pneumatic signal that the valve can work with. The type of transducer used is a function of the type of input signal and the range of the pneumatic signal to the valve. If the valve is to be used with a pneumatic controller due to system considerations, the type selected will depend on the inputs and outputs around the controller.

The types of limit switches and solenoid valves to be used will depend on system, safety, and logic requirements. A volume booster might be selected if the stroking time requirements can't be met without it. The selection of the airset depends on the capacity required and the inlet and outlet pressures. Gages and filters are normally recommended for these devices. Position sensors are usually used at the request of the end user. The type and output will depend on what type of output signal the end user needs.

10.2.7 Performance

This set of requirements goes right back to the performance demanded from the control system. The tighter the *process variable* has to be controlled, the better the valve assembly has to respond to the input signals from the controller. The discussion in Chap. 9 can be used to translate the process requirements into demands on the valve. Depending on what is listed here, you may have to select a completely different collection of equipment for the actuator and accessories to make the loop perform the way it should. The following set of questions on valve construction and performance needs to be addressed before the right choice can be made for a given application:

1. What kind of lost motion might be present in the drivetrain, all the way from the plug, ball, or disc to the actuator? For instance, on a rotary valve, how is the ball fastened to its shaft? How is the lever attached to the shaft? Is there a linkarm connecting the lever to the actuator stem? What do the rod-end bearings in the linkarm look like? How loose are they? Excessive lost motion in any of these parts will contribute to control problems.
2. What kind of lost motion is there in the positioner linkage? How is valve position sensed by the positioner? How close does it come to giving true ball, plug, or disc position? What is the operating principle behind the positioner? How much error might be contributed by the positioner itself?
3. How much force is required to maneuver the valve? Seal, packing, and bearing friction and flow forces all contribute to the total force required. Each of these elements should be examined and minimized whenever possible.
4. What does the drivetrain look like? Wind-up will be a function of the forces described above, coupled with the stiffness of the drivetrain. Things like shaft sizes and actuator mounting methods should be reviewed with an eye toward picking a valve assembly where wind-up is minimized.
5. What kind of positioning resolution will the valve provide? Resolution is generally a very good indication of how well the valve can find the “ideal” position at any point in time. It is a function of actuating forces, drivetrain stiffness, and positioner characteristics for a given assembly, and, as a result, can be difficult to predict without extensive testing. Low positioning resolution can sometimes lead to limit cycling because the valve never reaches the correct position but continues to bounce back and forth across the setpoint.

10.2.8 Maintenance

Like performance, maintenance has not received the attention it deserves when it comes to initial valve selection. If the valve is going to be hard to get to or is only scheduled for maintenance every 3 years, the choice for the valve may need to focus on reliability versus up-front cost. The same could be said if the valve is extremely important to the process but has no bypass, so a failure could bring the process down.

10.3 Summary

As can be seen from the two previous sections, good *valve selection technique* consists of getting the right information together regarding the application and then asking a series of questions that leads to the best choice for each of the elements that makes up the control valve assembly. The input to the process has already been summarized in **Fig. 10.1. Figure 10.2** is a generic example of a valve data sheet

CONTROL VALVE SELECTION DATA SHEET

DATE______
REV.______

A. General

1. Customer ______
2. Plant Site ______
3. Unit ______
4. P.O. Number ______
5. Contract Number ______
6. Specification # ______
7. Drawing # ______
8. Item # ______
9. Tag # ______
10. Serial # ______
11. Quantity ______
12. Application ______

B. Flow

1. Rated CV ______
2. Characteristic: ______
3. % Travel at Required CV: Norm. ______ Min. ______ Max. ______
4. ☐ 3-Way, % Travel for 2nd Flow Path: Norm. ______ Min. ______ Max. ______
5. ☐ Cavitating ☐ Flashing ☐ Noise Level ______

C. Valve Body Assembly

1. Type ______ Material ______
2. Size ______
3. Config.(In-Line, Angle, 3-Way) ______ Mfr./Model ______
4. End Connections: (Mat'l ____) Special Facings? Inlet ______ Outlet ______
5. Rating ______
6. Type of Bonnet: ______ Bolting: ______
7. Packing Material: ______
8. Packing Style: ______ ☐ Single ☐ Double ☐ Leak-off
9. Estimated Pkg. Friction: ______ lbs.
10. Stem Size: ______ in.
11. Allowable Leakage: Pkg: ______ ppm Gaskets: ______ ppm
12. ☐ Direct ☐ Reverse Acting

D. Trim

1. Materials
   - Plug/Ball Disc ______
   - Seat Ring ______
   - Seals ______
   - Guiding ______
   - Gaskets ______
   - Hardfacing ______
   - Stem ______ Size ______ in.
2. Port Size/Unb Area ______ in./ ______ in.^{2}
3. Shutoff Class ______
4. Shut-off ΔP ______ psig
5. Seat Load Req'd ______ lbs.
6. Flow Direction ☐ Up ☐ Down
7. Travel ______ in.
8. ☐ Balanced ☐ Unbalanced
9. ☐ Cavitation Trim ______
10. ☐ Noise Trim ______

E. Actuator ☐ Throttling ☐ On-Off

1. Type ______
2. Size ______ /Eff. Area: ______ in.^{2}
3. Mfg. Name/Model ______
4. Fail. Action ☐Open ☐Close ☐Last Pos.(Time ______ sec.)
5. ☐ Spring? Spring Rate: ______
6. Air Supply Req'd ______ psi ______ scfh
7. Push down to ☐ Open ☐ Close
8. Max. Air Press. Permitted ______ psia
9. ☐ Positioner? Type ______
10. Orientation ☐ Horz. ☐ Vert.
11. Benchset: ______ to ______ psig
    - Stroking Range
    - Open: ______ to ______ psig
    - Closed: ______ to ______ psig
    - Oper. Stroking Range
    - Open: ______ to ______ psig
    - Closed: ______ to ______ psig
12. Clearance Above Pipeline ______ in.

(a)

Figure 10.2 Valve selection data sheet.

CONTROL VALVE SELECTION DATA SHEET
PAGE 2

DATE_____
REV._____

F. Accessories

1. NEC Class ________ Div. ________ Group ________
2. ☐Positioner: ☐Dir. ☐Rev.;Input________ ☐Cam? Type________ ☐Gages? ☐High Perf.; Mfr. ________ Model # ________
3. ☐ Transducer: Input____ Output____; ☐Rack ☐Valve; Mfr.________ Model # ________
4. ☐ Pneumatic Controller; Input_____ Output_____ ☐Rack ☐Valve; Mfr.________ Model # ________
5. ☐ Lim. Switches; Qty.____ Rating____ ☐Open ☐Closed; Throws____ Poles____ Mfr.________ Model#________
6. ☐ Volume Booster; Capacity ____scfh; Input____ Output____ ☐Tunable?; Mfr.________ Model #________
7. ☐ Airset; Capacity_____scfh; ☐Filter ☐Gages ☐Rack ☐Valve; Input_____ Output_____ Mfr.________ Model#________
8. ☐ Position Sensor; ☐Remote ☐Local; Type________; Output________; Mfr.________ Model #________
9. ☐ Solenoid Valve; Qty.______; Flow Cap.:____scfh or cv; Input Signal________ Mfr.________ Model #________
10. ☐ Other ________________________

G. Performance (Valve Assembly)

1. Hysterisis plus deadband ________%
2. Linearity ________%
3. Accuracy ________%
4. Positioning Resolution ________%
5. Input Resolution ________%
6. Freq. Resp. ________%
7. Stroking Speed

	10% Step		100% Step	
	Open	Closed	Open	Closed
(Sec.) T63				
(Sec.) T99				
Overshoot %				

H. Special Tests/Handling

☐ Hydro, Press.________psia Time________
☐ Leakage
☐ Noise________________dba
☐ Material Cert's
☐ Performance
☐ Packing/Shipping
☐ Oxygen Cleaning
☐ Inspections
☐ Other ________________

I. Paperwork

☐ IM's
☐ Dimen. Dwgs.
☐ Parts List
☐ Spare Parts List
☐ Special Test Reports
☐ Mat'l Certs.
☐ QA ________________
☐ Other ________________

J. Miscellaneous

BY ________________ DATE ________________

(b)

Figure 10.2 *(Continued)*

showing the decisions that have to be made as a construction is settled on. It is a hybrid of several existing sheets and includes some additional information that can be helpful in maintaining the valve.

Figure 10.3 is a matrix that attempts to summarize how the various inputs identified in **Fig. 10.1** relate to the decisions made during the valve selection process. It also helps to illustrate the relative importance of the input data since it's easy to see how many different selection elements are affected by each input data point.

Inputs	Parameters																			
B. Flow	1	2	3	4	5	6	7	8	9	10	11	12	13	14	15	16	17	18	19	20
1.Flow Rate	X	X	X				X	X	X		X		X		X				X	X
2. P_1	X	X	X				X	X	X	X	X	X	X		X				X	X
3. P_2	X	X	X				X	X	X	X	X	X	X		X				X	X
4. T_1	X		X				X	X	X			X					X	X	X	
5.Fluid Type	X	X	X				X	X	X		X	X					X		X	
6.Fluid Density	X		X				X	X	X		X								X	
7. Viscosity	X		X				X	X	X		X	X							X	
8. PV	X	X	X				X	X	X		X								X	
9. Tc	X	X	X				X	X	X		X								X	
10. Pc	X	X	X				X	X	X		X								X	
11. Erosive	X	X	X				X	X	X		X	X								
12.Two-Phase	X	X	X				X	X			X								X	
13. 3-Way	X	X																	X	
C. Rating	X	X	X	X		X														
1. T max.																				
2. T min.																				
3. P max.																				
4. Mat'l.																				
D. Pipeline	X	X	X	X	X															
E. Materials			X				X	X				X					X			
F. Leakage							X	X		X		X					X			X
1. Class																				
2. Δ P																				
3. Limits																				
4. EPA	X																X			
G. Actuator							X						X					X	X	X
1. Type										X	X		X	X	X					X
2. Spring													X	X	X					X
3. Fail													X	X	X					X
4. Supply													X	X	X					X
5. Orientation													X							
H. Accessories																				
1. Positioner													X	X	X	X			X	X
2. I/P or E/P																X				X
3. Controllers													X		X	X				X
4. Limit SW.																X				
5. Booster																X				
6. Pos.Sensor																X				

(a)

Figure 10.3 Valve selection matrix.

Inputs	Parameters																			
	1	2	3	4	5	6	7	8	9	10	11	12	13	14	15	16	17	18	19	20
7. Sol. Valve																X				
I. Performance																				
1. Throttling	X								X				X		X	X				
2. Loop									X				X		X	X	X			
3. Accuracy									X				X		X	X	X			X
4. Speed													X		X	X	X		X	X
5. Linearity									X				X		X	X	X			X
6. Resolution									X				X		X	X	X		X	X
7. Freq. Resp.													X		X	X	X		X	X
J. Maintenance	X		X				X	X				X	X			X	X			

MATRIX-VERTICAL LISTING
VALVE PARAMETERS

1. Valve Type
2. Valve Size
3. Valve Material
4. Valve End Connections
5. Facings
6. Valve Rating
7. Valve Trim Type
8. Valve Trim Material
9. Valve Trim Characteristic
10. Balancing
11. Flow Direction
12. Soft Parts
13. Actuator Type
14. Spring
15. Actuator Size
16. Accessories
17. Packing Style/Material
18. Bonnet Style
19. Travel
20. Actuator Settings

(b)

Figure 10.3 *(Continued)*

References

1. Anderson, G. D., *ST/FT - 12 Control Valve Selection,* Fisher Controls, Marshalltown, Iowa, 1982.
2. *ISA S20.50 Specification Form for Process Measurement and Control Instruments, Primary Elements & Control Valves,* ISA, Research Triangle Park, N.C., 1983.

Part

2

Installation and Setup

Chapter

11

Installation

This chapter will cover recommended practices associated with the installation of control valves into a chemical processing plant. It will deal specifically with the installation of the control valve itself and will not cover such things as piping layout and the location of the valve in the pipe run. These items are considered to be part of good piping and system design and are outside the scope of this book.

The importance of proper installation cannot be overestimated. Improper installation can cause major problems at start-up or can keep a valve from ever having a chance to work properly during its normal service life. **Figure 11.1** illustrates the extent of the problem. This is a summary of a pre-start-up audit done on 31 valves in a chemical plant on the Gulf Coast. The bar chart shows, by problem type, the percentage of valves that exhibited less than optimum performance. You can see that the percentages are relatively high and that many of these problems would have either delayed start-up or, if not detected at that point, would have affected overall process control performance for some time. It's evident from these data that there is plenty of room for improvement in this area, and the following paragraphs are aimed at helping you eliminate these problems and drive the percentages to zero.

11.1 Who Should Do It?

Before covering recommendations on how a valve should be installed, it's appropriate to talk about who should be doing this type of work. Most unit start-ups do not go very well. The typical scenario involves multiple attempts to bring the unit on-line, with each attempt ending in the discovery of another operational problem that surfaces and has

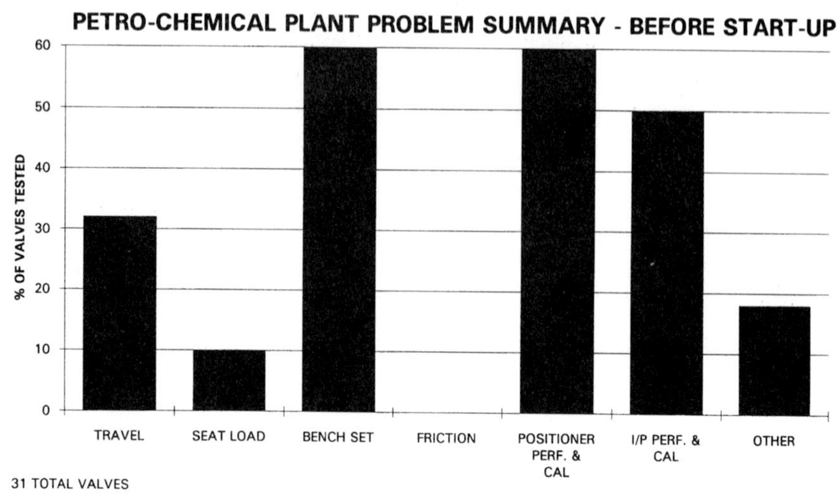

Figure 11.1 Start-up audit summary.

to be dealt with before proceeding. In many cases, the source of the problem is traced to a control valve that was working correctly when it arrived on-site, but whose operation is now less than satisfactory.

Why would this happen? It can be traced to the common industry practice of using low-bid piping contractors to install and set up the control valves. A control valve is an engineered product that needs to be handled with some care during installation. It is strongly recommended that personnel with an I&C background be used to at least supervise the installation of control valves so that you can be sure that they will work properly when the unit is brought on-line and throughout their service life. As the following sections illustrate, there are a lot of things that can go wrong because good installation procedures are not followed, resulting in delayed start-up and even damaged equipment. Keeping the right people involved from day 1 can help to avoid these problems and, once again, save the plant money in the long run.

11.2 Recommended Installation Practices

11.2.1 Before installation

There are a number of things that can be done in anticipation of installation that will help eliminate problems. First of all, *always* read the instruction manuals. Most good ones include product-specific precautions to be taken before and during installation that can help ensure proper operation once the valve is installed and operating. It is amazing how many problems could have been avoided if the installation crew had access to and had read the information that the manufacturer provided.

Most valves arrive on-site some time before they are to be installed. *Proper storage and protection* techniques need to be employed to avoid damage to the equipment before it ever sees the pipeline. Make sure that any openings on the valve and/or accessories are covered and sealed to prevent moisture or debris from entering into internal passages. This is especially critical for the accessories because of the small orifices involved and their propensity to plug if debris enters into the pneumatic circuits. Preferably, the valves should be stored inside and away from high-traffic areas. The flange faces on the body need to be protected so that no leak paths are created due to physical contact with other equipment. The same goes for the pneumatic tubing. Most valves are equipped with ¼-in copper tubing that can be bent and crimped fairly easily, resulting in flow restrictions or even leakage. Any open exhaust vents on the actuator or things such as solenoid valves also need to be sealed to keep debris and humidity outside of the system.

As for the pipeline that the valve is going into, it also needs to be kept clean and the flange faces protected. Preferably, it should be flushed and hydro tested before the valve is installed to minimize the chances of debris such as weld slag collecting in the valve and affecting operation and to guard against damage to the valve caused by the high pressures associated with the line hydro. The latter is particularly critical when dealing with valves with bellows since the bellows can be easily damaged if overpressured.

11.2.2 Installation into the pipeline

Assuming that the above recommendations are followed, the valve should be in good shape when the time comes to put it into the pipeline. The protective coverings need to be removed from the ends of the valve and the pipe flanges. The ends should then be inspected for any sign of damage, and corrective action should be taken if any damage is found. The ends should then be cleaned. A light coat of lubrication or gasket compound can be applied to the faces where the gasket is to be placed if this is compatible with the process and gasket material used. This lubrication will help the gasket to seal, will facilitate disassembly, and can also be used to hold the gaskets in place during installation.

The valve and pipeline ends need to be examined to be sure that they are of a consistent design. You should not try to use a flat-face flange design on the valve body with a raised-face flange design on the pipeline, for example. Leakage could result.

Once this inspection stage is complete, the pipe needs to be properly supported so that the piping ends can be adjusted to bring the faces parallel and in-line with one another. The face-to-face dimension for the piping then needs to be measured and compared to the face-to-face dimension for the valve body, including the two gaskets, to be sure that the body will fit in the space and that excessive force will not have to be used to draw up the flanges onto the body. If the spacing between the body and the pipeline flanges is too high, there is a good chance that the joint will leak due to line loads. If more than about one-third of the flange bolt torque is required to draw the joints up snug against the gasket, the line loads could be excessive, and other measures should be taken to get a better fit between the valve and the space in the pipeline.

The *gaskets* need to be selected based on the expected service conditions and the design of the gasketed joint. Take a moment to make sure that the gaskets properly fit on both the piping and the valve. Chapter 7 covers the selection of line gaskets in detail. Check the flange thickness for both the valve and piping to make sure that the studs selected are long enough to go through the flanges with about one nut thickness on either end. A good rule of thumb is that one thread should be exposed at either end of the studs, once the nuts have been installed.

Once the gaskets have been selected and the piping has been properly adjusted, the valve is ready to go into the line. The valve should be picked up and supported using either properly rated nylon straps around the actuator or a lifting lug, if it is built into the actuator design (see the instruction manual for lifting details). Care should be taken not to damage the pneumatic tubing or accessories during this lifting operation. This can be done by making sure that the nylon straps are not resting on or pushing against any of the peripheral equipment found around the actuator structure.

In most cases in the chemical process industries, *bolted joints* are used to hold the valve in the pipeline. If, however, the valve is to be *welded* into the line due to concerns about line flange leakage, special precautions need to be taken. In particular, if the valve contains soft parts that could be affected by the high temperatures associated with welding or heat treatment, it needs to be dissembled and the soft parts removed while the welding is done on the body ends. Once the welding and heat treatment are completed, reassembly can take place. Experience has shown that if the soft parts are left in the body, they can be damaged or can relax to the point where leakage can result once the valve is brought on-line. Seals can be reused, but gaskets should be replaced upon reassembly.

If *bolted joints* are used, good bolting practices need to be employed to avoid leakage. The studs and nuts need to be well lubricated, including the nut faces, with a lubricant that takes into account the service conditions, especially temperature. The lubricant is important in controlling the torque-to-stress relationship in the joint and will help to make disassembly easier. For a standard globe-style valve, the body assembly should be properly positioned between the flanges, paying attention to the flowing direction for the valve assembly, and the studs should be inserted into the flange holes on the lower half of the body. The gaskets can then be put into position, using the studs to hold them in the proper location. All the studs should then be inserted and torqued in a crisscross pattern up to the recommended torque. See Chap. 7 for additional information on good bolting practices to ensure a tight joint. Make sure that the valve assembly is fully supported until the flanges have been drawn up completely. This is especially critical for valves with separable flanges since the valve body can sometimes rotate inside the flanges if the joints are not tight.

Although most manuals say that a control valve assembly can be installed in any orientation, it is highly recommended that the *standard orientation* be used whenever possible. For sliding-stem valves, this is with the stem vertical and the actuator above the valve body. For rotary valves, this is with the valve shaft horizontal and the actuator mounted vertically. Anything other than standard orientation puts unusual stresses and loads on the bearings and seals and will result in reduced service life. It also makes the valve much more difficult to service and can result in vent openings that point upward, collecting humidity and debris.

There are some precautions that need to be taken that are valve-type specific. For instance, on eccentric-disc valves, the travel needs to be closely checked before installation to be sure that the actuator stops are properly adjusted. If the disc rotates too far into the seat, it can damage it or the disc, and if it goes too far the other way, the disc can contact the body walls and be damaged. On all butterfly or eccentric-disc valves, line clearance needs to be checked to ensure that there is enough clearance inside the pipeline to permit the disc to rotate without hitting the inside wall of the pipe. You also have to be careful in centering the valve, or the same problem can result. Poor centering will also reduce the installed capacity of the valve due to increased turbulence. For these same valves, be sure to fully close the valve before line installation to guard against the disc being damaged by coming into contact with the pipeline. *Note: If the valve has a spring-to-open actuator, it may have been shipped with air locked in the actuator to keep the disc closed. Be careful when handling this type of valve, since inadvertent venting of this air could result in the valve stroking without warning.*

On some flangeless valves, overtorquing can result in damage to the body shell or seals, so the torque needs to be tightly controlled in these cases. Some flangeless valves feature centering holes or lugs to be used during installation to get the valve in the proper position.

One big problem with all rotary valves occurs when the actuator is removed for installation. If special care is not taken to mark the parts and get them back together correctly, you can end up with the ball, plug, or disc not being properly aligned with the shaft and lever. If this happens, the valve will no longer stroke correctly to the open and closed positions, and leakage or damage to the internal parts will result.

For rotary valves on slurry-type service, the ball, plug, or disc should always rotate open to a position above the shaft. This keeps any debris that may collect in the bottom of the valve from being pulled back through the seals and bearings during the valve stroke. If it appears that this orientation rule is not being followed, the engineering department should be consulted so that corrective action can be taken.

Final adjustments or connections that might be required once the valve is in place in the line include:

- Make sure that the packing has been properly tightened.
- Connect the packing leak-off to appropriate tubing, if present.
- Install any grounding straps recommended in the instruction manual if the valve is on oxygen service.
- If purged bearings are being used, attach the purge lines as recommended in the instruction manual.
- If a bellows seal is being used with a monitoring tap, attach appropriate tubing to the tap.
- If a line hydro is to be carried out, be sure to take precautions to protect valve parts, such as bellows, that could be damaged. See the instruction manual for appropriate measures.
- Install insulation, if appropriate, on the valve assembly, taking care not to insulate parts such as extension bonnets, that are designed to dissipate heat.
- Install heat shields, if necessary, between heat sources and susceptible parts, such as accessories.

- Inspect all accessories and their mounting parts for damage or looseness that may have occurred due to handling during installation. Many problems with valve operation can be traced to this stage.

11.2.3 Pneumatic connections

Once the valve assembly is in place in the pipeline, the pneumatic and electrical connections need to be made. First of all, the pneumatic supply lines should have filters (3 μm) and isolation valves in place. Before connecting them to the valve, they should be opened up and allowed to vent for 2 to 3 min to clear any debris from the line. It is also recommended that a large paper bag be used to collect any oil or humidity that is present in the line during this purging. This can be accomplished by directing the airflow into the bag. If excessive amounts of oil or humidity are present at this stage, a review of the pneumatic supply system should be carried out and the problem corrected. Poor air quality is one of the major causes of premature failure of the pneumatic accessories. In fact, in light of this, the ISA is currently in the process of revising the "Quality Standard for Instrument Air" to provide better guidelines for achieving air quality consistent with good equipment performance (see reference 3).

Some type of *pipe sealant* should be used on the male threads of the pneumatic lines before connecting them to the valve. A TFE paste or tape is most common, but care needs to be taken not to use too much of either, or the excess can enter the lines and block the flow of air. Some plants have outlawed the use of this tape, but if used prudently, it should not pose any problems and is easier to apply than the paste.

The *air lines* connected to the valve need to be of sufficient capacity to drive the valve at its required stroking speed. This capacity depends on the size and capacity of the actuator, but $^3/_8$-in tubing is normally recommended. If we starve the valve at the pneumatic supply line, the dynamic response will suffer and, as repeated throughout this book, poor dynamic response from the valve means poor process control. Many valves are operating at substandard levels simply because the supply line capacity is too low.

In some cases, the connection line should be in the form of a "pigtail" to help keep the stresses low as the valve and piping shifts and/or vibrates. If the positioner is a very high-flow model or if a volume booster is present, there may be a need for two separate supply lines, one for the transducer and one for the device feeding the actuator. If a single line is used and the supply pressure drops due to high demand from the actuator, this drop in supply pressure can actually change the transducer signal output.

The *transducer* is one of the accessories that can be *remote mounted*. This is usually done because of vibration problems on the valve or to move the I/P out of a highly corrosive area. If all the transducers are mounted on a single rack, calibration can also be done more quickly. The downside to remote mounting is that the volume of air that has to flow through the I/P increases as the line length to the valve increases. Excessive line length can end up building another lag into the valve response because most transducers don't have very high flow capacity. If dynamic response is important for a given valve, the transducer should be mounted as closely as possible to the valve and/or a positioner and centralized rack mounting should not be used.

The same comments can be made regarding *split-ranging* of two valves off of one transducer. The line length is a concern because of the distance between the two valves, and to this is added the need to fill two pneumatic signal lines to the two separate valves. There is no way around sluggish valve response if this control strategy is adopted. If valve response is important on a loop, this scheme needs to be avoided. A better approach would be to split-range two transducers fed off of a single controller.

Once the pneumatic connections have been made, the assembly should be checked one more time to make sure that all temporary plugs and coverings installed as protection are removed from the actuator and accessory ports. This includes exhaust ports on boosters, solenoid valves, actuators, and the like. If any of these open ports are positioned such that water or debris could collect inside them, protective tubing should be connected to them and the open end of the tubing pointed downward to help keep the water and debris out. Make sure that if tubing is added in this way, it is of sufficient size so that the exhaust is not restricted. There have been cases where the stroking time of a valve more than doubled because the exhaust line from the actuator was restricted due to this type of tubing.

The inside of the accessory enclosures also needs to be checked to make sure that any packing material that was put in place during shipment has been removed. The accessories' performance can be affected if this material is inadvertently left in place.

A final step can be carried out at this point that will save a lot of time down the road. *Quick-connects* can be installed in the pneumatic lines that permit the rapid hookup of pressure-measuring devices anytime the valve and accessories need to be checked out or recalibrated. With these in place, no lines need to be broken into to measure pressure, so the tests can go faster, and there is less chance of leakage after the fact due to tubing connections not being drawn up tightly after a test. A convenient spot for these quick-disconnects is the positioner gage block. The gages can either be replaced by these

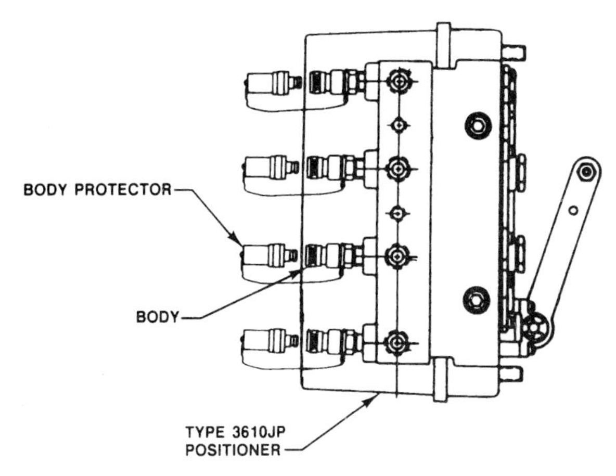

Figure 11.2 Quick-disconnect in positioner gage block. *(Courtesy of Fisher Controls International, Inc., Marshalltown, Iowa.)*

fittings, or they can be fitted with special couplings that permit the gages to be removed if more precise measurement of the pressures is desired. An example of one such arrangement is shown in **Fig. 11.2,** where four quick-disconnects have been installed in the gage block connections of a double-acting positioner. The four ports provide access to measure the supply pressure, the input signal, and the top and bottom piston pressures.

11.2.4 Electrical connections

Electrical connections can become very complicated, especially if the area the valve is going into is considered hazardous from an electrical code standpoint. Common electrical connections include things such as the transducer, solenoid valves, limit switches, and position feedback sensors. Each of these will have to be connected following the procedure outlined in the individual device's instruction manual, along with whatever guidelines apply from a hazardous location standpoint for a particular site. See Chap. 5 for a more complete discussion of hazardous area classifications.

11.2.5 Performance tests

The final step to be taken after the valve assembly has been installed and all connections made up is to check for proper operation. This can be done in a number of different ways, but the best approach is to temporarily hook up an adjustable input to the valve assembly and run at least a five-point calibration test, starting at 50 percent input and stepping to 75, 100, 75, 50, 25, 0, 25, and back to 50 percent. The valve position, along with all intermediate pressure readings, should be collected for each input point and then plotted to check for correct readings of the following operating characteristics:

Travel

Transducer zero and span

Transducer linearity

Transducer hysteresis plus deadband

Positioner zero and span

Positioner linearity

Positioner hysteresis plus deadband

Actuator stroking ranges: open and closed

If any of these values don't fall within normal ranges, they need to be corrected before the valve is put into service. Chapter 12 discusses how corrections can be made and also suggests what the acceptance criteria should be in each case.

It must be emphasized that this operational check is very important to ensuring that valve performance will be as expected during start-up and that every valve is given the maximum chance to provide good process control throughout its service life. Time spent at this stage in correcting problems is much cheaper than trying to do the same thing while on-line. That's why operations and maintenance departments should take an active role during the installation phase. They're going to have to deal with the problems sooner or later. Why not make a preemptive strike and correct them before they occur?

One final note. The *operational check* described above is normally a manual operation involving a lot of time taken to measure and manually record operational parameters. There is now equipment on the market that can hook up rapidly to a typical control valve and automatically generate a preprogrammed input signal while recording the valve pressures and travel. A "birth certificate" can then be generated for the given valve showing the correct as-left condition for the valve just before start-up (**Fig. 11.3**). This equipment can greatly simplify this final checkout and makes data storage and retrieval much easier. See Chap. 13 for a much more detailed description of how these systems work and what they can do, both from a start-up mode as discussed here and as a diagnostic troubleshooting tool.

FlowScanner Quick Report

Tag#: McK-Test Serial#: 10792793 01-17-1994
Test Type: 50 Second Scan Test: 08090836 08:26:40
Valve Style: ES / 1 Trim: Microform Actuator: 657 / 34

Test Parameter	Spec'd	Measured	Comments
OVERALL VALVE CONTROL			
Travel (IN)	.75	.756	OK
Dyn.Zero Travel (ma)	4	4.56	OK
Dyn.Max. Travel	20	19.55	OK
Dyn.Err.Band(%)	N/A	2.029	3% AND LESS IS NORMAL
Dyn.Lyn.(%+/-)	N/A	.408	1% AND LESS IS NORMAL
VALVE AND ACTUATOR DATA			
Friction Av. (LBS)	39	21	OK_ CLOSE ENOUGH
Maximum	39	32	VERY CLOSE TO SPEC
Minimum	N/A	17	
Seat Load - Test (LBS)	N/A	565	
- in Service	31	554	NO PROBLEM
Spring Rate		1087	OK
BenchSet(PSIG)	3-14.9	3.06-14.88	VERY CLOSE TO SPEC
POSITIONER DATA	Type: 3582		
Dyn.Set(PSIG)	3-15	3.34-14.76	WITHIN SPEC.
Dyn.Err.Band(%)	N/A	1.359	OK
Dyn.Lin.(%+/-)	N/A	.816	OK
I/P TRANSDUCER DATA	Type: 3311		
Dyn.Set(PSIG)	3-15	2.73-15.07	WITHIN SPEC.
Dyn.Err.Band(%)	N/A	1.214	OK
Dyn.Lin.(%+/-)	N/A	4.5	OK
SEAT EVALUATION			GOOD SEATING PROFILE

RECOMMENDED Valve Actuator Positioner I/P
REPAIR: _ _
ADJUST:
VALVE IS IN VERY GOOD CONDITION; READY FOR START-UP

Figure 11.3 Start-up birth certificate.

References

1. Bauman, H. D., *Control Valve Primer,* ISA, Research Triangle Park, N.C., 1991.
2. Hutchison, J. W., *ISA Handbook of Control Valves,* ISA, Research Triangle Park, N.C., 1971.
3. *ANSI/ISA-S7.3-1975* (*R1981*), ISA, Research Triangle Park, N.C., 1981.
4. *Design EWD EWS & EWT Instruction Manual, Form 2376,* Fisher Controls, Marshalltown, Iowa, 1981.
5. *Design CV500 Instruction Manual, Form 5302,* Fisher Controls, Marshalltown, Iowa, 1993.
6. *Design V100 Instruction Manual, Form 5061,* Fisher Controls, Marshalltown, Iowa, 1993.
7. *Design 8510 Instruction Manual, Form 5174,* Fisher Controls, Marshalltown, Iowa, 1992.
8. *Design CE Instruction Manual, Form 5192,* Fisher Controls, Marshalltown, Iowa, 1993.

Chapter

12

Setup and Calibration

This chapter will not go into detailed product-specific directions on how to set up and calibrate a control valve. These can normally be found in the instruction manual for the equipment. Instead it will examine common, generic procedures that need to be followed in order to avoid problems with valve operation. These procedures include things such as travel adjustment, packing adjustment, the infamous "benchset," zero and span on positioners and transducers, booster tuning, and limit switch settings.

The approach will be to first cover each of the procedures for a sliding-stem, globe-style valve since this is the most common type of valve in use in the chemical industry. Then, other valve types will be covered if there are any significant differences in the procedures.

12.1 Valve Setup

12.1.1 Benchset

We'll start with *benchset* because it is easily the most misunderstood concept associated with control valves, and it is usually the first thing that should be done when getting a valve ready for service. *Benchset* is the term used to describe the adjustment on an actuator with a spring that determines how much restoring force the spring will provide once the actuator is mounted on the valve. It is expressed as a pressure range for the rated travel of the valve and is calculated based on service loads inside the valve that must be overcome if the valve is to stroke properly (see Chap. 8 for more details). While it is calculated based on service loads, it is not meant to be set or checked with any friction or valve forces present. With these points in mind, the following procedure can be used to avoid the typical pitfalls encountered when trying to set the benchset on a control valve:

1. Consult the valve data sheet or the valve name tag to determine what the benchset setting should be. Note that for a benchset to be specified for an actuator, there must be a spring present since the specified benchset determines the spring adjustment. However, there are actuators with springs that have no provision for spring adjustment. The spring tension is determined by the spring characteristics and the space available in the assembly when it is put together. In this case, there is no adjustment possible, but there may or may not be a benchset specified, depending on the philosophy of the manufacturer. If the benchset is specified, it can be checked using the following procedure, but it cannot be adjusted if it is determined to be outside of the tolerance. The only remedy is to replace the spring or springs. Examples of this type of actuator are shown in **Figs. 12.1** and **12.2.**
2. Make sure that the actuator is disconnected from the valve or that there are no flow forces present and that the packing is completely loose, so there is no friction.
3. An adjustable pressure source with a gage accurate to $\pm$ 0.1 psi over the benchset range needs to be connected to the pressure inlet of the actuator. The positioner, if present, should be bypassed at this stage. Note that this procedure is normally carried out on single-acting actuators where pressure pushes the actuator in one

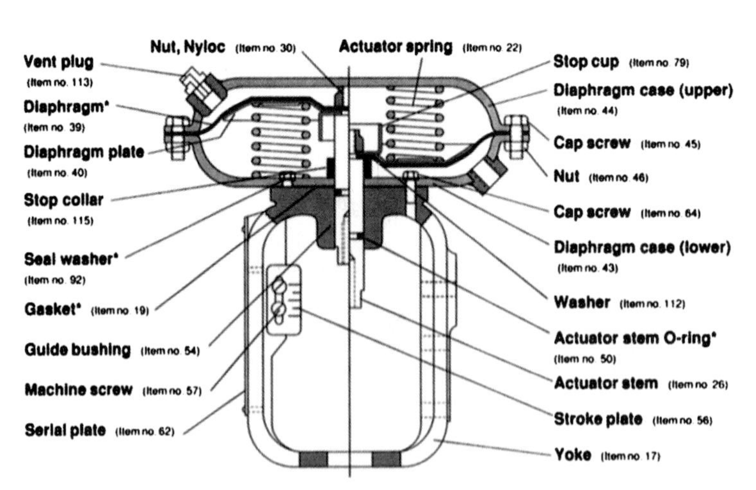

Figure 12.1 Spring and diaphragm actuator with no spring adjustment. *(Courtesy of Valtek International, Inc., Springville, Utah.)*

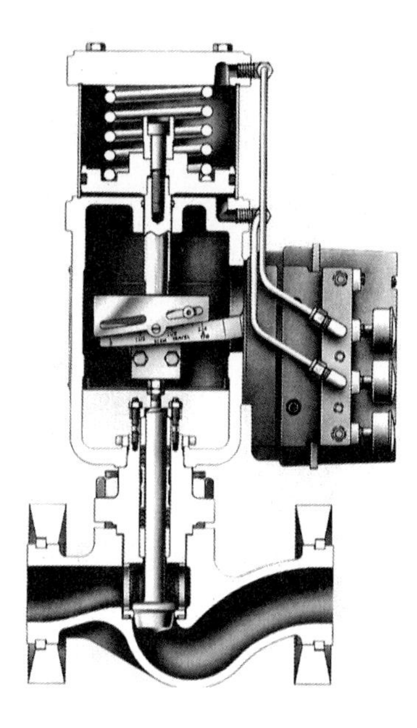

Figure 12.2 Piston actuator with spring but no spring adjustment. (*Courtesy of Fisher Controls International, Inc., Marshalltown, Iowa.*)

direction, and the spring supplies the restoring force. Springs can be supplied on double-acting actuators (**Fig. 12.2**) where pressure is acting to move the valve in both directions, but normally these are piston actuators with no provision for adjusting the spring tension (benchset). Essentially, the spring tension is set when the piston is assembled and can't be adjusted after the fact, so no benchsetting procedure is required. If a double-acting actuator does employ a method for adjusting benchset, or if you simply want to check the spring tension in an actuator with no benchset adjustment, the adjustable pressure source needs to be connected to the actuator inlet side opposite the spring, and the other side should be open to atmosphere. With this arrangement, the benchsetting technique is identical to that for a single-acting actuator.

4. Once the pressure source is connected, it should be used to drive the actuator up against its upper stop, either by increasing or decreasing the pressure depending on whether the actuator is fail closed or fail open, respectively. If the spring is opening the valve, it may need to be tightened to reach the upper stop. If this is the case, the benchset adjustor needs to be located on the actuator and the spring tightened until the actuator reaches the upper travel stop as illustrated in **Fig. 12.3.** This figure shows the upper stop as well as the benchset adjustor for a typical spring and diaphragm actuator with spring-to-open action. A spring-to-close model is shown in **Fig. 12.4.** The reason for always driving the

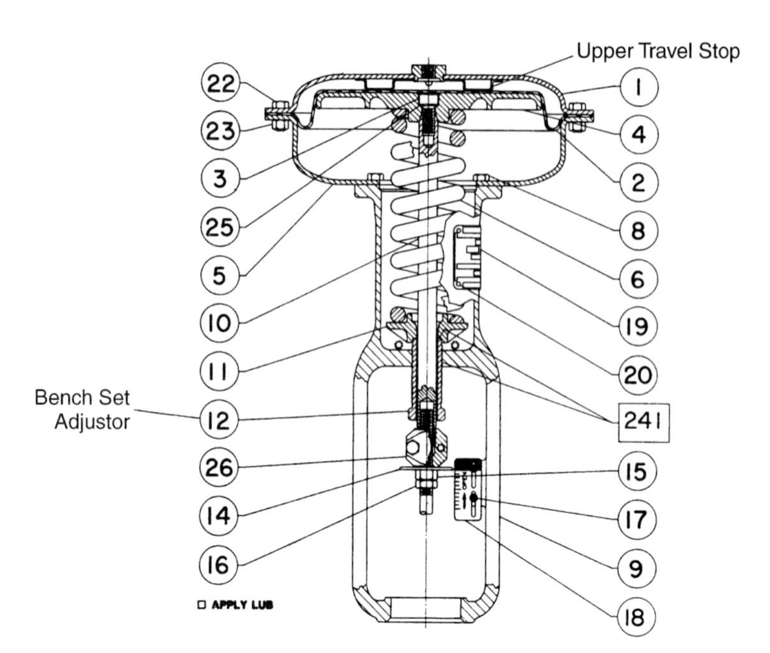

Figure 12.3 Spring-to-open actuator. (*Courtesy of Fisher Controls International, Inc., Marshalltown, Iowa.*)

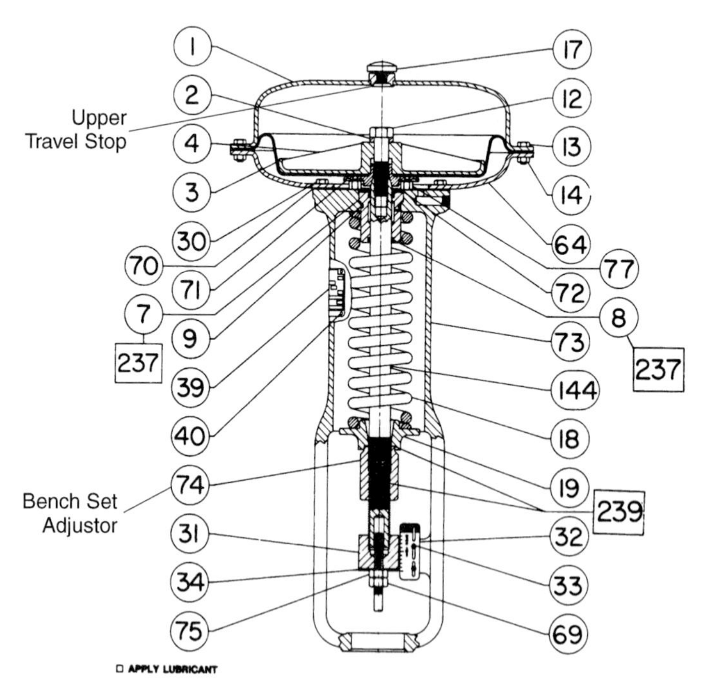

Figure 12.4 Spring-to-close actuator. (*Courtesy of Fisher Controls International, Inc., Marshalltown, Iowa.*)

actuator to its upper stop during the benchsetting procedure is that the upper stop is one of the reference points in setting travel, and we want to be sure that we set the benchset and the travel over the same stroking range. This will be covered again in more detail in the section on setting travel.

5. Let's assume that we're working with a spring-open actuator and we now have the diaphragm against its upper stop. We can check the spring tension, or benchset, by increasing the air pressure to the point where the actuator just begins to move and then further increasing it until the actuator has moved over the rated travel for the valve. These two pressures define the benchset for the actuator and can be compared to the values specified on the valve data sheet or the name tag. For a spring-to-close model, the pressure needs to be reduced until the actuator begins to move off of the upper stop and then further reduced until the actuator has stroked over the rated travel range for the valve. The measured benchset for a valve does not have to agree exactly with the specified values. Normally the benchset value that corresponds to the end of the stroke where the valve seats is deemed to be the most critical since it will determine the seat load for the valve. It should be equal to the specified value ±0.2 psi. The range of the benchset will have a relatively wide tolerance band because of the manufacturing tolerances on the spring rate and changes in the actuator effective area with travel. If the range (the difference between the two pressure values measured above) is accurate to within ±10 percent, it is considered to be close enough. If the range is outside of this limit, the spring needs to be changed since this is an indication that the spring rate is incorrect. If any adjustments need to be made based on the above tolerances, the spring can be loosened or tightened and the values checked again until they fall within the acceptable range.

This procedure applies equally as well to *rotary valves,* but for *reverse-acting* valves, it needs to be slightly modified. For this type of valve, the plug moves down to open and up to close. This means that the lower travel stop should be the reference point since valve travel will be defined by stroking from the lower travel stop up to the seat. As a result, the beginning point for the procedure should be with the actuator on the lower stop, and the critical reading that needs to be adjusted to within ±0.2 psi is the opposite reading since the valve now seats at the top of the actuator travel.

12.1.2 Setting valve travel

Sliding-stem valves. For sliding stem valves, the travel is defined by the valve seat on one end and the opposing actuator travel stop at the other. The following procedure can be used to make sure that the travel is set correctly and that the travel is set over the same range as the benchset:

1. Check the nameplate or valve data sheet for the specified travel.
2. With the actuator mounted on the valve but not connected at the stems, connect an air supply to the inlet of the actuator.
3. Placing an object such as a crescent wrench handle between the valve and actuator stems, use the actuator to drive the plug into the seat. This procedure assumes that the plug stays on the seat once it is placed there. If there are pilot springs or a bellows in the valve, the plug may push back from the seat due to the resistance of the springs or bellows. If this happens, measure the amount the plug moves back and subtract this amount from the rated travel figure used in step 5 of this procedure. This will ensure that the total travel is still correct.
4. Check to be sure that the actuator travel is at least 10 percent more than the rated travel in the valve. This will ensure that the valve can be adjusted so that the travel will be limited by the valve seat and one actuator stop as described above. If the actuator travel is not larger than the rated travel, the travel will be limited by the actuator at both ends and the valve may never reach the seat. If the actuator travel is too small, check to make sure that it is not being limited by an adjustable travel stop or manual operator. Assuming that this is not the case, the actuator will have to be replaced or disassembled and the travel corrected. The only other option is to reduce the rated travel for the valve.
5. With the actuator on the upper stop, stroke it down the rated travel for the valve by changing the pressure and then connect the actuator and valve stems. Normally this is done using some type of split-nut arrangement (**Fig. 12.5**).
6. Once the connection is made, stroke the valve several times and check the actual travel against the travel specified. It should be within the rated travel, $+\frac{1}{16}$ in, -0. If adjustments need to be made, pull the valve off the seat, loosen the stem connector, and turn the valve stem into the connector for more travel or out of the stem connector for less travel. *Never turn the valve stem with the plug on the seat.* This can damage the seat and result in leakage.

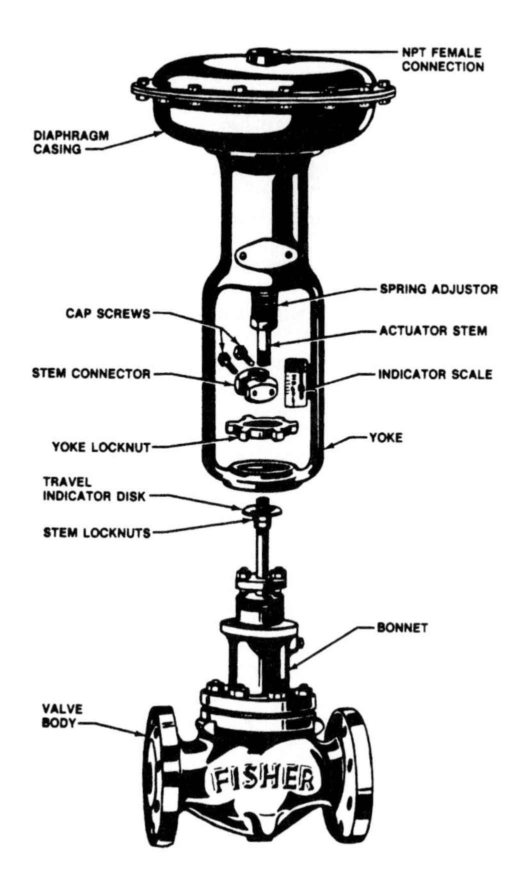

Figure 12.5 Actuator showing split-nut stem connector. (*Courtesy of Fisher Controls International, Inc., Marshalltown, Iowa.*)

7. Some valves do not use a split-nut stem connection. An example of this is shown in **Fig. 12.6.** In this case, the valve stem needs to be screwed into the actuator stem as far as it will go and then the valve stroked down toward the seat. The travel will be limited by the lower stop in the actuator in this first stroke. The valve should then be stroked to the open position and the valve stem screwed out of the actuator by about one-half turn. Stroke the valve closed again and note the stem position. This procedure should be repeated until the valve stroke has dropped due to contact between the plug and the seat and is equal to the rated travel for the valve. The valve stem clamp then needs to be tightened to prevent further stem rotation.
8. Adjust the travel indicator accordingly.

This procedure will assure that the benchset and travel are set over the same range since they were both set using the upper travel stop as

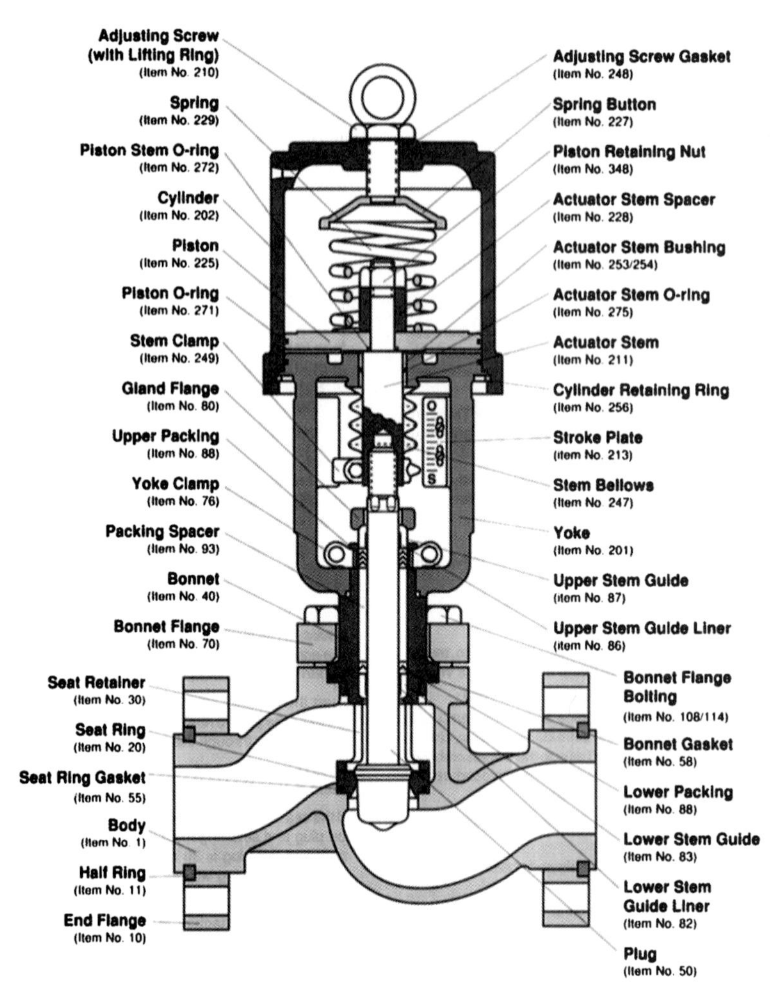

Figure 12.6 Actuator without split stem connector. (*Courtesy of Valtek International, Inc., Springville, Utah.*)

a reference. This is important because the actuator will normally have as much as 25 percent overtravel. If the benchset is mistakenly set from the lower actuator stop, the installed spring load will be incorrect by as much as this percentage, which could result in the valve not closing properly or not getting fully open. Always set the benchset and travel over the same stroking range to avoid this type of problem.

For reverse-acting sliding-stem valves, the valve seat is actually hit on the upstroke of the actuator, so the reference point for this procedure needs to be changed to the lower actuator stop for this type of valve. This is also a little more complicated because the valve plug needs to be pulled up into the seat rather than driven down into the seat as described in step 3 above. The other difference with this type of valve is that screwing the valve stem into the actuator stem decreases travel instead of increasing it.

Rotary valves. Rotary valves are different in that they normally use both actuator stops to determine travel. This is because, on most models, there is no positive stop inside the valve body like there is in a sliding-stem valve.

There are two principal ways of setting the *travel range* on a rotary valve. In the first method, the range is determined by the linear travel between the top and bottom actuator stops. This is usually not adjustable and is determined by the proper dimensioning of the related parts in the actuator. An example of this type of arrangement is shown in **Fig. 12.7,** where the travel of the diaphragm plate is not adjustable. The linear movement is transformed into the rated rotary travel through the lever arm, with most valves either stroking 60 or 90°. The tolerance on the stroke is relatively wide, given the manufacturing tolerances. In most cases, ±5° seems to be standard.

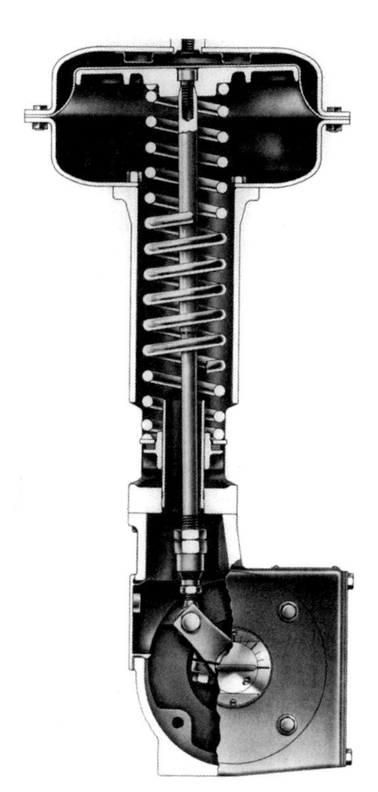

Figure 12.7 Rotary actuator without adjustable travel stops. (*Courtesy of Fisher Controls International, Inc., Marshalltown, Iowa.*)

While the travel range is fixed in the actuator, the *zero position* of the disc, ball, or plug does need to be adjusted so that the flow element comes into proper contact with the seat. If it doesn't stroke far enough toward the seat, the valve won't shut off, and if it strokes too far, the seat can be damaged. The "zero" position for the flow element is normally adjusted at the connection between the actuator stem and the lever. With the lever on the valve shaft, the linkarm can be turned into or out of the actuator stem to change the relative position of the flow element until it is in the proper closed position. On ball valves, this adjustment can be fairly rough since the ball merely rotates past the seal, so stroking too far will not result in damage to the seal. On conventional or high-performance butterfly valves, the adjustment needs to be precise, due to the risk of damaging the seal with overtravel.

Eccentric plug valves are a different story altogether. Because of their unique seal design, there is a significant amount of actuator torque that is required to put the plug into the seat far enough to shut off. At the same time, if all of the available actuator torque is transmitted to the plug, it might push in too far and damage the seat. Essentially, the actuator load ends up being shared between the seat and the travel stop in the actuator. This is done by very carefully adjusting the linkage to the point where the plug contacts the seat and the actuator travel stop at the same time. This can be tricky depending on valve design, and the instruction manual for the valve in question needs to be followed very closely in order to avoid problems.

The other type of rotary actuator that might be encountered has one or more *adjustable travel stops*. A standard arrangement is shown in **Fig. 12.8.** This type of design is more convenient and more flexible since both ends of the travel can be adjusted for a given valve. Adjusting the travel involves simply turning the travel stop bolts in or out until the travel is where it needs to be and then tightening the locknuts. Whenever possible, the travel in the upward direction for an actuator with a spring should be limited by the stop at the top of the actuator since this is the reference point that would be used in setting the spring tension. If the actual travel stop is very far from this point, the spring tension will not be correct for the assembly and operational problems could result.

For either actuator design, the best approach is to adjust the travel with the valve out of the line so that you can be sure that the valve is being properly positioned. The major cause of rotary valve problems is incorrect travel adjustment leading to excessive leakage, insufficient flow capacity, or damage to internal parts.

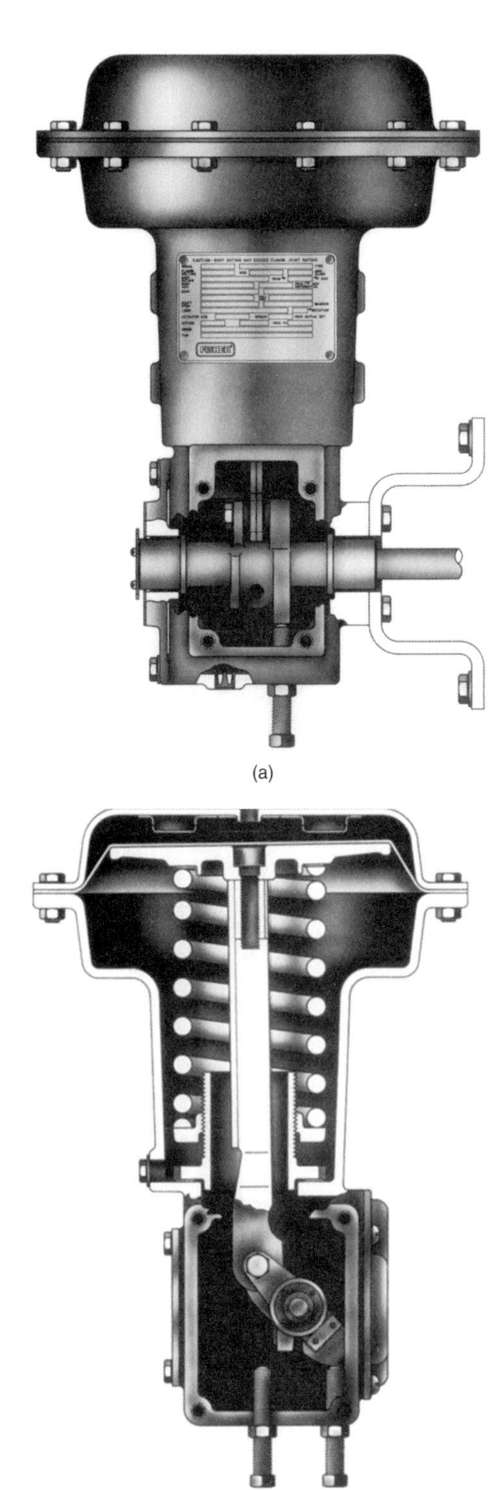

Figure 12.8 Rotary actuator with two adjustable travel stops. (*Courtesy of Fisher Controls International, Inc., Marshalltown, Iowa.*)

12.1.3 Packing adjustment

This step probably causes more problems with valve performance than all the other setup procedures combined. The packing needs to be tight enough to seal but not so tight as to begin to interfere with valve operation. With TFE-based spring-loaded V-ring packing, this is easy to achieve. The follower should be drawn down tight against the top of the bonnet. This will apply the proper amount of preload to the packing by compressing the spring inside the packing box. Additional sealing load will come from the pressure-assist design of the packing.

On all standard *jam-style packings* (graphite or TFE) without springs, the instruction manual should be consulted and the specified torque levels applied. If no torque levels are listed, the vendor should be contacted and asked to supply the correct values. **You should not guess at the appropriate values unless you are prepared to deal with leaks or sluggish response once the valve is on-line.** Before applying the torque, make sure that the studs and nuts are well lubricated so that the correct load is developed in relation to the applied torque. Always make sure that the nuts are tightened evenly to avoid problems with the packing follower binding on the shaft and not transmitting the load to the packing. Once the torque has been applied, the valve should be stroked 20 to 30 times to break in the packing and then it should be retorqued. If stick-slip action is observed, a small amount of lubricant may be applied to the stem and worked into the packing by stroking the valve. Of course, this technique won't work on a rotary valve since the lubricant won't be drawn down into the packing box. The only way to lube the shaft for a rotary valve is to pull the packing out and then apply an acceptable lubricant before putting the packing back in.

On *high-performance packing* sets, follow the instruction manual very closely if there is to be any hope of meeting published leakage performance. These systems usually include live-loading, meaning that the adjustment of the load is done by controlling the compression on a set of packing springs rather than using torque levels. Break-in cycling is not required for this type of design since the live-loading will react to any change in packing height to keep the load constant.

12.2 Calibration

Normally, the calibration that needs to be carried out on a valve assembly is on the transducer and the positioner.

12.2.1 Transducer calibration

The *transducer* takes the controller output and converts it into a pneumatic signal that the valve can work with. It can be direct or reverse acting, but in most cases, it will take a 4- to 20-mA input and convert it into 3- to 15- or 6- to 30-psi output. These devices can be adjusted to change the zero and the span so that the proper output is provided to the valve for a given input. For instance, for the example just cited, with the input set at 4 mA using a calibrated input device, the output should be a nominal 3 psi read with a calibrated gage. If it isn't, the zero should be adjusted until the appropriate signal is obtained. Once the zero is right, the input should be changed to the other end of the range and the output checked there. If the output is not 15 psi for an input of 20 mA, the range needs to be corrected until it is.

This is a simple process and only two or three comments need to be made before moving on. First, on most instruments, the zero and span are interactive, so the above procedure needs to be repeated several times before both values will be right. Second, the tolerances on the output need to take into account the next device in the line. If the output is going to a positioner, the tolerances should be 2.8 to 3.0 psi on the lower end and 15 to 15.2 for the upper value. This will ensure that the positioner will always be getting the full-open and full-closed signal at the endpoints of the 4- to 20-mA input range. If an actuator is being fed directly by the transducer, the tolerances should be centered on the nominal values and can be wider: 2.8 to 3.2 psi and 14.8 to 15.2 psi.

The last point on transducers is that the output on many models tends to drift with time, temperature, orientation, and the like. The end user needs to track this drift and change the initial setup tolerances in relation to the magnitude and direction of this drift to keep the output in the acceptable range.

12.2.2 Positioner calibration

Positioners act much like transducers in that they accept an input signal and provide an output that is a function of the input signal. The differences between the two devices include the facts that the input for a positioner can be either a pneumatic or electrical signal, the output for a positioner is valve position not pressure, and the relationship between the input and the output is not necessarily linear. The input-to-output function can be characterized by modifying the position feedback to give a linear, equal percentage, or quick-opening relationship. In most cases, the linear relationship is the best choice, and any characterizing that is required by the loop is done in the trim in the valve.

In setting up a positioner, adhere to the following procedure to avoid problems:

1. Check the data sheet to verify the input signal and the valve travel and to see whether it is direct or reverse acting and if any characterization is required.
2. Armed with this information, connect a calibrated electrical or pneumatic signal source to the input port of the positioner.
3. Check the *valve position feedback linkage* to make sure that it is secure and undamaged and that it has been properly positioned given the amount of travel for the valve (**Fig. 12.9**). Also, verify that the linkage is properly connected, taking into account the rated travel for the valve. Most linkages need to be adjusted with this travel in mind, and if this is done incorrectly, the position feedback can be adversely affected, or in the worst case, internal positioner parts can be damaged. A common problem is putting the linkage in the wrong hole on the feedback bracket and ending up with too much or too little feedback going into the positioner mechanism. Too much feedback can cause the cam follower on some models to roll off the cam and cause serious damage to the related parts.
4. Place some type of pressure-measuring device on the output port to the actuator. A gage may already be installed but be aware that most gages supplied with this equipment are not very accurate or reliable.

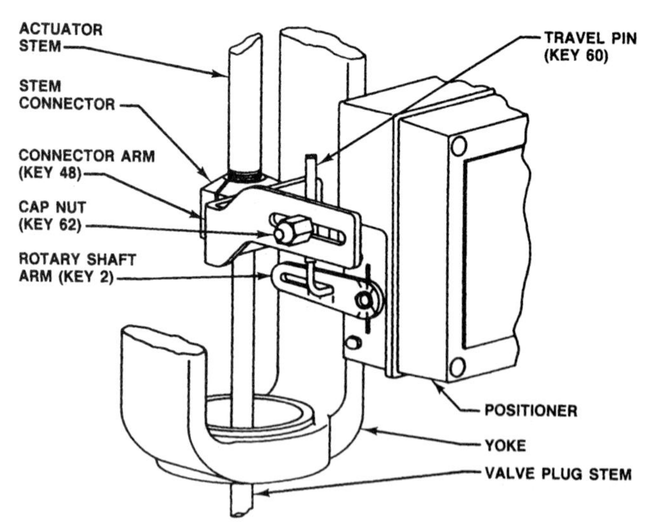

Figure 12.9 Details of feedback linkage. (*Courtesy of Fisher Controls International, Inc., Marshalltown, Iowa.*)

5. For a direct-acting positioner, reduce the input signal until it reaches the lower end of the input range and verify that the output from the positioner drops to 0 psi and that the valve takes whatever position corresponds to 0 psi input, either open or closed. If the output pressure does not drop to 0 psi, adjust the positioner zero until it does.
6. Now, increase the input signal until the upper range of the input is reached and verify that the output pressure goes to within 3 to 5 psi of full supply and that the valve reaches full travel. If the output pressure is not "maxing out," or "surging," as it is often called, the range needs to be changed until it does. If it cannot be adjusted to achieve this, the positioner is malfunctioning and should be scheduled for maintenance.
7. Repeat the above procedure several times until the performance is as described at the two endpoints for the input: Pressure drops to zero at one end with the valve reaching one endpoint in travel and goes to full supply at the other end with the valve stroking completely in the other direction. Repeating this is necessary because the zero and span on the positioner interact. Be careful adjusting the zero on positioners with nozzle-flapper designs. Sometimes locking the nozzle in place after adjustment can change the setting. The tolerances on the points where these two things occur should be as follows: nominal value −0, +5 percent (as a percentage of the range) at the lower end and nominal value +0, −5 percent at the upper end. This will ensure that the endpoints are reached at the nominal input points and that the usable input range for the positioner will be maximized, making it less susceptible to random changes in input.
8. Note that there are many things that can affect valve travel that aren't related to positioner performance. As a result, the positioner may be in perfect condition, but the valve travel may not look right. If the positioner is going from full supply to zero pressure at the endpoints, it is doing the best that it can, and if there are still problems with valve travel at that point, look elsewhere for the cause of the problem. The benchset may be wrong, the supply pressure may be too low, or the travel setting may have been done incorrectly. Get these problems corrected and then go back to finish the positioner calibration.
9. The *valve position* at intermediate points can be checked against the input signal using a deviation cycle as described in Chap. 11 to see if the linearity and hysteresis plus deadband are within the specified values for the device. These are normally inherent characteristics and can't be adjusted or corrected without repair or replacement.

12.3 Booster Tuning

Volume boosters are added to valves to increase stroking speeds by providing more air flow to the actuator. They act as pneumatic relays by taking a small-volume pneumatic signal and transforming it into the same pressure signal but with a much higher-volume flow so that the actuator and valve can move faster in response to the original signal.

The downside to all this stroking speed is that it has a tendency to make the valve assembly unstable with excessive overshoots and constant cycling as a result. To account for this, most boosters have *bypass adjustments* (**Fig. 12.10**) that permit the booster to be tuned so that small changes in input bypass the large relay and the valve remains stable. If adjusted properly, the booster should only be active for large step changes. In reality, most boosters are set up with the bypass completely closed, resulting in unstable operation that can sometimes be misinterpreted as sticking or jumping inside the valve.

To set a booster correctly, decide on the size of the step change that the booster needs to respond to, and then provide that step at the input to the booster. Starting with the bypass all the way closed, repeat the step change several times and back off on the bypass until the booster stops opening. Then repeat the step as the bypass is closed until the booster just begins to work again. Lock the bypass in place and place a sticker or tag on the booster that indicates that the bypass should not be changed without appropriate approvals.

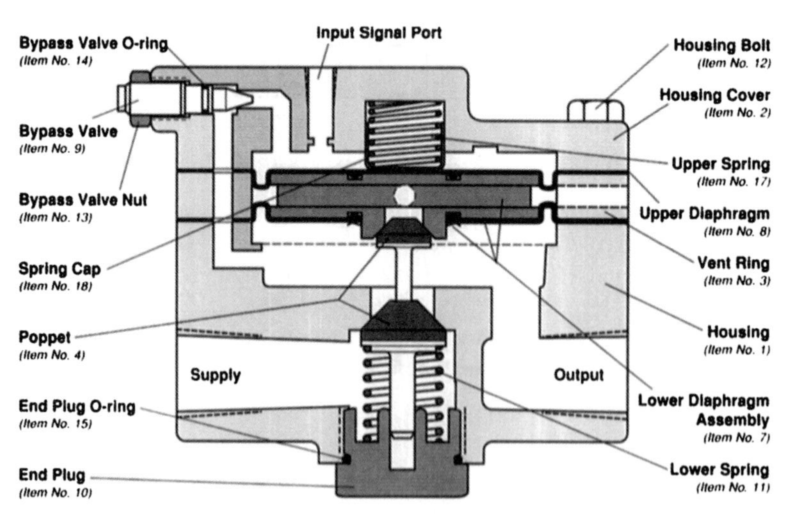

Figure 12.10 Volume booster with bypass. (*Courtesy of Valtek International, Inc., Springville, Utah.*)

12.4 Limit Switch Settings

There is nothing very complicated about limit switches on control valves. They should be set to activate as closely as possible to the actual endpoints for travel to avoid the chance of an open or closed indication when the valve has not fully stroked. In most cases, they can be set in the range between $\frac{1}{16}$ and $\frac{1}{8}$ in of the actual endpoints without too much trouble. Recognize that they have a certain amount of deadband and that there is a mechanical tolerance on the trip point, so the trip arm will contact the switch-activating arm well before the trip actually happens. Once they are set, they need to be locked securely in place on strong mounting brackets because the activating force is high for most switches.

Part

3

Maintenance and Operations

Chapter

13

Control Valve Maintenance

This chapter covers various aspects of recommended control valve maintenance procedures. Control valve maintenance is a very interesting subject. As has been said throughout this book, control valve performance is key to plant performance. If we accept this fact, it would seem to justify a relatively large investment in control valve maintenance since, along with proper valve selection, it is one of the key determinants of overall valve performance.

Surprisingly, in spite of the above, *valve maintenance procedures* have not changed significantly in the last 20 to 30 years and the hours dedicated to them have actually fallen off in the average plant, particularly in the last 15 years, as plants have downsized. Why this seeming dichotomy? There are a number of reasons. First and foremost, valves are robust. They continue to function, even if in poor condition. As a result, many plants can take the approach that no news is good news. "The valve still works, so leave it alone and concentrate on areas where we know we have problems." As a result, you end up with valves whose overall performance is less than it could be but which are judged to be good enough to get by.

Second, there is no *reference standard* to easily compare valve performance to. In other words, even though the valve is not performing at optimum levels, if there is no way for the end user to know this, it follows that no corrective action would be deemed necessary. Because there is no performance standard, end users do not know that they have a problem until complete failure takes place. This leads to a more reactive-type approach to valve maintenance (see the next section). Third, there is an *organizational structure* in most plants that keeps them from proactively addressing performance problems. This is covered in Sec. 13.2.

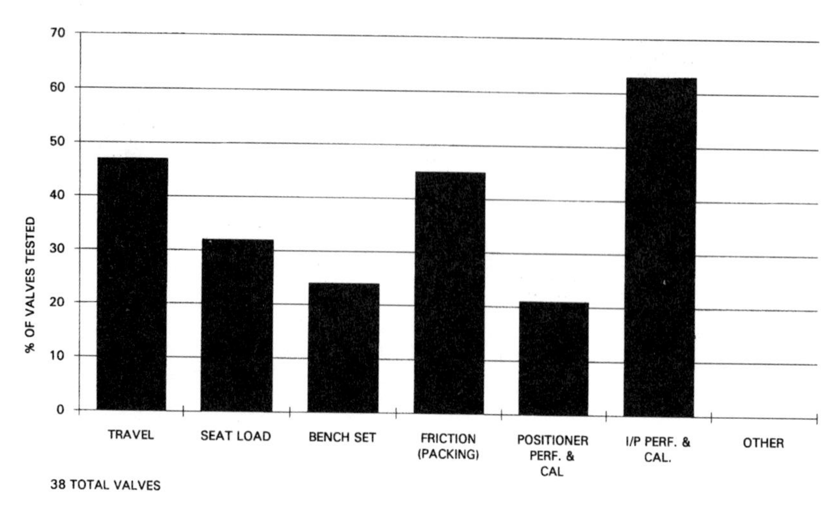

Figure 13.1 Problem summary—chemical plant.

What all this means is that the average plant is living with valve problems that are hurting overall process control and plant performance. **Figure 13.1** illustrates the results of a *control valve audit* at a typical chemical plant. The valves that were checked were valves that the plant considered to be in reasonably good shape (i.e., they were valves that were not scheduled for maintenance). As you can see from looking at the curves, the percentages of problems found are all high. The valves were still working but not working very well, which supports the argument just presented. The problems detected ranged from incorrect travel to I/P performance. Some further explanation is useful in interpreting the results of this survey:

Travel. Any valve that had a travel error greater than 10 percent of the specified value was included here. Travel problems can result in reduced capacity or poor shutoff.

Seat load. This category included valves with insufficient seat load or no seat contact. Both of these conditions result in excessive internal leakage.

Benchset. Any error outside the limits specified in Sec. 12.1.1 was part of this group. Benchset errors can affect stroking speed, seat load, and travel. If benchset was the primary cause of travel or seat load problems, it was not counted again in a second category.

Friction (packing). This included any valve with friction that was too high due to bearing or packing problems or friction that was judged to be too low, indicating a potential for packing leakage. Excessive friction could result in poor response, reduced travel, or inadequate seat load. Again, there was no double counting.

Positioner. This group included any problems related to the positioner, either calibration problems or poor dynamic response. Poor calibration could affect travel or seat load, and dynamic response relates to things like hysteresis, linearity, and stroking and frequency response. About 60 percent of this category was incorrect calibration.

I/P. Like the positioner, this included calibration and response problems, with the consequences being the same. For the transducer, the calibration problem percentage was close to 75 percent.

Other. This group had problems in areas such as spring rate, airset, boosters, and other accessories.

Figure 13.2 is a *composite* for four different types of plants and shows very clearly that there are major problems with travel settings, seat load, friction, and positioner performance. In fact, the average reading shows that over 40 percent of the valves tested might have all these problems. It's obvious from this kind of data that the valves, and therefore the plants, could be operating more efficiently. The good news is that a number of plants are beginning to realize the extent of the problem and are starting the search for a solution.

The change is coming about for a number of reasons. Increased competitive pressures are forcing end users to turn over new "rocks"

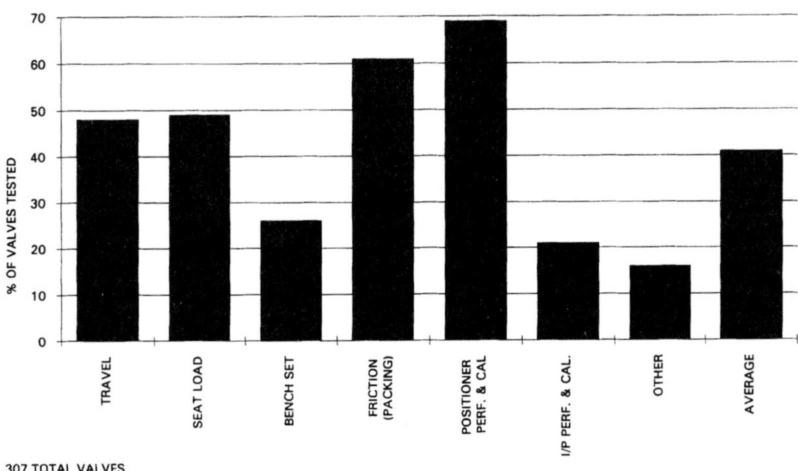

Figure 13.2 Weighted composite of problem summaries.

as they try to become more competitive. *Process audits* are being done that point to valve problems as the weak link in the process control chain. *Total quality* programs are emphasizing searching out even minor sources of deviations and eliminating them once and for all. And finally new "diagnostic" technology is permitting end users to get a better idea of how a valve is performing today, while providing a reference that can be used to determine how much better valve operation could be if all problems were corrected.

This new technology is discussed in Sec. 13.5. The balance of the chapter covers techniques and information that you should find useful as you try to transform your plant from one that just runs to one that performs at world-class levels.

13.1 Maintenance 101

Typical maintenance approaches include *reactive, preventive,* and *predictive maintenance. Reactive maintenance* is generally considered the least efficient method because, as the name implies, plant personnel take corrective action only after some type of breakdown has already occurred. In the process industry, allowing failures to happen usually means that the consequences and costs of the failure will be greatly magnified due to lost production associated with having equipment out of service. Since the failures tend to come with no warning, labor costs are also usually higher due to callbacks on second shift and the like.

While reactive maintenance does not seem to be very cost efficient, there is still a relatively large percentage of plants that utilize it when addressing control valves. This may be because there are a large number of valves in use and no one valve has the impact on operations that some of the other, larger pieces of equipment in a plant do. What we must not lose sight of is that, because of the large number of valves in use, the cumulative effect of multiple failures can be significant and justifies the consideration of more sophisticated approaches.

The next step up in addressing maintenance concerns is *preventive maintenance.* With this approach, the end user tries to prevent failures by scheduling regular interventions on equipment to repair, replace, or rebuild equipment or components. The timing for these interventions might be based on vendor recommendations or past history. This is better than a reactive approach because it cuts down on the on-line failures, but it does involve a higher level of maintenance activity, and some work may be done unnecessarily on equipment that was still in good condition. This is the most common approach used on control valves in the chemical process industry.

The third approach, which has gained a lot of favor over the last 10 years or so, is called *predictive maintenance* and involves trying to predict failure and schedule corrective action before it can occur. This is a little like "just-in-time maintenance" in that, at least in theory, there is no wasted effort since the interventions are based on impending failure. The predictions are the difficult part of this approach and are usually based upon the combination of some type of nondestructive examination and past history. This technique is currently not being used much for control valves due to the absence in the past of any way to easily check a valve to determine its current condition and remaining service life.

In general, this is considered to be the most cost efficient means of addressing maintenance concerns, and its application is the goal of most forward-looking maintenance personnel. As a result, there have been a number of efforts over the last several years to come up with new techniques that can aid in evaluating valve condition and to eventually predict useful life. These efforts are beginning to bear fruit with the introduction of new control valve diagnostic techniques that use pneumatic and valve position readings to determine the valve and accessory condition without having to disassemble the components. This technology will be covered in more detail later in this chapter.

To these three traditional maintenance techniques, a new technique called *preemptive maintenance* can be added. One of the messages conveyed in Chaps. 11 and 12 is that it is very important to get the valve set up correctly. Experience has shown that many problems that eventually require an intervention by the maintenance group can be traced to the fact that the valve wasn't set up properly to begin with. Preemptive maintenance is based upon the *total quality* concept that if one takes the time to do a job right the first time, many problems can be corrected before they occur. This can be a very powerful tool when working with control valves, and as a result, an entire section of this chapter is devoted to further explanation of this concept along with examples of how it can be applied.

13.2 Organizational Considerations

Before getting into the details of how to carry out maintenance on control valves, there are some changes in organizational structure that will help to make the maintenance efforts more effective. The typical plant organization is not one that lends itself to optimizing plant performance. Normally, there is an *operations group* charged with generating product, whose goals revolve around the following:

Increased yield (i.e., more product)

Reduced energy use (improved efficiency)

Reduced chemical use on the input side (improved efficiency)

Reduced cost: Labor, material, and capital

Better quality (more on-spec product; tighter specifications)

Safer working environment

Meet government regulations for emissions

Operations runs the plant and is supported by a number of other groups within the organization. The support groups that work with control valves include *mechanical maintenance, instrumentation and control,* and *systems engineering.* The primary problem revolves around the fact that the operations group is faced with the bottom-line goals listed above but does not usually have the detailed skills necessary to optimize the operation of the equipment itself. Operations personnel who are generalists must know a little bit about a number of different things. They rely on the support groups listed above to help them keep the plant running. But in most cases these support groups are not working to the same list of global goals. The organizational structure conspires to keep them focused on their own departmental goals, which are not always consistent with the overall goals for the plant.

The problems that can result from this type of structure are best illustrated with an example. As noted, operations wants to produce as much on-spec product as possible. To do this requires that the process control system be operating at optimum levels, including the control valves. Operations wants this to happen but must count on the support groups to make it happen and, worse yet, has no way of knowing if it *is* happening.

The support groups such as maintenance are not given, or evaluated on, the same set of goals. Maintenance is told to keep the plant running at minimum cost and with reduced staff. Anyone in this group who would suggest that they try to improve the performance of a system or valve that is already getting by would be labeled as a fanatic and encouraged to seek psychiatric help. This problem is further compounded by the fact that control valves are robust and can continue working under less than ideal conditions. They don't work very well, and the process can suffer, but they still work. The end result is that operations settles for a plant that continues to run, instead of the world-class performance that is really needed to stay competitive.

The only real solution to this dilemma is for management to take the blinders off and give the same goals to everyone in the organization. Everyone from the plant manager to the I&C techs working on the valves should have the same appreciation of increased efficiency in plant operations and be rewarded for it. Plant performance would improve dramatically with this approach and employee morale would improve at the same time as they began to see how their actions could have a bottom-line effect on making the plant more successful.

13.3 Preemptive Maintenance

As mentioned earlier, if care is taken to properly install and set up a control valve, many future problems can be eliminated. This new emphasis on up-front procedures is being dubbed *preemptive maintenance* to tie into the more common approaches already employed such as reactive, preventive, and predictive.

The key to making this technique work is identifying those up-front practices that can have a significant effect on future valve performance, either positive or negative. Each of the items covered in Chaps. 11 and 12 will be examined below in light of the above to illustrate the importance that should be placed on getting the particular activity done right the first time:

Activity	Consequences
Protecting the valve assembly before installation.	Prevents external leakage at the flange connections, tubing leaks, slow response due to tubing crimps.
Covering openings on accessories.	Prevents accessory malfunctions, slow response.
Flushing the pipeline, conducting hydro tests before valve installation.	Guards against trim damage, internal leakage, bellows damage.
Proper positioning of pipe.	Prevents external leakage at flange connection.
Proper flange gasket selection.	Prevents external leakage at flange connection.
Using straps at proper points to support valve during installation.	Guards against accessory malfunction, tubing leaks.
Using proper bolting technique on flange connections.	Prevents external leakage at flange connection.
Using the standard valve and actuator orientation.	Guards against premature bearing and seal failure, makes maintenance and service easier.
Adequate line clearance for both types of butterfly valves.	Permits the valve to fully stroke and avoids possible disc damage.
Properly centering butterfly valves in the pipeline.	Ensures that valve will fully stroke without hitting pipe and helps to ensure full flow capacity.

Activity	Consequences
Keeping butterfly valves fully closed during installation in pipeline.	Prevents damage to disc, problems with stroke, and seat leakage.
Controlling torque on flangeless valves.	Guards against damage to body and/or seal due to excessive loads.
Making sure that rotary valves open with the flow element above the shaft.	Helps to keep debris that collects in the bottom of valve from being dragged through the seal and bearings.
Protecting bellows during installation and line hydro.	Prevents external leakage past the bellows.
Installing insulation on correct parts.	Keeps the packing cool, helps to prevent external leakage past the packing.
Flushing pneumatic airlines.	Prevents debris from affecting accessory performance and slowing valve stroking speed.
Providing adequate air supply, pressure, and volume.	Capacity will affect valve stroking speed, inadequate pressure can result in poor shutoff, shortened travel, or reduced stroking speed.
Mounting the I/P as close to the valve as possible.	If the I/P is too far from the valve, it can add to lag and reduce the stroking speed.
Staying away from split-ranging off the I/P.	Split-ranging can add lag and reduce the stroking speed.
Making sure that all covers and plugs are removed from the accessories.	Can affect accessory performance and valve stroking speed.
Making sure the pneumatic exhaust lines have adequate capacity.	Can slow valve stroking speed.
Properly adjusting benchset.	Can reduce travel, hurt shutoff, and slow the valve stroking speed if not set correctly.
Properly adjusting valve travel.	On butterfly valves, overtravel can result in seal or disc damage. Improper zero setting for travel on rotary valves can result in excessive internal leakage. Undertravel can hurt valve flow capacity. Incorrect procedure on sliding-stem valves can result in excessive internal leakage.
Properly adjusting packing.	Can result in reduced travel, poor dynamic response, excessive internal leakage, or external leakage past the packing.
Properly calibrating the I/P and positioner.	Can result in reduced travel or excessive internal leakage.

As we review this list, it can be seen that 25 different activities need to be performed correctly while installing and setting the valve up or problems with valve performance will result. The other interesting point that results from this type of analysis is that these 25 activities, if done incorrectly, can result in a cumulative total of 55 different problems with valve performance. This is why preemptive maintenance is so important. We can eliminate 55 sources of valve problems before they occur if we practice the preemptive approach. To practice preemptive maintenance, the plant needs to emphasize the above activities in internal procedures and then make sure that the procedures are carried out by using the right people to handle setup and installation (see Sec. 11.1), and making sure that they have the proper training.

13.4 Preventive Maintenance

Preventive maintenance for control valves is not now particularly sophisticated because there has not been enough data collected regarding failure rates, time between failures, and failure types. In most plants, the critical valves are identified as the ones that can bring the process down if the valve fails. Failure, as it is defined here, is usually confined to things like excessive external leakage, excessive internal leakage, and severe stroking problems such as reduced travel or sticking and jumping. It does not include more subtle problems such as some of those identified above.

Once these critical valves are identified, they are put on a maintenance schedule that usually ties into the outage schedule on the unit in which they are found. The preventive maintenance consists of pulling the valve from the line, tearing the valve apart, and putting it back together with new or repaired parts, depending on the extent of the damage found upon disassembly. The teardowns generally coincide with unit shutdowns that might occur quarterly, semiannually, or once every 18 months. The hope is that by rebuilding the valves on this schedule, no catastrophic failures will occur between now and the next outage. Other troublesome valves are added to this teardown list, as time permits, based on feedback from operations. While this does seem to work in terms of reducing what are defined as on-line failures, there are a number of problems with this traditional approach:

- Many of the valves scheduled for overhaul don't really need it. For the critical valves, the teardown schedule is based more on the consequences of failure and on when one can get to them than on their actual condition. On the valves that have been selected based on feedback from operations, in many cases the problem reported is actually due to some other problem in the loop. The valve may have been operating correctly.

- Many of the repairs could be done with the valve left in the line. If a problem can be traced to an accessory, it is much simpler to leave the valve in place and fix only the external component that needs it.
- Major problems are usually detected, but some of the more subtle conditions go unnoticed, still costing the plant money in terms of lost performance.
- Many valves are repaired over and over again for the same condition without any thought given to determining the root cause and preventing the recurrence through corrective action.
- No formal records are kept on the as-found condition or on the corrective action taken.
- Many problems are the result of a misapplication of a valve or trim type or changing service conditions. An application review could help determine the root cause but is seldom carried out.
- Simply pulling a valve and rebuilding it does not guarantee that the valve will be going back into the line in any better shape than it left. If setup and installation are not done as outlined in the previous section, the teardown ritual may actually be causing as many problems as it solves.
- Pulling the valve means breaking the line flange joints, and getting the valve back into the line without problems is not always as simple as it sounds. Pipelines can move, making realignment difficult, and an assortment of other problems can surface, including installing the valve backward, using incorrect gaskets, and/or installing the insulation incorrectly.
- Non-valve-mounted accessories such as airsets and I/Ps may not even be checked if they don't go back to the shop with the valve.
- Dynamic performance of the valve, while critical to good process control, is very seldom checked. Normally, if any troubleshooting is done in this type of program, it is a static check where the valve performance is verified at several discrete points. No thought is given to how quickly a valve is responding to a change in input from the controller.

It's clear from a review of the problems with the traditional preventive approach that there is a pressing need within the industry for a better way to determine which valves need work and what needs to be done. An approach that can help do this is covered in Sec. 13.5. But before turning to that subject, some general guidelines need to be reviewed for the teardown and reassembly of a typical control valve. Regardless of how we determine which valves need work, the teardown and rebuild process is critical to getting valve performance problems corrected. To be sure that the process is carried out correctly, the following checklist should be referred to. Note that it relates to a sliding-stem globe-style valve, but the same general techniques could be used for other types of valves, taking into account the differences in construction.

1. Consult the material safety data sheet (MSDS) for the fluid being handled by the valve to determine what safety measures are appropriate for handling the valve, including decontamination, if necessary. Find the instruction manuals, cross-sectional drawings, and spare parts list. Make sure that the spare parts required are on hand before beginning disassembly.
2. Set the valve up in a sturdy vise and leave the strap loosely connected to valve assembly to support it in case the vise lets go.
3. Perform a detailed visual inspection of the valve, noting any watermarks, damaged linkages, and the like.
4. If possible, perform an as-found diagnostic test as described in the next section before disassembly. This can help to document problems found with the operation of the valve.
5. Center punch a mark on the actuator, bonnet, and body so that the orientation can be kept track of as the valve is disassembled. Sometimes this can provide clues to the source of problems found during disassembly.
6. Pressurize the actuator and accessories and use a soap solution to check for air leaks.
7. Check tightness of the packing nuts.
8. Once preliminary work is completed, relieve tension in the actuator spring, if present, and disconnect the valve stem from the actuator stem. Be careful not to rotate the valve stem while it is on the valve seat. This could damage the seating surface.
9. Disconnect the actuator assembly from the bonnet and lift it off.
10. Loosen the bonnet bolting but leave several nuts in place on the studs to guard against the bonnet popping up due to internal pressure when it is pried loose from the body. Remove the bonnet. Pull packing and packing parts out of the bonnet.

11. Leaving the internal parts in place and noting the orientation, look for any unusual marks and traces on the assembly. Take particular care when examining the gasket faces.
12. Pull the stem and plug out and repeat the above inspection.
13. Pull the cage out of the body and remove the seat ring following the guidelines in the instruction manual. If the seat ring is screwed in, be extremely careful when applying the torque to back it out. If the seat ring suddenly breaks loose or if one of the parts slips or fails, there is a serious risk of injury.
14. Inspect the inside of the body for any signs of leakage or erosion.
15. Clean and grit-blast the body parts and inspect them again. Repair and replace as necessary, based on the inspection. Chase the body studs with a die.
16. As far as the topworks are concerned, if the operation checked out properly and none of the elastomeric parts or bearings are scheduled for replacement, it can be left assembled.
17. Note that the elastomers and bearings in the actuator should be replaced about every 4 years of standard service to guard against leakage and poor response due to increased stiffness. If the actuator is taken apart, the diaphragm should not be reused.
18. To reassemble, begin by inserting the seat ring into the body. The seat ring and plug sealing faces should be recut if necessary to clean up any marks or wire-drawing. The seats should be lapped if the valve is a class IV shutoff or better. Lapping is explained in Sec. 13.6.
19. Using all new gaskets, reassemble the body assembly, and torque the bonnet bolts to the recommended settings. Reinstall the packing per the instruction manual. Take special care in adjusting the torque.
20. Reinstall the actuator assembly onto the bonnet and set the benchset, if applicable, per the instructions in Chap. 12.
21. Set the travel and connect the two stems. Tighten the packing.
22. Recalibrate the accessories, tune the booster, and set the limit switches, all as outlined in Chap. 12.
23. Rerun an operational check using a diagnostic scanner. If possible, check for seat leakage per the ANSI specification. See Chap. 2 for details.
24. Reinstall the valve per the guidelines given in Chap. 11.

Before turning to predictive measures, it should be pointed out that there are some additional preventive activities that can be done periodically to keep valves in good condition and that don't involve complete teardown. Some suggestions are shown in **Table 13.1.**

TABLE 13.1 Preventive Maintenance Practices

Item	Activity
Elastomer O-rings and diaphragms.	All elastomers age with time, temperature, compression, and the like. Aging causes lack of elasticity and results in leakage and poor dynamic response. Verify hardness on a regular basis and replace very 4 years.
Diaphragm casing bolts.	Retorque each outage to maintain seal.
Benchset.	Review actuator sizing and verify spring adjustment.
Fittings and tubing.	Regular bubble checks to look for leakage.
Accessories.	Regular diagnostic surveys to trend performance signatures, bubble checks for leakage.
Stem connector.	Retighten each outage. Check to make sure it's properly aligned.
Bearings.	Check for wear every 2 years, especially those with integral O-ring seals, like in the actuator.
Travel.	Verify the actuator travel is slightly greater than rated valve travel. Recheck valve travel each outage.
Spring rate.	Verify that benchset range has not changed and that the spring is not wound too tightly. Check linearity against specs.
Travel stops, manual handwheels.	Ensure that they're not interfering with valve travel.
Zero and span on accessories.	Recheck to make sure valve is getting full-open and full-closed signals with standard input ranges.
Air supply.	Check to make sure it's 5 psi higher than air-to-diaphragm range but less than the maximum rated pressure to diaphragm. Look for restrictions that would slow valve travel.

13.5 Predictive Maintenance and Diagnostics

As mentioned in the previous section, even those plants that have tried to address valve reliability and performance through a preventive program have been frustrated with the amount of wasted effort involved in trying to determine when a valve needs to be serviced. Up until several years ago, the only practical way to determine if a valve needed work was to take it out of service, inspect it, and at least partially disassemble it. Once all this is done, you might just as well do a complete teardown and rebuild because of the amount of time already involved.

Recent advancements in *computer-based technology* have made checking out a valve's operation much simpler and less time consuming. Using this equipment will make the preventive approach much more efficient since preventive measures can be concentrated on those valves that really need it. It also provides a standard to which current valve operation can be compared to so that you can know whether you should expect better. The industry is not there yet, but eventually, as more data are gathered with these systems regarding correlations between today's performance and remaining life, true predictive techniques can be employed that will result in keeping valves operating at peak efficiency with a minimum of maintenance effort. Now, let's take a look at how these systems work and how they can help to improve a plant's approach to control valve maintenance.

The approach to be outlined here is analogous to vibration surveys on rotating equipment where accelerometers are used to "look inside" pumps and compressors to determine the condition of internal parts such as bearings and take corrective action before failure occurs. Instead of using vibration readings, this new diagnostic equipment for valves records pneumatic pressures, along with valve travel, and develops correlations between these readings that help to check the operating condition of the valve assembly and accessories and compares it to base signatures that show how the valve should be operating.

The technology in question is based on a *customized data acquisition system* aimed at air-operated valves, and a typical hookup is shown in **Fig. 13.3.** It measures valve position along with pneumatic pressures in the actuator and accessories as the valve is stroked through all or part of its stroke. Analysis of these data can determine critical operating characteristics for a valve and can help determine what needs to be done so that the valve can adequately respond to the input from the controller.

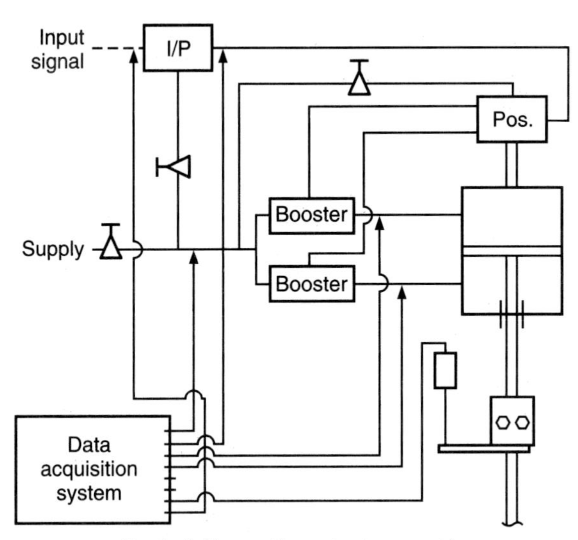

Figure 13.3 Typical diagnostic system connections.

Figure 13.4 shows the background data that are used to establish a baseline for the valve in question. **Figures 13.5** through **13.10** are examples of signature curves for a control valve assembly and its individual accessories that show the type of things that are checked. These characteristics are verified while stroking the valve at a speed that approximates actual service, so it gives the user a better idea of how a valve responds under dynamic conditions.

As already noted, hysteresis, deadband, and linearity all contribute to control problems. As we review these signatures, it can be seen that these properties are measured directly and by further investigation, the root cause of excessive readings can be determined and corrected. For instance, let's say that hysteresis is high. By looking at each of the accessory curves, we can determine which one is causing the problem and fix only the component that needs it, saving time and money. Or maybe the accessories check out, but the friction reading in the valve is 4 times what it should be. We can also do things like verifying proper packing torque or checking bearing or seal condition so that, once again, the real cause of the problem is identified with a minimum of effort. Another common problem discussed earlier is insufficient air supply volume. By looking at **Fig. 13.10,** we can immediately see if the valve is being starved during operation and can correct the situation to enable the valve to respond quickly to a change in input. The name-tag

```
Fisher Diagnostic Services   FlowScanner       Valve Nametag

Tag#: McK-Test         Serial#: 10792793          02-08-1994
Description: Sample Valve for Test                  13:24:24
Plant Site : Austin

Body Style: ES          Body Size: 1     IN        Class: 150

Trim Style: Microform

Flow Direction: Up                  Flow Pressure: OPENS

Port Diameter: 1/4     IN               The valve is UNBALANCED
UnBalanced Area: .0500 Square IN

Stem Diameter: .394  IN         Packing Type: TFE / Single
Spec. Packing Friction: 39      Total Stem Friction: 39   LBS

Leak Class: IV                  Seat Type: Metal
Spec. Inlet Pressure: 99.93     Outlet Pressure: 80.06 PSIG

Required Seat Load: 31     LBS

Valve Travel: .75    IN         Stroking Time:  Open:  10 SEC
                                                Close: 10 SEC

Actuator Type: 657              Bench Set : 3-14.9    PSIG
Style 34                        Effective Area 69        Sq IN
Air Pressure Closes the Valve

ACCESSORIES NOTED
I/P Type: 3311  ; Resistance: 176  Ohms; Output: 3-15      PSIG

Positioner Type: 3582       Zero Control signal = OPEN

Other:
Other:
Other:

Noted Comments:
Valve Tested from display stock
```

Figure 13.4 Background valve data.

information in **Fig. 13.4,** if properly filled out, also provides an opportunity to review for application problems.

Figures 13.5 through **13.10** deal with sliding-stem valves, but rotary models can also be analyzed with similar results **Figure 13.11** shows the actuator/valve signature for an eccentric-plug valve. We can examine the dynamic response characteristics for this type of

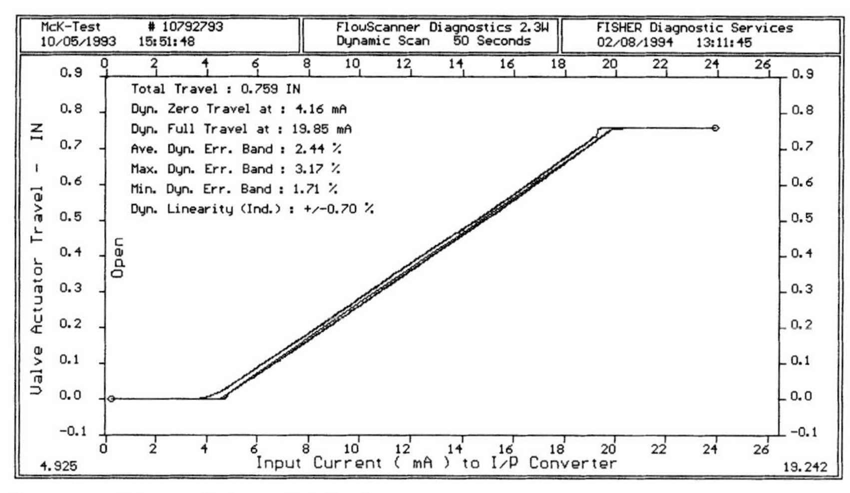

Figure 13.5 Diagnostic trace: Total valve response.

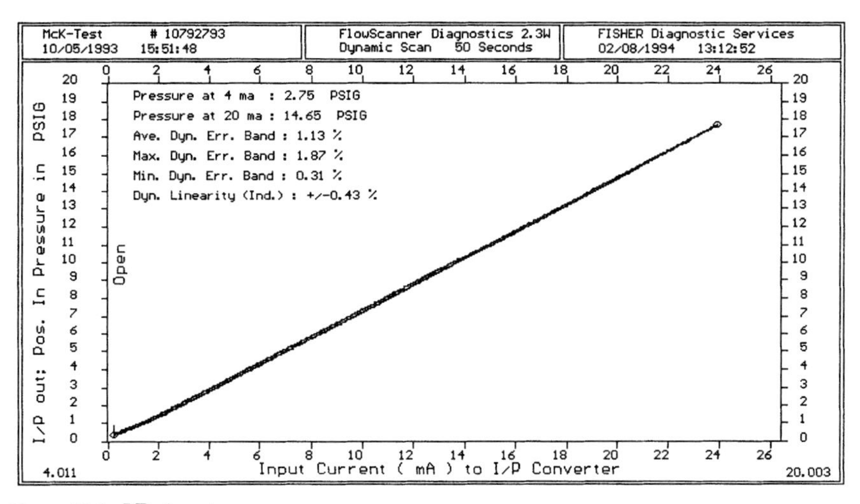

Figure 13.6 I/P signature.

valve just like we did for the sliding-stem valve in **Figs. 13.5** through **13.10.** We can also examine things such as seating characteristics by checking the shape of the curve as the valve seats out (the lower left portion of the curve). In this way, we can actually check to see that the plug is making proper contact with the seat and will shut off. Although shutoff is not part of the process control considerations for

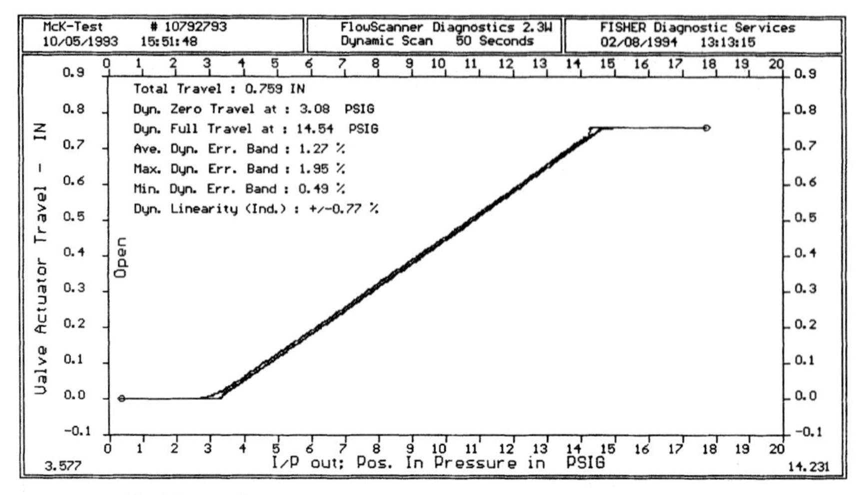

Figure 13.7 Positioner signature.

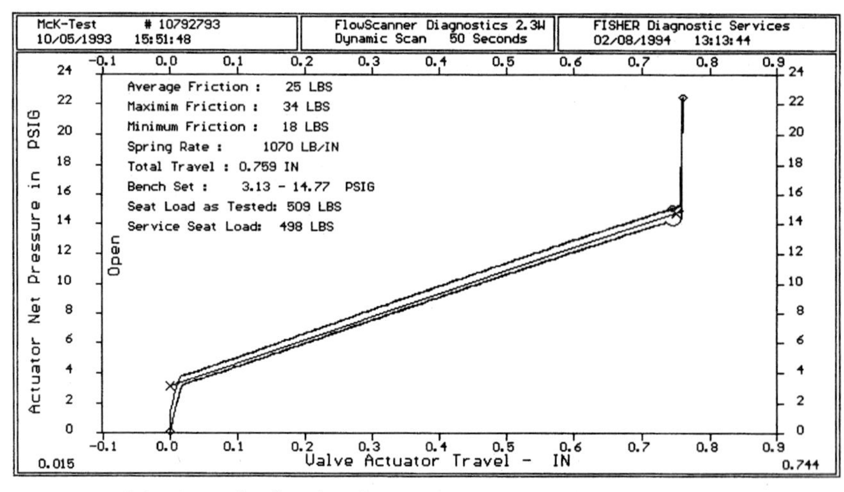

Figure 13.8 Actuator and valve signature.

a valve, there are a number of applications where it is important to the process itself. **Figure 13.12** shows the summary report where the field data can be compared directly to the specified data and recommendations made for corrective action.

Not only does the use of this equipment make maintenance simpler, it also improves valve dynamic response. To help illustrate the

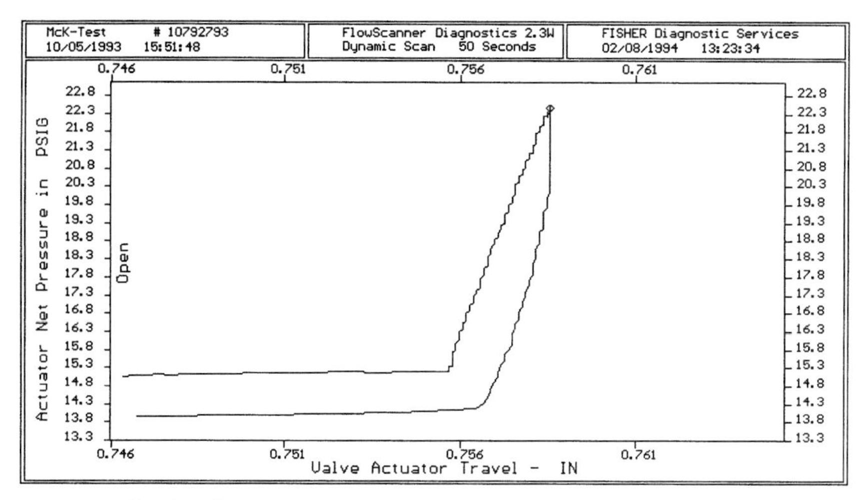

Figure 13.9 Seating signature.

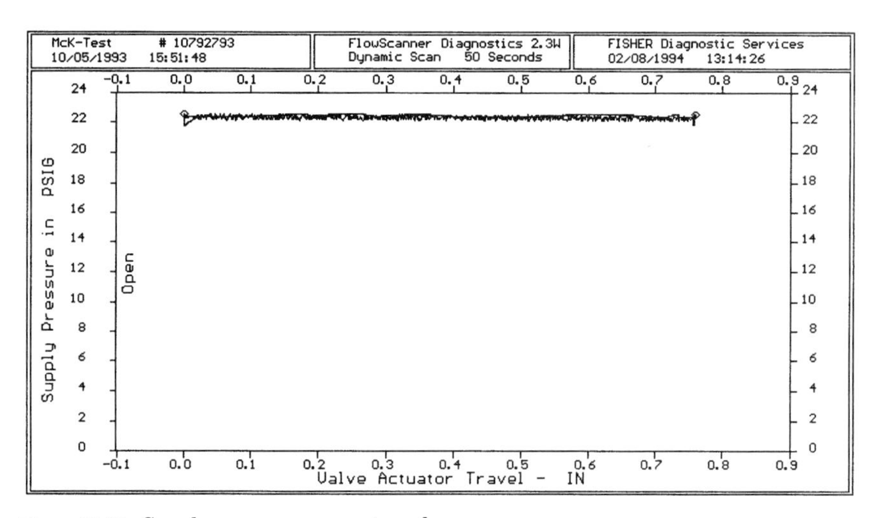

Figure 13.10 Supply pressure versus travel.

kind of *process control* improvement possible utilizing this type of maintenance approach, some typical problems were simulated with a valve. The valve was then installed in a pressure control loop with the controller settings adjusted as they might be in a process plant. **Figure 13.13** shows that the resultant loop variability is very high. The problems were detected and corrected using control valve diag-

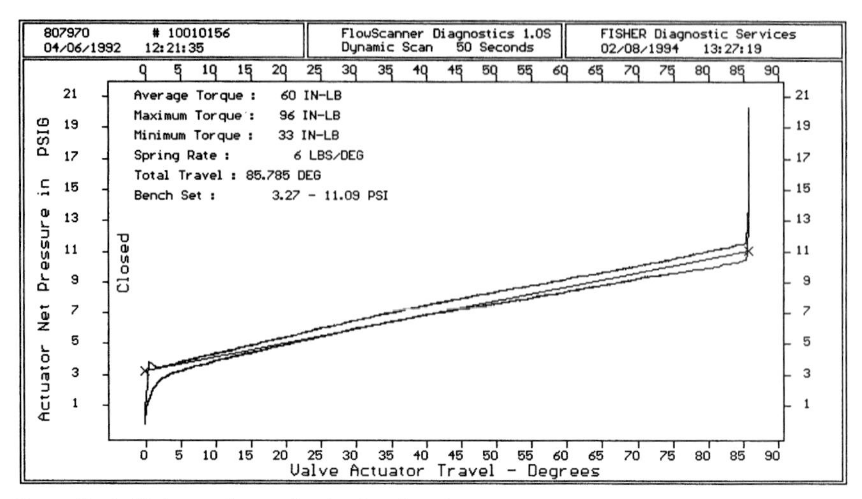

Figure 13.11 Rotary valve and actuator signature.

nostics and the loop was retuned using automated loop tuning (see Chap. 17). **Figure 13.14** shows the results, and the improvement is obvious.

This equipment can be used in a *maintenance* or *troubleshooting mode* as just described. But it can also be used as a *calibration* and *setup tool* to enable the end user to ensure that a valve is set up and operating properly before it is put into service. This is especially important with control valves since so many problems can be introduced at this initial stage (see Sec. 13.3). Once this as-built set of signatures is run, it can be stored in computer memory and can serve as an auditable calibration record. It can also serve as a baseline measurement and be compared to any future curves that are generated to see if valve performance has changed and why.

As stated earlier, regular use of this type of technology can begin to address the problems associated with the traditional approach to valve maintenance. The system can be used to inspect a valve without disassembly to determine if there is an application problem, if the end user should expect better performance, and the real cause of the problem, if there is one. Although not yet available, the advent of smart valve-mounted devices, along with fieldbus communication standards will enable this same type of testing and evaluation to be done on-line from the control room. Computer-based maintenance programs will even allow automated diagnostics to take place to determine the prob-

Fisher Diagnostics FlowScanner Quick Report

Tag#: McK-Test Serial#: 10792793 02-08-1994
Test Type: Test: 10051551 13:21:50
Valve Style: ES / 1 Trim: Microform Actuator: 657 / 34

Test Parameter	Spec'd	Measured	Comments
OVERALL VALVE CONTROL			
Travel (IN)	.75	.759	CLOSE ENOUGH
Dyn.Zero Travel (ma)	4	4.16	OK
Dyn.Max. Travel	20	19.85	OK
Dyn.Err.Band(%)	N/A	2.443	OK; LESS THAN 4%
Dyn.Lyn.(%+/-)	N/A	.704	OK; LESS THAN 1%
VALVE AND ACTUATOR DATA			
Friction Av. (LBS)	39	25	NORMAL
Maximum	39	34	VERY TO CLOSE TO NOMINAL
Minimum	N/A	18	
Seat Load - Test (LBS)	N/A	509	
- in Service	31	498	MORE THAN ENOUGH
Spring Rate		1070	OK
BenchSet(PSIG)	3-14.9	3.13-14.77	VERY GOOD
POSITIONER DATA	Type: 3582		
Dyn.Set(PSIG)	3-15	3.08-14.54	VERY GOOD
Dyn.Err.Band(%)	N/A	1.266	OK; LESS THAN 3%
Dyn.Lin.(%+/-)	N/A	.771	OK; LESS THAN 1%
I/P TRANSDUCER DATA	Type: 3311		
Dyn.Set(PSIG)	3-15	2.75-14.65	SPAN COULD BE A LITTLE MORE
Dyn.Err.Band(%)	N/A	1.129	OK; LESS THAN 2%
Dyn.Lin.(%+/-)	N/A	.429	OK; LESS THAN 1%
SEAT EVALUATION			GOOD CONTACT SIGNATURE

RECOMMENDED	Valve	Actuator	Positioner	I/P
REPAIR:	-			
ADJUST:				

Figure 13.12 Diagnostics summary report.

lem and then check stock for replacement parts or automatically cut a purchase order if the parts aren't found. The applicable section of the instruction manual could even be automatically generated and attached to the work order to explain what needs to be done and how long it will take.

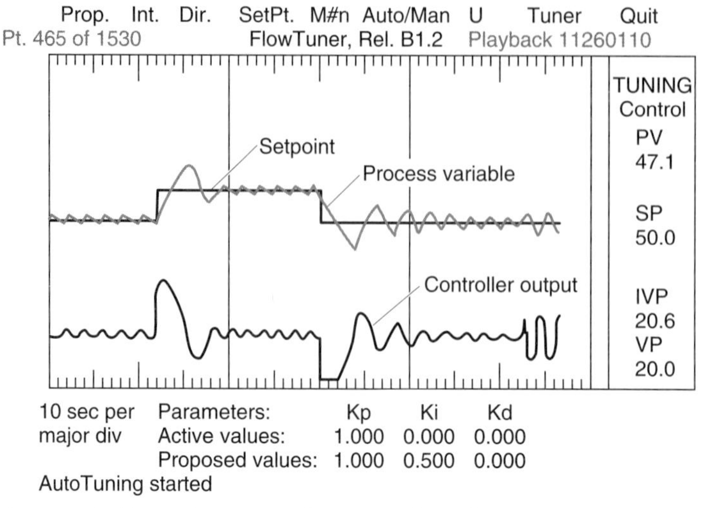

Figure 13.13 Dynamic response for valve with problems.

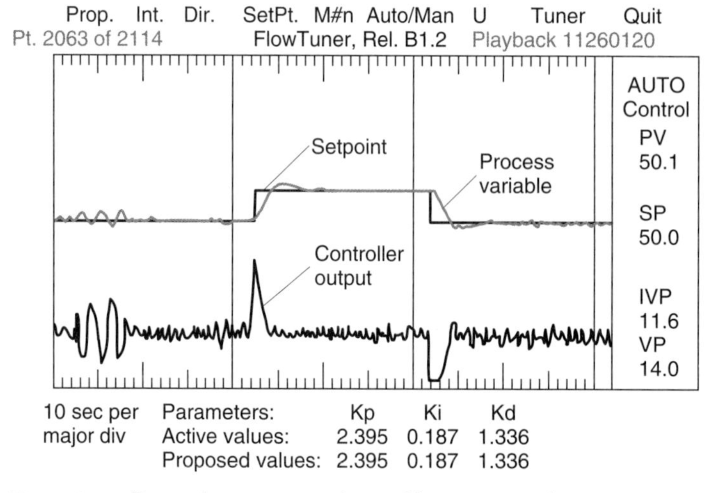

Figure 13.14 Dynamic response, valve problems corrected.

13.6 Control Valve Troubleshooting

Table 13.2 is a listing of a number of different valve problems or symptoms, with potential root causes, and recommended corrective action. It is presented in the form of a troubleshooting diagram that references certain common procedures used in valve maintenance. These procedures are explained in greater detail in the next section.

TABLE 13.2 Troubleshooting Diagram

Problems and symptoms	Causes	Solutions
1. Body erosion.	1*a*. Velocity. 1*b*. Particulates in flowstream. 1*c*. Cavitation and flashing.	1*a*. Increase valve trim size to slow fluid. 1*b*. Switch to streamlined design to reduce fluid impingement. 1*c*. Switch to C5 body material. 1*d*. Switch to torturous-path trim to slow fluid. 1*e*. Switch to low-recovery valve and trim to control cavitation. 1*f*. Repair by welding up with stainless material.
2. Trim erosion.	2*a*. Velocity. 2*b*. Particulates in flowstream. 2*c*. Cavitation and flashing.	2*a*. Increase valve and trim size to slow fluid. 2*b*. Switch to hardened trim. 2*c*. Switch to torturous-path trim to slow fluid. 2*d*. Switch to low-recovery valve and trim to control cavitation. 2*e*. Switch to streamlined design to reduce impingement.
3. Seat ring-to-plug leakage.	3*a*. Low load (benchset, calibration, friction, etc.). 3*b*. Poor surface condition (lapping, materials).	3*a*. Use proper surface preparation (lapping). 3*b*. Correct actuator and valve setup (benchset, calibration, friction, etc.).
4. Seat ring-to-body leakage.	4*a*. Low load (inadequate torque, parts stack-up, improper gasketing). 4*b*. Surface condition (cleanliness, finish). 4*c*. Porosity in body.	4*a*. Correct bolt load, parts stack-up, gasketing. 4*b*. Recut, clean up gasket face. 4*c*. Porosity in casting can sometimes result in leakage around gaskets. Check for porosity. Grind out and weld up.
5. Packing leakage.	5*a*. Stem finish/cleanliness. 5*b*. Bent stem. 5*c*. Low packing load. 5*d*. Wrong packing type or configuration. 5*e*. Excessive packing stack height (graphite). 5*f*. Corrosion and pitting (graphite). 5*g*. Seized or cocked packing follower.	5*a*. Clean up and polish stem to 4 rms finish. 5*b*. Straighten stem to within 0.002 in over stroking length. 5*c*. Retorque bolting or use live-loading. 5*d*. Check packing type and configuration against application. Repack as necessary. 5*e*. Install spacers to minimize packing height. Repack valve. 5*f*. Use sacrificial washers. Remove graphite packing if valve is to be inactive for more than 2 to 3 weeks. 5*g*. Inspect and replace any damaged parts such as flanges, nuts, and followers. 5*h*. Switch to high-performance packing system.

TABLE 13.2 Troubleshooting Diagram (Continued)

Problems and Symptoms	Causes	Solutions
6. Sliding wear.	6*a*. High cycling (unstable loop?). 6*b*. Excessive contact stress. 6*c*. Misalignment. 6*d*. Surface finish not to specification. 6*e*. Incorrect materials choice.	6*a*. Tune loop; reduce friction to reduce instability. 6*b*. Increase bearing size. 6*c*. Remachine parts to correct alignment. 6*d*. Polish surfaces. 6*e*. Review materials choice in light of application. 6*f*. Switch to sliding-stem globe-style valve because of better guiding.
7. Bonnet-to-body leakage.	7*a*. Low load from bonnet bolting (torque, internal parts, stack-up, spring rate in gasket). 7*b*. Surface finish. 7*c*. Stud leaks.	7*a*. Retorque bolting. Check parts stack-up against drawings. 7*b*. Retouch, clean up gasket faces. 7*c*. Sometimes casting porosity can let process fluid seep into bottom of stud holes. Leakage around studs looks like leakage past bonnet gasket (see **Fig. 13.15**). Grind out and weld up porosity.
8. Loose stem connection or broken stem.	8*a*. Improper torque or pinning. 8*b*. Vibration or instability.	8*a*. Purchase the stem and plug as an assembly. 8*b*. Review trim-style application. 8*c*. Reduce clearances between cage and plug. 8*d*. Switch to a welded plug or stem connection.

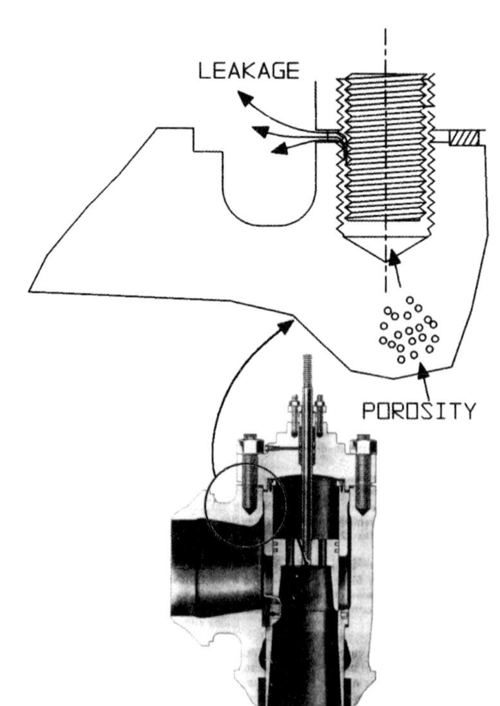

Figure 13.15 Porosity in body masquerading as bonnet joint leakage.

Problems and symptoms	Causes	Solutions
9. Excessive leakage past piston seal.	9*a*. Cage finish too rough. Cage I.D. too large. 9*b*. Improper installation: graphite rings, omniseal. 9*c*. Is leakage normal for the type of seal? 9*d*. Exceeding temperature limitations for seal. 9*e*. Seal simply worn out due to cycling.	9*a*. Polish cage bore, check I.D. against drawings. 9*b*. Replace seal, follow installation instructions. 9*c*. For some seals such as graphite piston rings, high leakage is normal. 9*d*. Change to high temperature design. 9*e*. Replace seal. Address loop stability if cycling is caused by this.
10. Valve will not respond to signal.	10*a*. No air supply or low air supply. 10*b*. Leaks in actuator. 10*c*. Solenoid closed on inlet lines. 10*d*. No controller input signal. 10*e*. Crimped, broken air lines. 10*f*. Leaking air fitting. 10*g*. Incorrect flow direction causing excessive loads on plug. 10*h*. Incorrect air line connections. 10*i*. Packing parts binding on stem or shaft. 10*j*. Defective positioner or I/P. 10*k*. Packing overtightened. 10*l*. Trim is seized. 10*m*. Plug stuck in seat.	10*a*. Check the system in accordance with the P&IDs. Verify that all air supply valves are open. 10*b*. Measure and verify sufficient air supply pressure. 10*c*. Listen for blowby at the seals or diaphragm. Repair or replace defective parts. 10*d*. Actuate solenoid valve. Replace if defective. 10*e*. No controller input may indicate a fuse has blown. Replace. 10*f*. Check all air lines to see they are not crimped or broken. Repair or replace. 10*g*. Check fittings for leaks. Tighten or replace. 10*h*. If the valve was just installed, check the flow arrow to ensure the process is flowing in the proper direction. Flow above the seat can add pressures the actuator may not be able to overcome. Reverse flowing direction, if appropriate. 10*i*. Check the air to and from a piston actuator to ensure the supply is not connected to the exhaust and vice versa. Check all connections. 10*j*. Check the packing gland. Improper gland configuration is a primary cause of rod binding. Replace parts and polish trim. 10*k*. Check the positioner and/or the I/P to see if the output can be changed manually. If not, it is defective. Repair or replace. 10*l*. Overtightened packing or binding in guides can cause excessive friction that blocks valve. Loosen, lubricate, cycle, and retorque. 10*m*. Replace or repair seized trim. Damage may be polished out. 10*n*. Pull or machine plug out of seat. Repair or replace affected parts.

TABLE 13.2 Troubleshooting Diagram (Continued)

Problem and symptom	Causes	Solutions
11. Valve will not open to rated travel.	11*a*. Insufficient supply pressure. 11*b*. Leaks in the actuator or accessories. 11*c*. Incorrect positioner or I/P calibration. 11*d*. Incorrect travel adjustment. 11*e*. Incorrect actuator spring rate. 11*f*. Incorrect benchset. 11*g*. Bent stem or shaft. 11*h*. Damaged valve trim. 11*i*. Debris in trim. 11*j*. Incorrect flow direction. 11*k*. Actuator is too small. 11*l*. Excessive packing friction. 11*m*. Incorrect position of manual operator on travel stop.	11*a*. Verify adequate supply pressure. 11*b*. Stop all leaks in actuator, air lines, fittings, and accessories. 11*c*. Correct positioner and/or I/P calibration. 11*d*. Readjust valve travel. 11*e*. Change actuator spring. 11*f*. Adjust benchset. 11*g*. Replace bent stem or shaft. 11*h*. Replace damaged trim. 11*i*. Clean out valve trim. 11*j*. Reverse flowing direction. 11*k*. Replace actuator. 11*l*. Loosen packing, cycle, lubricate and retorque. 11*m*. Readjust manual operator or travel stop.
12. Valve travel sluggish or slow.	12*a*. Excessive packing friction. 12*b*. Stem or shaft bent. 12*c*. Inadequate supply pressure. 12*d*. Inadequate supply volume. 12*e*. Undersized accessories. 12*f*. Excessive friction in piston-type actuator. 12*g*. Bearing friction. 12*h*. Poor positioner response.	12*a*. Readjust or replace packing. 12*b*. Replace bent shaft or stem. 12*c*. Increase supply pressure. 12*d*. Go to bigger supply line or add capacity at valve. 12*e*. Increase flow capacity of accessories. 12*f*. Clean out, polish cylinder I.D., remove excess lubricant. 12*g*. Repair or replace defective bearings. 12*h*. Repair or replace positioner.
13. Valve travel is jumpy.	13*a*. Stick-slip action in packing seals or bearings. 13*b*. Volume booster bypass may need to be adjusted. 13*c*. Positioner may be defective. 13*d*. Positioner gain may be too high.	13*a*. Loosen, lubricate packing. Replace or repair seals and bearings. 13*b*. Adjust booster bypass. 13*c*. Repair or replace positioners. 13*d*. Adjust positioner gain. Replace with lower gain model.
14. Rotary valve will not rotate	Rotary valves have some unique problems. In addition to those items already covered in items 10 and 11: 14*a*. Actuator stops set wrong, stopping the valve mechanically before it fully rotates. 14*b*. Broken shaft. 14*c*. Overtravel can cause severe damage to eccentric valves; valves can jam. 14*d*. Dirt or corroded valve seats can cause broken stems or valve can jam. 14*e*. Changing service conditions, higher pressures and greater pressure drops may stop the valve from rotating due to insufficient torque, high bearing loads. 14*f*. Overtightened line bolting can increase friction between the ball and seal.	14*a*. Readjust actuator stops. 14*b*. Replace shaft. 14*c*. Replace damaged parts, readjust travel. 14*d*. Replace or clean parts. 14*e*. Recheck actuator sizing and valve service limits. Change valve and/or actuator, as appropriate. 14*f*. Loosen line bolting.

Problem and symptom	Causes	Solutions
15. Poor flow control (rotary and sliding stem).	See items 12 and 13 relating to sluggish response and "jumpy" travel. Other causes include: 15*a*. Deformed cage. 15*b*. Damaged piston rings. 15*c*. Erosion, corrosion, and cavitation can alter trim profile. 15*d*. A twisted shaft will indicate a position that is untrue in regard to disk and seat. The valve may indicate full open or full closed and may really be mid-range. 15*e*. Valve may be installed backward. 15*f*. Incorrect selection of flow characteristic. 15*g*. Low performance valve package.	15*a*. Replace cage. 15*b*. Replace piston rings. 15*c*. Resolve sources of damage. Replace parts. 15*d*. Replace shaft. 15*e*. Reverse valve in line. 15*f*. Correct flow characteristic. 15*g*. Select valve assembly with control requirements taken into account.

13.7 Common Valve Maintenance Procedures

13.7.1 Packing maintenance

Valve packing, as we've already noted, is one of the more troublesome elements of control valve operation. As a result, the end user is often faced with the prospect of pulling it out and installing a new set. The best way to do this is to take the bonnet off of the valve and then push the old packing out from the bottom, using the following procedure. Note that this procedure covers a sliding-stem globe-style valve, and, as such, it can be done in-line. For rotary valves, the procedure differs in that there is no bonnet, so the valve has to be taken from the line to extract the packing as indicated below:

1. Apply enough air pressure to the actuator to put the valve in an intermediate position so that there is no residual stem load. Disconnect the actuator and valve stems. Relieve the air pressure, and disconnect the actuator supply and any leakoff piping.
2. Remove the yoke coupling, yoke locknut, or the yoke bolting, and remove the actuator from the bonnet.
3. Loosen the packing flange nuts so that the packing is not tight on the valve plug stem. Remove any travel indicator disk and stem locknuts from the valve plug stem threads. *Safety note:* When lifting the bonnet, be sure that the valve plug and stem assembly

remains on the seat ring. This avoids damage to the seating surfaces as a result of the assembly dropping from the bonnet after being lifted part way out. The parts are also easier to handle separately. Use care to avoid damaging gasket sealing surfaces. If the cage cannot be held in the body due to gasket adhesion, control it so that it will not cause equipment damage or personal injury should it fall unexpectedly.

4. Unscrew the bonnet bolting and carefully lift the bonnet off the valve stem. If the valve plug and stem assembly start to lift with the bonnet, use a brass or lead hammer on the end of the stem and tap them back down. Set the bonnet on a cardboard or wooden surface to prevent damage to the bonnet gasket surface.
5. Remove the valve plug, the seat ring, and the cage. *Note:* All residual gasket material must be removed from the cage gasket surfaces. If the gasket surfaces are scored or damaged during this process, smooth and polish them by hand, sanding with 360-grit paper and using long, sweeping strokes. Failure to remove all residual gasket material and/or burrs from the gasket surfaces will result in leakage.
6. Clean all gasket surfaces with a good-quality degreaser. Remove any residual tin or silver from all gasket surfaces.
7. Cover the opening in the valve body to protect the gasket surface and to prevent foreign material from getting into the body cavity.
8. Remove the packing flange nuts, packing flange, upper wiper, and packing follower. Carefully push out all the remaining packing parts from the body side of the bonnet using a rounded rod or other tool that will not scratch the packing box wall.
9. Clean the packing box and the related metal packing parts: packing follower, packing box ring, spring or lantern ring, special washers, etc.
10. Inspect the valve-stem threads for any sharp edges that might cut the packing. A whetstone or emery cloth may be used to smooth the threads if necessary. They can also be chased with a die.
11. Remove the protective covering from the body cavity, and install the cage using new top gaskets. Install the plug and then slide the bonnet over the stem and onto the studs. Lubricate the stud threads and the faces of the hex nuts. Replace hex nuts and torque the nuts in a crisscross pattern to no more than one-quarter of the nominal torque value specified. When all the nuts are tightened to that torque value, increase the torque by one-quar-

ter of the specified nominal torque and repeat the crisscross pattern. Repeat this procedure until all the nuts are tightened to the specified nominal value. Apply the final torque value again and, if any nut still turns, tighten every nut again.

12. Install new packing and the metal packing box parts according to the appropriate arrangement in the instruction manual. If desired, packing parts may be prelubricated for easier installation. Slip a smooth-edged pipe over the valve stem, and gently tamp each soft packing part into the packing box.
13. Slide the packing follower, wiper, and packing flange into position. Lubricate the packing flange studs and other related parts and the faces of the packing flange nuts. Replace the packing flange nuts. *For spring-loaded TFE V-ring packing,* tighten the packing flange nuts until the shoulder on the packing follower contacts the bonnet. *For other standard packing types,* tighten the packing flange nuts to the recommended torque. For high-performance packing sets, adjust the live-loading springs as indicated in the instruction manual.
14. Mount the actuator on the valve body assembly, and reconnect the actuator and valve stems according to the procedures in the appropriate instruction manual.
15. Cycle the valve 20 to 30 times and recheck packing load.

Packing can be replaced with the valve in the line, but it is not recommended due to the increased risk of stem or packing box damage. If it must be attempted, follow the above procedure with the changes noted below:

1. Remove the packing loading parts so that the top of the packing rings can be seen.
2. Very carefully insert a corkscrew packing extraction tool into the packing box and twist it into the top of the packing until it can be used to pull the top packing ring out.
3. Repeat this procedure until all the upper packing has been removed. If there is a spacer or bushing below the packing or between the upper and lower packing sets on a double arrangement, it usually has some type of slot or extraction hole. If it does not, it will have to be left in place. Assuming that it can be extracted, pull it out, and continue the above process with any packing left below the spacer.
4. Once all the packing and internal parts have been removed, do your best to clean the box out and inspect for any signs of damage.

This cleaning and inspection will be very difficult to accomplish with the bonnet in place.

5. Normally you should remove the stem connector and the actuator so the rings can be slid down over the stem. If this is not possible, split rings can be used, and they can be forced onto the stem by twisting them until the opening is large enough to slide over the stem. Split rings are not recommended due to their propensity to leak. If they are used, make sure to stagger the splits to reduce the potential for leakage.
6. If any damage is found, the valve should be disassembled and the situation corrected at the first opportunity. Effective corrective action cannot be taken with the bonnet on the valve, and repacking with the bonnet on the valve will improve packing performance for a limited time, at best, if the stem or box is damaged in any way.
7. Repack and reassemble as noted above, using split rings if the stem connector was not removed.

13.7.2 Lapping the seats

Lapping is a procedure used to provide a better fit and surface finish between the valve plug and the mating seat. It applies only to metal-to-metal seating and is normally used for class IV or V shutoff on control valves. Classes I, II, and III don't require it and class VI nearly always requires soft seats. The plug and seat in their as-machined state do not always fit together perfectly around their circumference. Imperfections in fit result in excess leakage, so lapping is required to eliminate these imperfections and to make sure that the two parts fit together as closely as possible. Lapping should be carried out as follows:

1. Lapping should be done with the standard guiding in place to make sure that the parts are lapped in the positions that they will be in once the valve is fully assembled. For this reason, it is normally done with the bonnet in place.
2. With the seat ring in place in the body, apply a light coating of coarse grinding compound (600 grit) to both the seat ring and the plug. If the seating surfaces are made of stainless steel, use some white lead in the grinding compound to keep it from tearing or galling. Insert the plug and stem into the body and assemble the bonnet onto the body opening. The bonnet does not have to be bolted into place as long as the guiding simulates actual service.

3. Lapping requires that a very light load be applied to keep from tearing the metal, so if the plug is heavy, a spring should be used to support some of the load. The spring can be inserted over the stem and then a piece of strap iron can be locked into place on the stem and used as a grinding handle (**Fig. 13.16**).

4. Gently rotate the plug and stem four or five times, over about a 45° arc. Pick it up and move to a new position and repeat. Continue this procedure, lapping over the entire circumference at least once. Pull the assembly apart, clean the surfaces, and look for a fine continuous lap line on both the plug and the seat. Using a mirror will make the line easier to see on the seat ring inside the valve body.

5. If the lap lines look good, reassemble and repeat the procedure with a fine grit compound. If the lap lines are not continuous, repeat with the course compound. If they are still not continuous, try coining the surfaces by hitting the top of the stem two or three times with a heavy, but soft hammer, and lap again. If this still doesn't provide the desired results, the plug and seat should be remachined to provide a better initial fit and the process restarted.

6. When the fine grinding is done, thoroughly clean the surfaces and reassemble, torquing the bonnet in place. If possible, a seat-leak test should then be carried out to ensure tight shutoff.

7. High-temperature valves should be heated, if possible, before beginning this process to better duplicate actual guiding and fit in service.

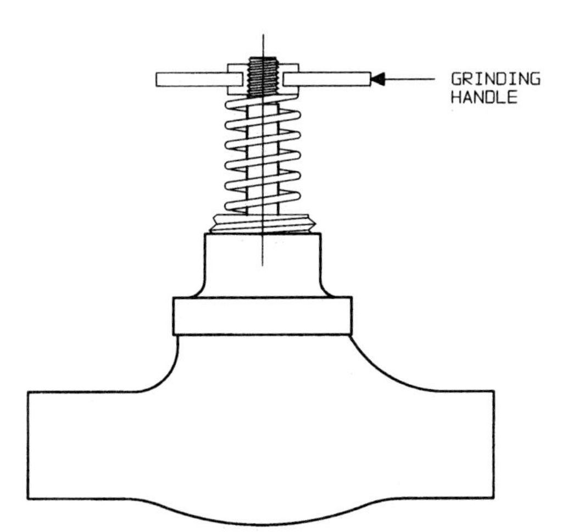

Figure 13.16 Lapping tool used with spring.

8. Double-ported bodies will never seal as well as a single-port design, but they can still be lapped to improve shutoff. Special considerations for these valves include:

   The top seat grinds faster than the bottom. Use a coarser grit on the bottom ring to help correct for this.

   Never leave one seat dry while grinding the other one. This will tear the metal and hurt the shutoff.

   Heavy grinding on one seat may be required to get the two seats to contact at the same time.

9. Note that despite claims to the contrary, blue-lining to check for seat contact will not provide the same tight shutoff seen with lapping. Tests have shown that there can still be relatively large imperfections present even though the blue-line shows continuous contact between the two surfaces.

13.7.3 Replacing the actuator diaphragm

After isolating the valve assembly from all pneumatic and/or fluid pressures, relieve *spring compression* in the spring, if possible. (On some spring and diaphragm actuators for use on rotary-shaft valve bodies, spring compression is not externally adjustable. Initial spring compression is set at the factory and does not need to be relieved in order to change the diaphragm.) Remove the upper diaphragm case. On direct-acting actuators, the diaphragm can be lifted out and replaced with a new one. On reverse-acting actuators, the diaphragm head assembly must be dismantled to change the diaphragm.

Most pneumatic spring and diaphragm actuators utilize a *molded diaphragm* for control valve service. The molded diaphragm facilitates installation, provides a relatively uniform effective area throughout the valve's travel range, and permits greater travel than could be possible if a *flat-sheet diaphragm* were used. If a flat-sheet diaphragm is used in an emergency repair situation, it should be replaced with a molded diaphragm as soon as possible.

When reassembling the diaphragm case, tighten the cap screws around the perimeter of the case firmly and evenly to prevent leakage. Be careful not to tear the diaphragm in the area of the bolt holes during reassembly. Avoid reusing a diaphragm since they are prone to leak if reused.

13.7.4 Replacing threaded-in seat rings

Threaded-in seat rings, as noted in Chap. 2, are no longer the preferred design for control valves in the chemical process industry. Nevertheless, this design is encountered fairly often due to its popularity in the past. The main reason this design has fallen from favor is that the seat rings can be very difficult to get out. Adhering to the following recommended practice should help extract the seat ring with a minimum of effort and risk to personnel:

1. Before trying to remove the seat ring(s), check to see if it has been tack-welded into the body. If it has, grind out the weld.
2. To make disassembly easier, soak the ring and threads with penetrating oil and allow them to sit for some time so that the oil can do its job in loosening up the threads.
3. Insert a seat ring puller like that shown in **Fig. 13.17** against the lugs or in the slots of the ring. Be careful to hold the puller down against the ring while applying torque, and any rounded edges on the lugs or slots should be corrected to keep the puller from slipping past the lugs or slots.
4. The torque can be applied manually or with the aid of a hydraulic torque wrench. If the power wrench is used, be extra careful to avoid slippage due to the high torques and the safety risk to personnel if something slips or breaks. If the valve has been pulled from the line, a lathe or boring mill may be the easiest way to apply the torque to back the ring out.

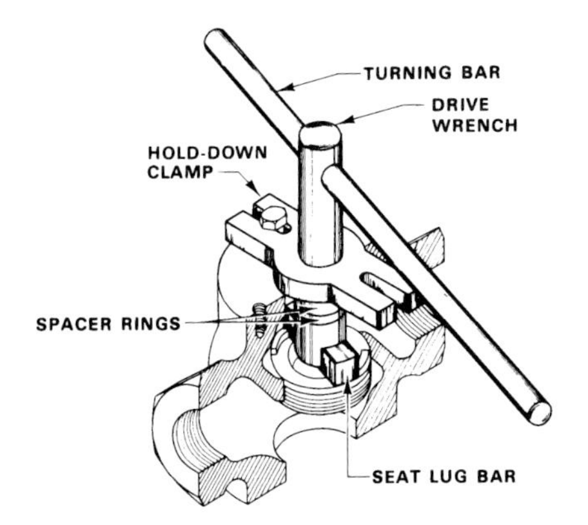

Figure 13.17 Seat ring puller. *(Courtesy of Fisher Controls International, Inc., Marshalltown, Iowa.)*

5. The bonnet bolting can be used as a reaction point for the torque and to hold the puller down into the body.
6. On particularly stubborn rings, using an impact wrench can help to break them loose.
7. As the ring starts to come out, the bolts holding the puller in the body must also be loosened to permit the ring to move up.
8. Once the ring is out, thoroughly clean and chase all threads.
9. Apply a heavy coat of lubricant or pipe compound to all threads and reinstall and torque to specified levels. The ring may be tack-welded in place, as necessary.
10. On double-ported valves, the port the farthest distance from the actuator is the smallest and needs to be installed first.

13.8 References

1. Preckwinkle, S. E., *Maintenance Guide for Air Operated Valves, Pneumatic Actuators & Accessories,* Electric Power Research Institute, Palo Alto, Calif., 1991.
2. Ozol, J., "Experiences with Control Valve Cavitation Problems and Their Solutions," *Proceedings of EPRI Power Plant Valves Symposium EPRI,* Palo Alto, Calif., 1987.
3. McElroy, J. W., *Light Water Reactor Valve Performance Surveys Utilizing Acoustic Techniques,* Philadelphia Electric Co., Philadelphia, Pa., 1987.
4. Fitzgerald, W. V., "Automated Control Valve Troubleshooting: The Key to Optimum Valve Performance," *ISA Proceedings,* ISA, Research Triangle Park, N.C., 1991.
5. Ferguson, Brian, "Air-Operated Valve—Preventive Maintenance Program," *Proceedings of the 2d NRC/ASME Symposium on Pump & Valve Testing,* Washington, D.C., 1992.
6. Hutchison, J. W., *ISA Handbook of Control Valves,* ISA, Research Triangle Park, N.C., 1971.
7. *Control Valve Handbook,* 1st ed., Fisher Controls, Marshalltown, Iowa, 1977.
8. *Instruction Manual, EHD, EHS, & EHT, Form 5163,* Fisher Controls, Marshalltown, Iowa, 1985.

Chapter

14

Operational Concerns and Process Optimization

The operations group doesn't care what kind of valve is in the line or how it is maintained as long as they can produce the maximum amount of on-spec product at the minimum cost. In reality, the choice of valve and the maintenance program in place has a tremendous effect on how the plant operates. True *process optimization* is achieved when the process control system is permitted to do its work. This means keeping the process variable as close as possible to the setpoint at all times. To better understand how a control valve affects process control performance, let's examine a typical single-loop control system. The discussion to follow is not meant to be a treatise on process control; there are a number of other books (see the references) already available that cover this in much more detail and with a lot more math. The emphasis here will be relating valve performance to loop performance on a basic level.

14.1 Control System Basics

An example of a *single-loop system* is shown in **Fig. 14.1.** There are three primary components: the control element, the final control element, and the measurement element. The control element, in this case a standard proportional, integral, derivative controller, continuously monitors the process variable and then compares it to the setpoint. Assuming a constant setpoint, the load changes in the process conspire to drive the process variable off the setpoint. The measurement element, a flow transmitter operating off of an orifice plate, measures the flow and sends a signal back to the controller, where it is compared to the setpoint.

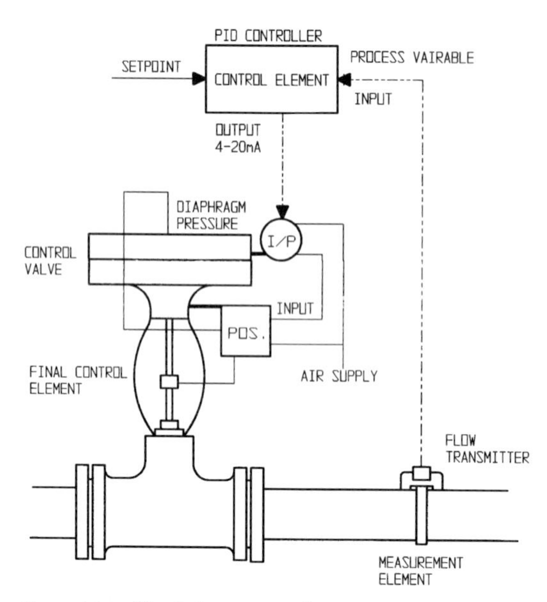

Figure 14.1 Single-loop control system.

Through the balance of this discussion, it will be assumed that the flow transmitter accurately represents the flow through the pipe at all times. This is not really the case. *Measurement elements* do exhibit static error and also have frequency response limits that mean that the actual flow may sometimes be different from the signal sent back to the controller. However, the error is usually small, and the dynamic response is better than the final control element, so the assumption of "perfect" performance is not too far off the mark. We also want to concentrate on control valve performance and by assuming perfect transmitter performance, it permits us to avoid getting caught up in analyzing how the transmitter enters into the equation.

The *error* between *setpoint* and *process variable* is computed and analyzed and then the controller output is adjusted accordingly, based on the error and how the controller is tuned. Tuning the controller involves adjusting the proportional, integral, and derivative settings. These settings are explained in more detail in Sec. 5.13, but in short, they determine how a controller reacts to an error between the process variable and the setpoint.

Once the process variable has been measured and transmitted and the error analyzed, the controller output to the valve changes and, in theory, results in a correction by the valve that brings the process variable closer to the desired setpoint. In reality, the controller output has to pass through and be transformed by a number of physical elements before any change in position occurs at the valve plug. It is this change in position that actually causes the flow change that results in a correction or change in the process variable. Until the valve plug moves, the controller has had no effect on the loop. This is why valve performance is so important.

The *controller output* is normally in the form of a 4- to 20-mA signal when it leaves the controller. This has to be transformed into a pneumatic signal before the valve can work with it, and this is done at the transducer (I/P). This is the first opportunity for error to be introduced based on how well the I/P can make this transformation. The pneumatic signal from the I/P is sent to the positioner, which acts just like a controller in that it compares this input signal (like the setpoint) to the actual valve position and then generates a pneumatic output to the actuator that attempts to eliminate the error between the position desired based on the input signal and the real position of the valve. The positioner-actuator combination also introduces error, as covered in the next section.

The point is, controllers have become very sophisticated in how they look at and react to errors, but the only corrective action they can take is through the valve assembly. With this in mind, the next section will address how differences in valve performance and construction can affect the way the loop performs.

14.2 Valve Performance and Its Effect on Process Control

With the preceding section as background, let's look at how a valve responds to the output from the controller and relate this response to how well the system will perform as a whole. This discussion is aimed at demonstrating why you cannot afford to accept poor valve performance on critical loops.

Starting with the system in steady-state equilibrium, we introduce some type of *disturbance* that results in a change in flow. Let's say the downstream pressure drops and, as a result, the pressure drop increases across the valve. If the pressure drop increases, the flow will increase, assuming that the valve is not in a choked or critical flow situation. The increase in flow is sensed by the flow transmitter and sent back to the controller where it is compared to the setpoint and an error calculated. The error will change as a function of time, so the absolute error, the derivative of error, and the integral of the error can be calculated at any point in time. The controller has an adjustment for the proportional, integral, and derivative modes, and each adjustment determines the size of the correction as a function of the absolute error, the integral of the error, and the derivative of the error. In other words, the change in controller output due to the proportional mode will be a linear function of the magnitude of the absolute error and the proportional setting on the controller. In the same way, the integral and derivative corrections will be linear functions of the settings for these variables in combination with the magnitude of the integral and derivative of the error, respectively.

What this means is that if all three modes are used with a standard controller, the derivative correction will dominate just after the disturbance since the derivative of the error due to a load change will usually be larger than either the error itself or the integral of the error. The proportional mode will take over from the derivative once the system has begun to react and the magnitude of the error has stabilized. Then, as the size of the error begins to get very small, the proportional action gives way to the integral action, which makes the final corrections. This transition versus time for the three modes is illustrated in **Fig. 14.2.** (Note that derivative action is most beneficial on slow processes such as temperature or level loops and can be more trouble than it's worth on fast loops. Proportional and integral modes tend to be used on all types of loops.).

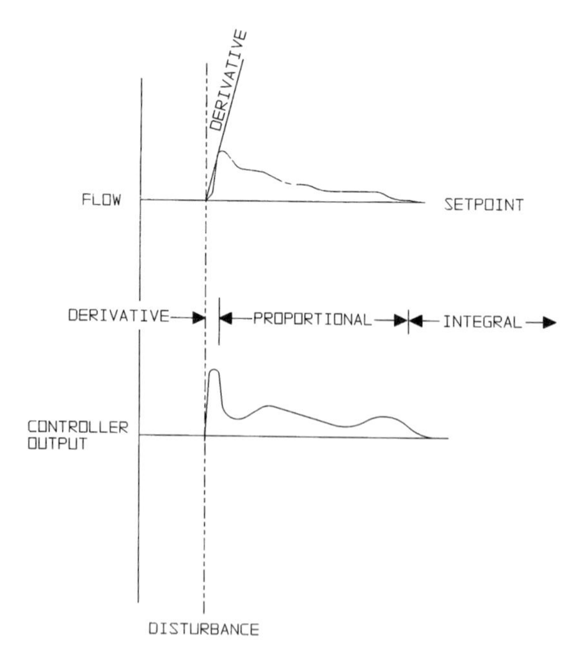

Figure 14.2 Three-mode controller action.

Now, let's see what happens at the valve as the controller input changes. The large change in controller output initially due to the derivative action goes to the I/P, which would like to change its output very rapidly as well but may be limited by a number of factors. If the output volume is large between the I/P and the positioner, the pressure will take some time to build up and a *lag* will be introduced. The magnitude of this lag will be a function of the length and cross-sectional area of the tubing that has to be filled, so mounting the I/P off the valve or split-ranging the I/P will mean slower response. In addition to tubing volume, the flow rate of the relay in the I/P and the air supply volume and pressure will also play a major role in determining how fast the I/P output can climb in relation to the change in controller signal. Lags are undesirable and result in poor process because they increase the correction time, meaning that the process stays off of the setpoint longer. To reduce lag to a minimum and improve I/P response, make sure that the I/P flow rate and air supply volume and pressure are consistent with the tubing volume to be filled. All things being equal, short, small tubing, high I/P flow capacity, and high air supply capacity and pressure will all work to keep I/P lag to a minimum.

I/P hysteresis plus deadband can also play a small role in increasing the lag, but it is usually very small in comparison to the above effects, so, in most cases, it does not deserve a lot of attention. The same comment applies for linearity. If it gets very bad, it needs to be corrected, but in most cases, it is not a primary concern. If the above problems are adequately addressed, the I/P should work well.

Assuming that everything is set up properly on the I/P, it should provide an output that follows the controller with a minimum of lag. This output is then transmitted to the positioner where there is even more risk of error being introduced. The *positioner* compares the input signal to the valve position feedback signal generated by the feedback linkage. When the I/P signal changes, it results in an error being detected between where the valve should be and where the positioner thinks it is. The error then initiates a change in valve position. We'd like this change in position to instantaneously and accurately reflect the changes in I/P output. Several things can inhibit its ability to do this, and if they are allowed to occur, they can hurt control performance.

One problem can be avoided if you make sure that the *air supply* has plenty of capacity and pressure to drive the actuator to where it needs to go. There should be no more than a 10 percent drop in supply pressure with a 100 percent step change to the valve. If it is greater than this, the size of the airset needs to be increased so that the positioner is not being starved. Another precaution that needs to be taken is to keep at least a 5-psi differential between the pressure required to stroke the valve and the minimum supply pressure. If the margin is less than this, it needs to be increased either by reducing the pressure required to stroke the valve or by increasing the supply pressure. The reason this is important is that the pneumatic relays in the positioner operate on a pressure drop principle. If the pressure drop gets too small, the flow through the relay will drop off and the stroking speed will be reduced. Slower stroking speed means more lag and more variation between the setpoint and the process variable.

Another problem that surfaces with many positioner-actuator combinations is poor *input resolution.* This characteristic is covered in Sec. 9.3 and reflects the size of the change in input signal required to initiate a change in valve position. If input resolution is poor, it takes a greater change in controller output to initiate the correction at the valve. This means more lag and more error. It's a little like driving an old car with a worn out steering gear. The wheel can be turned 10° before any change in the car's direction is seen. Once a correction is made in one direction, the wheel has to be swung all the way back through the play in the linkage to keep the car from wandering in the other direction. No matter how good a driver you are, you will find yourself wandering into the ditch or into oncoming traffic, neither of which would be considered a desired result. The loop behavior is very similar, with the process variable wandering from side to side across the setpoint.

This behavior is due in part to the inherent characteristics of the control valve assembly, and these can be corrected by repairing or replacing the positioner and/or the positioner linkage. Reducing the valve friction will also help to improve the input resolution, so things like switching to TFE packing or correcting binding or bearing problems in the valve and actuator will also help.

The stroking speed of the positioner-actuator combination is also a major determinant of process control performance. Again, if the positioner flow capacity is low in relation to the actuator volume and pressure range required to stroke the valve, the corrective action at the valve will be slow, and increased error will result. Increasing the supply flow capacity, reducing the actuator volume, and/or reducing the pressure range required to stroke the valve are all steps that will increase stroking speed and reduce lag. At the same time, be careful not to increase the speed to the point where the valve overshoots the setpoint and begins to cycle due to instability. **Table 9.1** illustrates some general guidelines regarding stroking times that seem to be reasonable based on equipment currently available (see reference 1).

Most of what we've dealt with so far relates to the ability of the valve to quickly respond to changes in input so that the correction takes place quickly and the process variability is minimized. Speed is important, but what about accuracy? If the valve can't get to the right position, the variability will be high even during the steady state. Input resolution was mentioned earlier. If a valve does not respond to a change in input of 1 percent or less, it follows that the loop will not be able to control to within 1 percent, so input resolution is an important measure of the potential for loop accuracy.

Positioning resolution is another good measure of overall loop accuracy. All valve assemblies have friction, and static friction will always be slightly higher than dynamic levels. This results in a tendency for valve assemblies to move in a stick-slip mode for small corrections, and there is minimum size to this jump that defines the positioning resolution. The size of this jump seems to vary with friction, the ratio of static-to-dynamic friction, and the dynamic characteristics of the actuator and positioner. If the minimum jump is 5 percent of the span, for example, it follows that the best you could do is to get within about 2.5 percent of the correct position. The overshoot would result in a correction back the other way, and a limit cycle would be set up whereby the valve and the loop would bounce back and forth over the correct setpoint. Not only does this result in a significant amount of error, but it also causes premature failure in parts such as the packing and bearings due to the high cycling. Try to select a valve with positioning resolution characteristics that are consistent with accuracy requirements for the loop.

14.3 Performance Summary

Having just reviewed valve performance and its implications for loop performance, you may be ready to concede that more attention needs to be paid to selecting a valve with good performance characteristics and then setting up a maintenance program to make sure that performance levels are sustained through the life of the plant. But, at the same time, you may be asking what measures should be taken to make sure that this gets done. The recommendations are really very simple and can be summarized as follows:

1. Take the time to identify the performance characteristics required for a particular valve in a given application (i.e., analyze required performance levels for each loop).
2. Define and document these requirements.
3. Include them as one of the key elements in the purchasing decision.

4. Use the material covered in Sec. 14.2 to ensure that the valves will meet the requirements and take corrective action with either the valve equipment or the supporting hardware in case things such as stroking time and positioning resolution are not up to standards.
5. Test the valves in-line, before start-up, to verify that they meet the requirements defined above.
6. Set up a regular troubleshooting and testing routine to correct problems as they develop so that performance levels can be maintained.
7. Routinely review loop requirements to see if any adjustments need to be made to valve performance to provide a better fit and improved overall process control.

14.4 References

1. Bialkowski, Bill, *The Entech Report,* vol. 5, issue 2, Entech Control Engineering, Inc., Toronto, Canada, 1993.
2. Murrill, P. W., *Fundamentals of Process Control Theory,* ISA, Research Triangle Park, N.C., 1981.
3. Lloyd, Sheldon, and Anderson, Gary, *Industrial Process Control,* Fisher Controls, Marshalltown, Iowa, 1971.
4. Kallen, H. P., *Handbook of Instrumentation and Controls,* McGraw-Hill, New York, 1961.

Part

4

General Topic and New Developments

Chapter

15

Financial Considerations

As mentioned early on in this book and reinforced on several occasions, the *true cost* of a control valve has very little to do with the *initial purchase price*. Nevertheless, up-front price is the easiest and most obvious method of cost comparison, so most end users still use it as the primary consideration in determining which control valve to select for a given application. A better method would be to compare the total *life-cycle return on investment* for a control valve that would take into account not only up-front cost but other bottom-line aspects of control valve performance such as those identified in the following two-part list:

1. Acquisition
   a. Purchasing and engineering costs
   b. Delivery integrity
   c. Inspection and QA costs
2. Commissioning and start-up costs
3. Post-start-up
   a. Reliability or up-time
   b. East of repair and troubleshooting
   c. Frequency of repair
   d. Cost of repair
   e. Safety considerations
   f. Dynamic performance
      (1) Stroking speed
      (2) Frequency response

(3) Linearity
(4) Hysteresis plus deadband
(5) Input resolution
(6) Positioning resolution
(7) Load sensitivity

g. Leakage

(1) Lost process fluid and energy
(2) Safety exposure for personnel
(3) Damaged trim
(4) Regulatory compliance

h. Required inventory

(1) Parts
(2) Documentation

i. Support staff and skills

j. Timely support

The next few paragraphs will go over each one of these elements and relate it as best as possible to the initial cost of a typical valve. The true cost cannot always be estimated in a generic approach like this because the process that the valve is being used on may be a key factor. In this case, you will have to make your own estimates, but at least a framework will be presented that will help to guide the thought process.

To help set the stage for the balance of this discussion, assume that a *typical control valve* in a chemical industry application might be a 4-in, 600-lb sliding-stem globe style with stainless body and trim and with a spring and diaphragm actuator, positioner, airset, and I/P. This type of assembly might sell for an average of $8000, and if we assume a ±10 percent range to cover the various prices in the marketplace, this means that you would be faced with selecting from valves whose up-front price range from $7200 to $8800, or a difference of $1600. Now, let's see how this *$1600 pricing difference* stacks up against the other life-cycle costs and benefits just identified.

15.1 Life-Cycle Costs

15.1.1 Pre-start-up costs

Four *cost elements* have been identified here, including purchasing and engineering, delivery integrity, inspection and QA, and commissioning and start-up. Essentially, this grouping covers the preparatory work required up until the point where the plant is actually considered to be in full production. It is a labor-based cost to the end user (or contractor) and includes the work necessary to:

1. Specify the valves
2. Communicate the requirements to four or five vendors
3. Review the proposals
4. Negotiate the price
5. Select a vendor
6. Enter the order
7. Expedite delivery
8. Inspect the equipment
9. Coordinate delivery
10. Stage and store the equipment
11. Check documentation
12. Take corrective action on mistakes found
13. Install valves
14. Set up and calibrate valves
15. Run start-up testing
16. Take corrective action on start-up problems

Based on the above list of activities, it has been estimated that an average valve would require about 26 h of this type of support from the time the application is identified to the point where the valve is installed and operating correctly. If we assume that the burdened cost per hour for this type of work is going to be approximately $50/h, we arrive at a total investment of $1300. A review of this list shows that there are only two or three items that are really adding value to the process. The rest of them are either duplication of effort or activities designed to detect and correct problems that shouldn't be there in the first place.

If we changed the rules and decided that the end user and vendor should be partners rather than adversaries, the time related to items 2 through 5 could be almost eliminated. If we utilize only vendors that are committed to *total quality* and who have been audited against purchase process requirements, items 7, 8, 9, 11, and 12 can either be eliminated or performed by the vendor. If proper attention is paid to installation, setup, and calibration as detailed in Chaps. 11 and 12, item 16 should also be minimized. By effectively managing the pre-start-up activities and limiting them to value-added labor only, we should easily cut the total support in half, saving $650 in the process.

15.1.2 Post-start-up costs

Once the plant is running, the opportunities for reducing true costs or increasing return on investment are even greater. The first item, reliability, relates directly to process up-time. Fewer problems means that the process can stay on-line longer and more product can be generated. As a result, this is more of a return-on-investment item than a cost savings. What kind of effect this can have on operational returns is hard to estimate because it depends on the application and on the relative importance of the valve to the overall process. Nevertheless, end users who are committed to world-class performance need to address this aspect of control valve performance if they want to make the best overall valve choice.

This will mean another break from doing business the same old way. Valve reliability indices will need to be determined based on past end-user experience coupled with vendor input. A combination of mean time between failures and a factor accounting for the type and severity of failures would be a good start. If this type of performance index is used in combination with the valve criticality rating recommended in Chap. 10, you can begin to come up with guidelines like a recommended minimum reliability index that is a function of the valve criticality rating. In this way, you would end up with better-performing valves in those applications where they are really needed, and process up-time would be maximized. Noncritical applications could be filled with cheaper, less reliable designs. As mentioned earlier, the advantages associated with reliability-centered valve selection are difficult to quantify, but given the revenue associated with the generation of incremental end product, they should be significantly larger than the $1600 figure that we've established as our reference point.

In addition to reliability, you also need to be concerned with how easily a problem with a valve can be detected, how often a valve breaks down, and how much time and material will be needed to correct the problem. Once again, this may be a qualitative measure based on a standard review of valve design features and parts costs, but it still needs to be part of the purchase decision.

As an example, a normal valve can have a service life of 20 years or more. Average maintenance practices might mean that the valve is repaired seven times over this period, or about once every 3 years. The industry average cost of repairing a valve, whether it is done in-house or contracted out, is about $900, so this means that $6300 will be spent over the life of an average valve just in repair costs. The net present value of these expenses are going to be around $4100, or 51 percent of the initial cost of the valve itself. If the design is such that the number of repairs done can be reduced along with the cost of each repair, the life-cycle cost for the valve can be reduced significantly in comparison to the initial price. The reliability index suggested above can be used here as well, along with an estimate of the average cost of repair for a particular valve. Through the proper application of this type of information, the total repair cost can be managed, resulting in a potential 10 percent reduction, or about a $400 savings in net present value terms.

Safety is also a principal concern for an operating plant, so the valve design should be reviewed for possible problems with injuries to personnel who are working on or around it. External leakage is one example that could result in harmful exposure to plant personnel or emissions beyond the federal mandate. The designs of seals and gaskets need to be effective and easily maintained to reduce this risk. Pinch points and the sudden release of spring energy are two other common sources of injury to personnel. The design chosen should minimize the risk associated with these two problems, while clearly marking any remaining safety hazards. The cost savings associated with "safe" designs is hard to quantify, but this does not mean that it can be ignored during the selection process.

Dynamic performance of the control valve and its effect on process control has been covered in a number of other areas in this book and is one of its main themes. In general, you would like to eliminate any deviation between the process variable and the setpoint, regardless of load or setpoint changes. In practice, this is impossible but is still the reference to which process control performance is compared. Extremely tight control is not required or desired on all loops, but on those judged to be critical to overall plant performance, it needs to be a prime consideration. That's why the control valve input data sheet in **Fig. 10.1** puts so much emphasis on including loop characteristics when specifying the application conditions for a valve. Using this information, the end user and vendor can select a valve that best fits the dynamic demands of the process and that will provide the level of control desired.

This will mean that for critical applications, you may end up spending more money up-front to get a high-performance valve with rapid stroking speed, good frequency response, low hysteresis plus deadband, low load sensitivity, and high input and positioning resolution. The pay-off from an operational standpoint will come with reduced steady-state variation between setpoint and process variable, and faster response to load or setpoint changes. Both of these will result in more on-spec product produced with less waste. While the financial benefit of reduced cost or increased production will depend on the application, it's safe to say that it will easily surpass the $1600 figure in less than a week in most critical applications.

Leakage is another of the principal problems that you want to avoid with control valves. External leakage results in personnel exposure or fines from the EPA, and the fluid that escapes is wasted. Internal leakage damages trim due to erosion and wastes process fluid or energy in steam applications. The costs for external leakage are hard to quantify and about the best that can be done is to review the design with an eye toward minimizing leakage and then monitor and document actual performance by valve type and design.

The cost of internal leakage, on the other hand, can be estimated by putting together the cost to repair the plug and seat along with the cost of the fluid lost due to excess leakage. This type of analysis is only required on valves that frequently shut off to class IV levels or better. With a valve of this type, experience has shown that to disassemble the valve, recut the seats, and put it back together again costs an average of $600. Recutting the seats on a valve that was torn down already for other reasons will only cost about $100. If the valve leaks and the trim is damaged four or five times over the life of the valve, this will add about $300 in new costs to the earlier estimate for net present value repair costs.

As far as lost fluid is concerned, if a class IV valve leaks to class III levels, it means a tenfold increase in leakage to 0.1 percent of valve capacity. Using our standard 3-in valve and assuming a shutoff pressure drop of 500 psi, this translates into an additional leakage amount of 2.2 gpm, or over 3000 gallons per day of wasted fluid. Once again, even with relatively cheap fluids that cost only 10 cents per gallon, this excess leakage costs more than $300 a day. Even if a valve is closed only 1 percent of the time, this still represents $1000 per year in waste.

Required inventory is used here in a broader sense. It does include parts as you would expect, but it also addresses the type of documentation and skills required to keep a valve operating as it should. Valves that are overly complicated in design carry a hidden cost of requiring additional skills from the staff that may or may not be available. The same can be said of the practice of using a number of different vendors. The plant ends up with multiple designs all requiring their own parts stock and documentation, and the staff has to be able to maintain the skills necessary to work on these different designs. Progressive companies in the chemical industry are beginning to realize this and are cutting down on the number of different designs and vendors used in their plants.

Another intangible part of the valve selection process is product support. How quickly does a vendor respond to an inquiry or problem? How near is the vendor to the plant? What kind of stock is carried locally? These are all questions that need to be asked up-front because they are the primary factor in determining how fast you can react to and recover from a random failure. If it's a failure that brings the process down, any unnecessary delay is costing money, and the costs will build up very fast in relation to our reference figure above.

15.2 Cost Summary

At the beginning of this chapter, we decided that a typical purchase range given market conditions today would be about $1600. What this means is that using the normal method of specifying and purchasing a control valve based on meeting minimum purchase order requirements at minimum costs, the potential savings in going with the bottom-dollar bid is $1600. This is the most that the end user will save in up-front costs.

Now, compare these savings with the real cost of selecting a control valve whose price is low but whose life-cycle performance is less than desired. The following rough estimates were made in the previous section:

Concern	Cost ($)
Pre-start-up	650
Reliability	1600+
Repair	400
Safety	?
Dynamic performance	1600+
Leakage	
Trim damage	300
Lost fluid	1000
Inventory	?
Timely support	?
Total	5550

Even ignoring all the intangibles and taking a very conservative approach, you can see that the net present value of the cost factors is much larger than the $1600 price range. The message is clear. End users and vendors should break out of the paradigm of selecting equipment based on up-front price and begin to collect data that will enable them to truly select the valve with the lowest overall cost to the process versus the value provided.

Chapter

16

Regulatory Concerns

One of the common trends for the chemical industry that was mentioned early in the book is the increasing effect that government regulations are having on an average plant's operations. This chapter will summarize four of the more recent acts of legislation and regulation and then cover the implications that they have for control valve selection and use in a typical chemical process plant.

16.1 Process Safety Management Rules (OSHA)

The official name for this OSHA safety standard is *Process Safety Management of Highly Hazardous Chemicals,* OSHA 29CFR1910.119. It was pulled together and enacted on May 26, 1992 by OSHA in reaction to the perceived high frequency of major accidents in the process industries. It greatly expands the number of chemicals considered to be hazardous and then applies new requirements for documentation of procedures and equipment used in handling these chemicals. This new list of hazardous chemicals is shown in **Table 16.1.** Each category of new requirements is addressed in the following sections with special emphasis on how they might affect control valves. It should be noted that if valves are selected, installed, and maintained as recommended in this book, the risk of chemical release due to a valve malfunction will be greatly reduced. This is one of the intangible benefits associated with adopting the philosophy of this book.

16.1.1 Scope

This regulation applies to any process facility that handles the chemicals listed in **Table 16.1** at or above the *threshold levels* listed in pounds. In determining whether the threshold limit is reached, the

TABLE 16.1 OSHA PSM 1910.119—List of Highly Hazardous Chemicals, Toxics, and Reactives

Chemical name	CAS*	TQ†
Acetaldehyde	75-07-0	2,500
Acrolein (2-propenal)	107-02-8	150
Acrylyl chloride	814-68-6	250
Allyl chloride	107-05-1	1,000
Allylamine	107-11-9	1,000
Alkylaluminums	Varies	5,000
Ammonia, anhydrous	7664-41-7	10,000
Ammonia solutions (>44% ammonia by weight)	7664-41-7	15,000
Ammonium perchlorate	7790-98-9	7,500
Ammonium permanganate	7787-36-2	7,500
Arsine (also called arsenic hydride)	7784-42-1	100
Bis (chloromethyl) ether	542-88-1	100
Boron trichloride	10294-34-5	2,500
Boron trifluoride	7637-07-2	250
Bromine	7726-95-6	1,500
Bromine chloride	13863-41-7	1,500
Bromine pentafluoride	7789-30-2	2,500
Bromine trifluoride	7787-71-5	15,000
3-Bromopropyne (also called propargyl bromide)	106-96-7	100
Butyl hydroperoxide (tertiary)	75-91-2	5,000
Butyl perbenzoate (tertiary)	614-45-9	7,500
Carbonyl chloride (see phosgene)	75-44-5	100
Carbonyl fluoride cellulose nitrate (concentration >12.6% nitrogen)	9004-70-0	2,500
Chlorine	7782-50-5	1,500
Chlorine dioxide	10049-04-4	1,000
Chlorine pentafluoride	13637-63-3	1,000
Chlorine trifluoride	7790-91-2	1,000
Chlorodiethylaluminum (also called diethylaluminum chloride)	96-10-6	5,000
1-Chloro-2,4-dinitrobenzene	97-00-7	5,000
Chloromethyl methyl ether	107-30-2	500
Chloropicrin	76-06-2	500
Chloropicrin and methyl bromide mixture	None	1,500
Chloropicrin and methyl chloride mixture	None	1,500
Cumene hydroperoxide	80-15-9	5,000
Cyanogen	460-19-5	2,500
Cyanogen chloride	506-77-4	500
Cyanuric fluoride	675-14-9	100
Diacetyl peroxide (concentration >70%)	110-22-5	5,000
Diazomethane	334-88-3	500
Dibenzoyl peroxide	94-36-0	7,500
Diborane	19287-45-7	100
Dibutyl peroxide (tertiary)	110-05-4	5,000
Dichloro acetylene	7572-29-4	250
Dichlorosilane	4109-96-0	2,500
Diethylzine	557-20-0	10,000
Diisopropyl peroxydicarbonate	105-64-6	7,500
Dilaluroyl peroxide	105-74-8	7,500
Dimethyldichlorosilane	75-78-5	1,000
Dimethylhydrazine, 1,1-dimethylamine, anhydrous	57-14-7 124-40-3	1,000

Chemical name	CAS*	TQ†
2,4-Dinitroaniline	97-02-9	5,000
Ethyl methyl ketone peroxide (also methyl ethyl ketone peroxide; concentration >60%)	1338-23-4	5,000
Ethyl nitrite	109-95-5	5,000
Ethylamine	75-04-7	7,500
Ethylene fluorohydrin	371-62-0	100
Ethylene oxide	75-21-8	5,000
Ethyleneimine	151-56-4	1,000
Fluorine	7782-41-4	1,000
Formaldehyde (formalin)	50-00-0	1,000
Furan	110-00-9	500
Hexafluoroacetone	684-16-2	5,000
Hydrochloric acid, anhydrous	7647-01-0	5,000
Hydrofluoric acid, anhydrous	7664-39-3	1,000
Hydrogen bromide	10035-10-6	5,000
Hydrogen chloride	7647-01-0	5,000
Hydrogen cyanide, anhydrous	74-90-8	1,000
Hydrogen fluoride	7664-39-3	1,000
Hydrogen peroxide (52% by weight or greater)	7722-84-1	7,500
Hydrogen selenide	7783-07-5	150
Hydrogen sulfide	7783-06-4	1,500
Hydroxylamine	7803-49-8	2,500
Iron, pentacarbonyl	13463-40-6	250
Isopropylamine	75-31-0	5,000
Ketene	463-51-4	100
Methacrylaldehyde	78-85-3	1,000
Methacryloyl chloride	920-46-7	150
Methacryloyloxyethyl isocyanate	30674-80-7	100
Methyl acrylonitrile	126-98-7	250
Methylamine, anhydrous	74-89-5	1,000
Methyl bromide	74-83-9	2,500
Methyl chloride	74-87-3	15,000
Methyl chloroformate	79-22-1	500
Methyl ethyl ketone peroxide (concentration >60%)	1338-23-4	5,000
Methyl fluoroacetate	453-18-9	100
Methyl fluorosulfate	421-20-5	100
Methyl hydrazine	60-34-4	100
Methyl iodide	74-88-4	7,500
Methyl isocyanate	624-83-9	250
Methyl mercaptan	74-93-1	5,000
Methyl vinyl ketone	79-84-4	100
Methyltrichlorosilane	75-79-6	500
Nickel carbonyl (nickel tetracarbonyl)	13463-39-3	150
Nitric acid (94.5% by weight or greater)	7697-37-2	500
Nitric oxide	10102-43-9	250
Nitroaniline (para nitroaniline)	100-01-6	5,000
Nitromethane	75-52-5	2,500
Nitrogen dioxide	10102-44-0	250
Nitrogen oxides (NO; NO_2; N_2O_4; N_2O_3)	10102-44-0	250
Nitrogen tetroxide (also called nitrogen peroxide)	10544-72-6	250
Nitrogen trifluoride	7783-54-2	

TABLE 16.1 OSHA PSM 1910.119—List of Highly Hazardous Chemicals, Toxics, and Reactives (Continued)

Chemical name	CAS*	TQ†
Nitrogen trioxide	10544-73-7	250
Oleum (65% to 80% by weight; also called fuming sulfuric acid)	8014-94-7	1,000
Osmium tetroxide	20816-12-0	100
Oxygen difluoride (fluorine monoxide)	7783-41-7	100
Ozone	10028-15-6	100
Pentaborane	19624-22-7	100
Peracetic acid (concentration >60% acetic acid; also called peroxyacetic acid)	79-21-0	1,000
Perchloric acid (concentration >60% by weight)	7601-90-3	5,000
Perchloromethyl mercaptan	594-42-3	150
Perchloryl fluoride	7616-94-6	5,000
Peroxyacetic acid (concentration >60% acetic acid; also called peracetic acid)	79-21-0	1,000
Phosgene (also called carbonyl chloride)	75-44-5	100
Phosphine (hydrogen phosphide)	7803-51-2	100
Phosphorus oxychloride (also called phosphoryl chloride)	10025-87-3	1,000
Phosphorus trichloride	7719-12-2	1,000
Phosphoryl chloride (also called phosphorus oxychloride)	10025-87-3	1,000
Propargyl bromide	106-96-7	100
Propyl nitrate	627-3-4	2,500
Sarin	107-44-8	100
Selenium hexafluoride	7783-79-1	1,000
Stibine (antimony hydride)	7803-52-3	500
Sulfur dioxide (liquid)	7446-09-5	1,000
Sulfur pentafluoride	5714-22-7	250
Sulfur tetrafluoride	7783-60-0	250
Sulfur trioxide (also called sulfuric anhydride)	7446-11-9	1,000
Sulfuric anhydride (also called sulfur trioxide)	7446-77-9	1,000
Tellurium hexafluoride	7783-80-4	250
Tetrafluoroethylene	116-14-3	5,000
Tetrafluorohydrazine	10036-47-2	5,000
Tetramethyl lead	75-74-1	1,000
Thionyl chloride	7719-09-7	250
Trichloro (chloromethyl) silane	1558-25-4	100
Trichloro (dichlorophenyl) silane	27137-85-5	2,500
Trichlorosilane	10025-78-2	5,000
Trifluorochloroethylene	79-38-9	10,000
Trimethyloxysilane	2487-90-3	1,500

*CAS = chemical abstract service number

†TQ = threshold quantity in pounds

end user must measure or estimate the maximum amount of the chemical in use in all vessels, containers, and piping at any one time.

In addition to the chemicals in the quantities listed above, this standard also applies to any *flammable liquid* or *gas* in quantities of 10,000 lb or more unless the material is used solely as an on-site fuel and is not used in any part of the chemical process. It also includes an exemption for flammable liquids stored in atmospheric tanks which are kept below their normal boiling point without taking any special measures such as chilling.

16.1.2 Employee participation

This section of the standard requires that the facility develop a *written plan* to meet the requirements and that the plan be reviewed and commented on by the employees. Any good management team would take this step, but the government saw fit to mandate it in true bureaucratic style.

16.1.3 Process safety information

This section relates to the requirement that the processing facility develop and maintain a *written record* of the hazards associated with the handling of the chemical or chemicals used in the process, the technology of the process itself, and the equipment used in the process. This involves the development of a relatively large database of information and is one of the big headaches associated with this legislation.

As far as the chemical hazards themselves, most of these could be covered by pulling together the appropriate MSDSs once the chemicals in use have been determined. Most plants should already have the MSDSs on file in response to the earlier implementation of the OSHA Hazard Communication Standard (29CFR1910.1200).

The information on the *technology of the process* itself includes things such as a flow diagram of the process, maximum amount of the chemicals on hand, and upper and lower limits on things such as temperature, pressure, flow, and mixtures. It also includes an evaluation of what might happen if these limits are exceeded, especially from a safety standpoint. Pulling together this kind of information could involve a considerable amount of reverse engineering if it was not included in the original process design or if it has been subsequently lost.

The third item deals with the equipment used, and this is where the control valve is treated. Any equipment used needs to have a file developed that includes the following information:

Materials of construction

Piping and instrument diagrams

Electrical classifications

Pressure relief system design and design basis

Ventilation system design

Applicable design codes and standards used

Material and energy balances

Safety system design

This information is aimed at ensuring that the equipment used has been properly designed so that it will stand up to the demands of the process and not fail prematurely, putting employees at risk. This also includes information on any special QA or performance testing intended to show that the equipment, as supplied or maintained, will perform as designed.

End users should look for vendors who are familiar with this part of the standard and who can supply the necessary documentation as part of the purchasing package, preferably in a format that the on-site database can accept without having to duplicate keystrokes. Cutting down on the number of vendors used will also help to reduce the work required to pull this information together.

16.1.4 Process hazard analysis

The process hazard analysis is considered the cornerstone of the standard and requires that the end user take the assembled information and perform a *written hazard analysis* that looks at any and all risks that might exist with a given process. This will take a considerable amount of engineering, and OSHA has recognized this by permitting a staged phase-in, spread over 4 years, with completion required in 1996.

The analysis must be conducted by a *knowledgeable team* of experts in the process and should include past accidents or incidents, all the information gathered in the previous section, along with related industry experience. It culminates in a report that summarizes the probability of an incident occurring, the consequences of the incident, and the type and number of personnel affected. It may also recommend corrective action to reduce the risk of an accident and requires that the facility take action on the recommendations and document any corrective action. In reviewing the requirements covered so far, it should be obvious that the documentation pulled together to support this standard will be horrendous. This review has to be redone every 5 years, or whenever the process changes appreciably, and documentation has to be retained for the life of the process.

As far as valves are concerned, the vendors can be a big help in this area by providing background on the types and frequencies of failures that might occur with a control valve in light of the service that it is in. The valve failure can then be evaluated in terms of how it might affect the process and the associated risks.

16.1.5 Operating procedures

Once the design review is done and the documentation pulled together on all the equipment used in the process, some type of assurance needs to be given that the process will be operated correctly. This section requires that *written operating procedures* be pulled together that take into account all the safety risks and limits identified in previous steps. The procedures will cover various stages in the process such as start-up, shutdown, normal operation, temporary conditions, and emergency conditions.

It should specifically address the function of any safety features, equipment, or systems. The procedures should be made easily available to anyone working in or around the process, and they should be reviewed on a regular basis or anytime that the process changes. All the traditional safety procedures such as lockout/tagout, confined space entry, breaking into pipelines or vessels, and access limitations should be part of, or at least referenced in, these operating procedures. Once again, knowledgeable valve vendors can help by providing appropriate operating and maintenance practices that will help to minimize any risks related to valve operation.

16.1.6 Training

This section recognizes that the massive effort spent in pulling together all the procedures and documentation required is all for naught unless they get used. The only way to ensure that they will be used is to put together a *comprehensive training program* that tells everyone involved what they should be doing to support the regulation. This section calls for an initial training program for all personnel involved, including management, lays out a schedule for refresher training, and covers the requirements for documentation of training.

Valve vendors can supply training material or even assist with the training as necessary. This is one area where vendor consolidation can help. Reducing the types of valve equipment in use will help in reducing the amount of different procedures and training that the site will have to address.

16.1.7 Contractors

This is a very important part of the standard since the majority of accidents and incidents in the process industry over the last several years have involved *contractor personnel.* This section is aimed at spelling out very clear requirements for both site and contractor management that will help cut down on the safety risk for contractors working at a particular site. For the end user, these include things like checking a contractor's past and present safety performance, going over hazards with the contractor staff, and providing adequate safety training.

The contractor needs to maintain written safety procedures, verify that employees have been properly notified of any potential hazards at a site, conduct appropriate in-house training, and verify that employees have been properly trained on-site.

For control valves, any contract work should be carried out by the OEM since it is most familiar with the design and any unique safety hazards associated with the valve. Any valve contractor used should have a good safety track record and be fully committed to safe operating practices. The end user should ask to see things such as written safety practices, training records, safety audit reports, and documented evidence of safety performance by OSHA standards.

16.1.8 Pre-start-up safety review

Anytime the process or the related equipment changes, the end user is required to perform a pre-start-up safety review before bringing the process on-line. The review will provide assurance that the equipment is properly designed, that procedures have been changed as necessary, that applicable areas of the process hazard analysis have been redone, and that training on any new elements has been carried out. This is to make sure that no changes are made without proper consideration of how the change might affect the overall safety.

The thing to remember here is that even on routine maintenance, if the control valve is not put back together exactly as it was, the standard considers this to be a change. The term *replacement in kind* governs here. This is defined to mean that any parts or equipment that are replaced must be replaced with material that is equivalent from a design and performance standpoint. There must even be a QA system in place to assure that this will happen as specified in the standard.

This was brought up in this section because of the common practice in some plants of using non-OEM parts in an effort to save money. Regardless of claims to the contrary, there is no assurance that these parts meet the original OEM's design and operating guidelines, so their use would be considered an official change and would have to be covered by a new safety review. The end user could certify that they do not affect the safety function through some type of design review, but if they do, they are assuming a tremendous *liability* in exchange for saving a relatively small amount of money. It is better to stay away from using anything but original parts.

16.1.9 Mechanical integrity

The intent of this section of the standard is to preserve the mechanical integrity of any process-handling equipment, thereby avoiding the *catastrophic release* of hazardous chemicals since it is this type of release that has the most potential for injury to personnel. The methods that OSHA prescribes to maintain mechanical integrity include:

- Operating the process within defined limits
- Having written maintenance procedures covering regular inspection and detection activities
- Giving maintenance personnel involved overview training that covers the hazards of the process and assuring that they can effectively carry out the tasks identified in the previous step
- Adopting state-of-the-art procedures in early detection and correction of problems that could lead to loss of mechanical integrity
- Covering this whole process by a written QA program that includes assurance that the equipment used is suitable for the process, that it is properly installed, and that it is properly maintained so that safety performance is not compromised

Note how much this last paragraph parallels the philosophy of this book: that to get optimum performance from a control valve, it must be correctly selected, installed, and maintained.

16.1.10 Hot work permit

OSHA included a hot work permit in the standard as a separate item because postaccident investigations showed that not all sites had an acceptable hot work permit system in place. Controlling hot work is considered to be critical to process safety because many of the most serious accidents have involved ignition of flammable gases or liquids, and the ignition could be traced to inadequate hot work permit systems, in many cases.

16.1.11 Management of change

This section of the standard sets out very broad guidelines for managing and controlling changes of any kind to the process, the chemicals in use, the processing equipment, or the facilities employed to support the process. Replacement-in-kind changes are, of course, exempted here because, by definition, they are not considered to have an effect on the safety performance of the process or the related equipment.

As with the other sections, this one requires a written procedure that includes a technical review of the proposed change, the impact of the change on safety and health, changes to procedures, and authorization requirements. Any resultant modifications to the process information described in Sec. 16.1.3 need to be carried out as well, and personnel need to be retrained, as necessary, given the consequences of the change.

This is another good reason to stick with OEM parts since they would normally be covered under the "replacement in kind" guidelines. End users should also keep valve vendors involved in any change related to their equipment since it may be easier for them to evaluate the consequences of a change as it relates to their equipment.

16.1.12 Incident investigation

The section of the standard requires that an *incident investigation* be carried out anytime a chemical release occurs or could have occurred given the circumstances (a near miss). The investigation must be carried out no later than *48 h* after the incident and includes such elements as a description of the incident, the factors that contributed to the incident, and any recommended corrective action. The results of the investigation must be reviewed with all personnel who could be affected by the findings and should be retained on file for at least 5 years.

16.1.13 Emergency planning and response

This section requires that the facility have in place an emergency action plan in accordance with 29CFR 1910.38(a), with the additional requirement to have a plan for handling small chemical releases.

16.1.14 Compliance safety audits

This section of the standard requires that an *audit* be carried out at least once every 3 years, verifying compliance with the standard. The audit team must include one person knowledgeable about the process. The facility must take prompt action in responding to the findings, and the last two audit reports should remain on file.

16.1.15 Trade secrets

This paragraph was inserted into the standard at the request of industry groups that were concerned about the misuse of *proprietary process information* that was pulled together to comply with the standard. Accordingly, this provision frees the end user to make this type of information available only to those with a *need to know* and also permits them to require that confidentiality agreements be signed, as necessary, to prevent the information from becoming public.

16.2 Clean Air Act—EPA

This act was passed in 1990 and is aimed at reducing the amount of hazardous chemicals being released into the atmosphere. The portion of the new regulation that has implications for control valves sets limits on what are called *fugitive emissions.* This category of chemical releases refers to small leaks from multiple sources that are relatively hard to locate and measure. Three of the primary sources of fugitive emissions are *line flange joints, valve gasketed joints,* and *valve packing,* with packing being the biggest contributor since it is a dynamic seal. The act, as passed, has provisions for a gradual phase-in as indicated in **Fig. 16.1.** The critical phase in this diagram is phase II, where the allowable leak rate drops from the existing 10,000-ppm limit down to a much more stringent 500 ppm.

There are 187 volatile organic chemicals (VOCs) that are officially covered by the act, but many end users are expanding the scope to cover other hazardous chemicals that could result in unsafe exposure to personnel. The act requires that potential sources of fugitive emissions be "sniffed" using a procedure called EPA Method 21. **Figure 16.2** shows a valve being tested using this procedure. The chemical concentration found using this test is compared to EPA limits to determine acceptability.

The way the standard is now set up, the valves in a plant that handle the offending chemicals are sniffed annually. If less than 0.5 per-

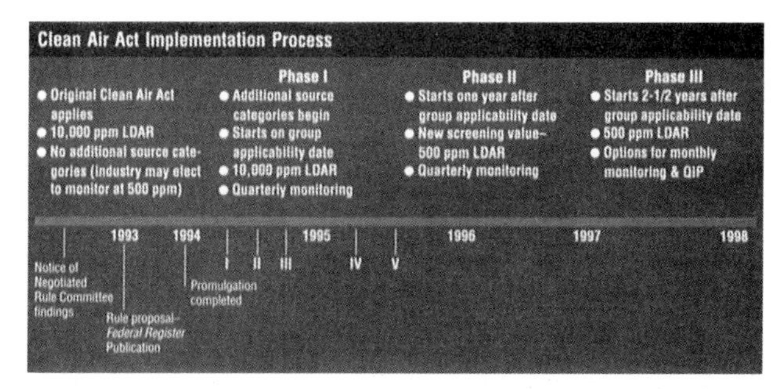

Figure 16.1 Clean Air Act implementation timetable. *(Courtesy of Fisher Controls International, Inc., Marshalltown, Iowa.)*

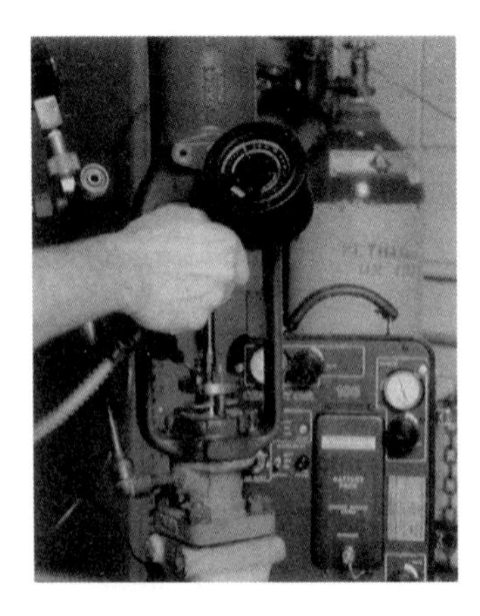

Figure 16.2 Valve being "sniffed" using EPA method 21. (*Courtesy of Fisher Controls International, Inc., Marshalltown, Iowa.*)

cent of the valves have leakage concentrations that are more than 500 ppm, the testing can continue on an annual basis. If the percentage is greater than 0.5, semiannual testing is required. Quarterly testing is required if the percentage of valves that don't pass rises above 1 percent. And monthly testing and/or a formal corrective action plan is required at sites where more than 2 percent of the valves don't meet the 500-ppm limit. This *progressive testing approach* is illustrated in **Fig. 16.3,** and it is obvious that it is structured to apply a severe financial penalty for poor performance, by increasing the frequency of testing as the percentage of leaking valves goes up. This is especially true if you consider that there may be 1000 or more valves covered by this act at a given plant site. Monthly testing on a population of this size can become very expensive. With this as background, how does a site go about making sure that it can stay in compliance?

The *gasketed joints* do not normally pose any problems if the recommended gasketing and bolting practices described in Chap. 7 are employed, Packing is another matter altogether. Consistently meeting and beating these limits normally requires that one of the high-performance packing systems described in Chap. 7 be employed. There are a number of systems available in the marketplace today. Before making a choice, ask the following types of questions about each of the packing systems under consideration:

- How is the packing load adjusted, and how is the load maintained over time? Live-loaded packing systems that incorporate a load indicator are highly recommended for systems that must remain extremely leaktight over long periods of time without having to be readjusted.

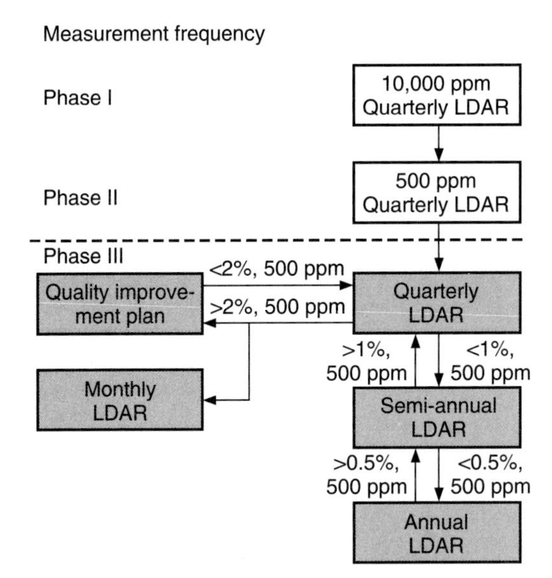

Figure 16.3 Progressive testing approach. (*Courtesy of Fisher Controls International, Inc., Marshalltown, Iowa.*)

- Does the system incorporate enhanced guiding techniques that help to minimize side loads on the packing?
- Is proper emphasis placed on elements that are not supplied as part of the packing system? Stem finish, for example, should be 4 rms or better if good long-term performance is going to be achieved.
- What kind of guarantees are given for maximum leak rates and service life? How were these limits established? Did proof of design testing include elements like high cycling (100,000 plus cycles), high temperature, thermal cycles, and high pressure? How close were the testing conditions to the service that the packing will see in your plant?

Proper selection and installation of a well-designed high-performance packing system will help to maximize the chances of a site being able to maintain compliance and keep testing frequency as low as possible.

As noted in Chap. 7, *bellows seals* can also be used to meet the EPA requirements but are much more expensive than the packing systems just described and require that special precautions be taken during installation and operation. It is better to try to make the packing systems work and go to the bellows seal only as a last resort.

16.3 Hazard Communication Standard

This regulation was implemented in 1986 and is aimed at assuring that workers are adequately informed about chemical hazards in the workplace. Its official designation is OSHA 29CFR1910.1200, but it is usually referred to as the *right to know* rules because of its underlying philosophy that employees working with hazardous chemicals have the right to know about potential health hazards. Its application has been very expensive for the industry and sometimes confusing since, if taken to the extreme, nearly every substance could be considered to pose some type of physical or health risk. Over the last several years, fortunately, there has been a trend to take a more reasonable approach by including only true hazardous chemicals.

This standard must be considered by anyone working with control valves because they may contain hazardous chemicals in service. Before the valves are disassembled in any way, the personnel carrying out the work need to be informed of any health risks they may be taking in working with the chemical in question. The risks, along with any special precautions, are listed on a MSDS. The standard specifies the type of information that needs to be on the sheet and requires that the facility provide adequate access to the sheets.

The personnel working on the valve can then consult these sheets to learn how best to protect themselves from the risk of harmful exposure. In most cases, it is recommended that the valve be cleaned before personnel begin the repair. This is most effectively done after at least partial disassembly because chemicals can remain trapped in some areas of the valve, such as the packing box and gasket grooves.

Compliance with this standard is particularly important when valve repair is *subcontracted* to an off-site facility because the end user can be held liable if one of the contractor's employees is inadvertently exposed to a hazardous chemical. Be sure that the valves are cleaned and that all MSDSs are supplied for the chemicals handled by the valves. Any valve repair company that is selected by an end user should be audited to be sure that it has an effective right-to-know program in place and that it is taken seriously.

Very briefly, the right-to-know standard breaks down into five sets of requirements:

1. *Identifying hazardous chemicals.* This section requires that all *physical and health hazards* be identified for any chemical used or manufactured in a chemical processing facility. The hazards need to be clearly *communicated* through the use of *labels* and the *MSDSs* mentioned above.

2. *Product warning labels.* This section covers the requirements for *labeling of containers* that handle hazardous chemicals. The label

must show the name of the chemical and any physical or health hazards. Exceptions include temporary containers, if the material in the container is going to be used immediately by the person who filled it, and piping.

3. *MSDSs.* This section covers the basic requirements for the information to be included on these *reference sheets.* The sheets have been a source of confusion for the industry because there is no standard format dictated by the guideline. As the industry has gained more experience with the regulation, this problem is gradually being solved as a de facto format has evolved. In any case, the information that must be covered includes:

   The common name and chemical name for the compound

   The name, address, phone number, and emergency phone number of the manufacturer

   The date of the form

   A list of hazardous ingredients in the compound

   Physical information about the compound

   Fire and explosion information

   Dangers of chemical reactions

   Measures to control the chemical's hazards

   Information about the compound's health hazards

   Information about how to deal with spills and leaks

4. *The written hazard communication program.* This section covers the requirements for a *written procedure* that completely describes the site's right-to-know program.

5. *Employee training.* This section describes the requirements for an effective employee training program aimed at ensuring that the procedures and practices covered in the site's right-to-know program are put to use by each and every employee.

16.4 ISO 9000

ISO 9000 is the generic name that refers to a series of standards that were put together by the International Organization for Standardization to define the requirements for a QA program covering the design, manufacture, and service of industrial products. They were first published in 1987 and were pulled together in an effort to define the minimum standards for products destined for use in the new *European Economic Community* (EEC). Initially, many companies pursued compliance with the idea of being able to supply products to the EEC. Recently, however, the use of the standards have evolved to the point where they are now being used to indicate that the manufacturer has a good, solid *QA program* in place that will assure customers, no matter what world area they are in, that the manufacturer will deliver exactly what was ordered.

For chemical processing plants, this means delivering the appropriate amount of chemical, on schedule, that meets whatever product specifications have been agreed to between the vendor and the customer. With this in mind, an ISO 9000 program needs to address anything in the process that could affect *quantity, delivery integrity,* and *product quality.* This is where control valves come into the picture. They would be included in the group of processing equipment that might have an impact on one of these critical factors. Each valve application has to be reviewed to determine how or whether it is considered critical to the delivery of a conforming end product. If it is judged to fall into this category, certain elements of its performance would have to be tightly controlled to ensure that it does not have a negative impact on the end product.

In most cases, this would mean verifying that things such as *process control performance,* and *shutoff* remain consistent over the valve's service life. The best way to confirm this is to make sure that the valve is properly selected, installed, and maintained, taking the service conditions into account. This means that a good QA program needs to verify that the following things happen for valves that are considered to be critical to end-product quality:

- That the *service conditions* are accurately identified and documented and then transmitted to the valve vendor so that the *valve design* is consistent with the demands of the process. Any differences between the purchasing requirements and the tender must be identified and resolved.
- That the equipment supplier has adequate *design control* and *manufacturing procedures* in place to guarantee that the equipment supplied is in conformance with the purchase order, as amended during the resolution stage mentioned in the preceding paragraph. A documented *vendor review* is part of the ISO requirements, and while it is not required that all vendors have ISO 9000 certification, it certainly makes life easier for the plant that is using the equipment.
- That the appropriate *practices* and *procedures* are utilized when installing and setting up the valve so that it is in good operating condition when the process comes on-line. Some type of documented in-line inspection should support this.

- That an effective *maintenance program* is in place to ensure that conditions cannot develop and go undetected that could affect the overall quality of the end product. This program stresses the application of *corrective action* aimed at a long-term solution of any problems that surface. Equipment with problems should be clearly marked to avoid inadvertently putting it back into service.

You can see in reviewing this list that it follows the recommended practices that are stressed throughout this book. What is new if ISO 9000 applies is that nothing can be taken for granted in the above elements. *Written procedures* must be established and followed for all steps identified, and then they must be assembled in a QA manual.

Any adjustments or calibration that are performed must be done with equipment that is guaranteed to be accurate to required levels, through the use of a *traceable calibration program.* Any interventions on equipment need to be documented so that no unauthorized actions could be taken that could affect valve performance. All work procedures that apply to equipment covered must be in written form and must be able to be audited. Any actions such as calibration and maintenance should be documented in written form, as mentioned, but should also be reflected through the use of some type of marking on the valve like a calibration sticker. All personnel working on the equipment must be appropriately trained, and records must be kept that verify conformance with the preceding requirements.

16.5 References

1. *ISO 9001, Quality Systems—Model for Quality Assurance in Design and Development Production, Installation and Servicing,* International Organization for Standardization, 1987.
2. Miller, M., *A Fisher Overview of Process Safety Management Rules,* Fisher Controls, Marshalltown, Iowa, 1992.
3. Cerrato, Jim, "Valve Packing Minimizes Leakage," *Intech,* vol. 40, no. 5, p. 40, ISA, Research Triangle Park, N.C., 1993.
4. Ritz, George, "Advances in Control Valve Technology," *Control Magazine,* March 1993.
5. Quernemoen, Bruce, "When Is Parts Substitution Okay?" *Hydrocarbon Processing,* May 1993.
6. *29CFR19b.119—Process Safety Management of Highly Hazardous Chemicals,* OSHA, Washington, D.C., 1992.
7. *EPA Clean Air Act,* EPA, Washington, D.C., 1990.
8. *29CFR1910.1200—Hazard Communication Standard,* OSHA, Washington, D.C., 1986.

Chapter

17

New Developments

There are a number of new developments that have either already arrived or are just on the horizon that will affect how control valves are used in the future. This chapter goes over a number of these and attempts to give you a better appreciation of how each one could affect the operation of a typical control valve.

17.1 Smart Valves and Instrumentation

This technology has tremendous potential for changing the valve industry. The term *smart* refers to the integration of *microprocessor* capability into the operation of the control valve. This can be done in any one of several ways.

One of the first commercial attempts to utilize "smarts" in a control valve package is illustrated in **Fig. 17.1.** This is a complete valve instrumentation package that features a smart module that includes provisions for monitoring and recording things such as:

- Inlet and outlet pressure
- Inlet temperature
- Stem position
- Top and bottom actuator pressure
- Positioner input signal

Using the information collected as shown in the schematic of **Fig. 17.2,** the microprocessor in the system can end up performing a number of tasks that enable the end user to better optimize the process. Typical features include:

Figure 17.1 Smart control valve system. (*Courtesy of Valtek International, Inc., Springville, Utah.*)

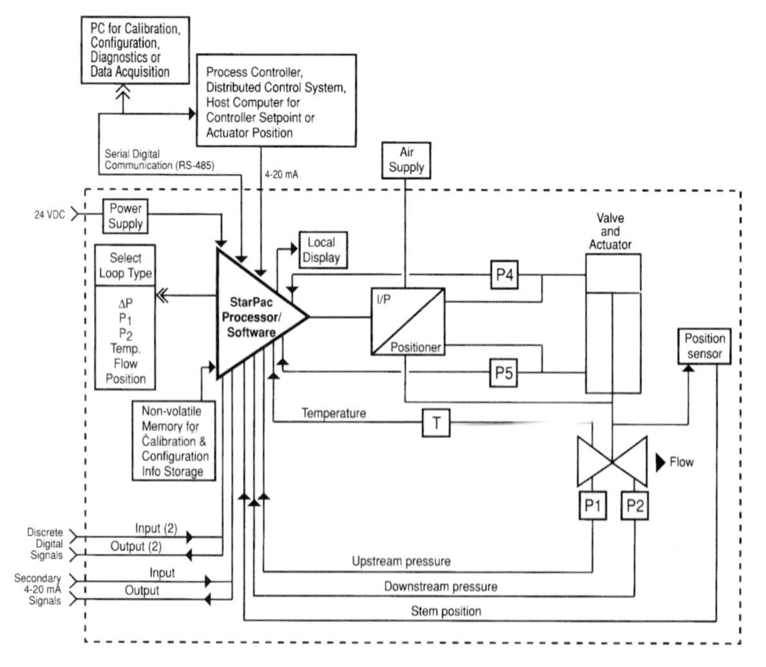

Figure 17.2 Smart control valve system schematic. (*Courtesy of Valtek International, Inc., Springville, Utah.*)

- *Instrument signature.* With this feature, the control room can identify the device that it is communicating with by checking an electronic signature that is stored in the memory of the device. This helps with commissioning activities since wires don't have to be traced.
- *Local process control.* The system measures pressure drop versus valve position, along with temperature. Armed with this information, the flow through the valve can be calculated and sent back to the microprocessor. The "brain" acts like a local controller by comparing the actual flow, temperature, or upstream or downstream pressures to whatever the defined setpoint is and making the necessary corrections. It can be tuned for better control just like a standard PID controller. This eliminates the need for additional sensors being mounted in the line since the process variable measurement is integrated into the valve itself.
- *Process diagnostics.* The initial process *signature* can be generated by examining the flow versus travel characteristics and then this signature can be compared to performance at a later date to see if any changes have occurred that would necessitate corrective action.
- *Built-in process transmitters.* As mentioned earlier, this system features built-in measurement devices that can record and transmit various process characteristics, eliminating the need to add these sensors somewhere else in the line.
- *Configurable failure modes.* Multiple failure modes can be configured so that different actions occur upon loss of power, air supply, command signal, or process signal.
- *Valve diagnostics.* Since travel and actuator pressures are measured, diagnostics can be performed on valve operation itself, looking for problems such as sticking in the trim and high packing friction. The operating principles behind these diagnostics are explained in more detail in Sec. 13.5.

These features have proven to be of interest to the industry, but the system does have some disadvantages that have prevented its widespread use. It is relatively expensive and requires an auxiliary 24-V dc power supply, which is not always readily available. Nevertheless, it does begin to show the kinds of benefits that smart devices can bring to the marketplace.

The other direction that smart devices have taken for control valve operation is the introduction of *smart positioners*. In this case, the microprocessor is integrated into the positioner package rather than requiring a separate module as with the preceding example. It is limited somewhat in the amount of computing power that is available, but it has a significant advantage in that it is loop-powered, so it doesn't require an auxiliary power supply.

Like the smart valve package, it can measure, record, and transmit things like valve travel and actuator pressure so that *diagnostics* on the valve can be performed locally, using a PC to communicate with the device, or the signals can be transmitted to the control room where the performance of the valve can be *evaluated on-line.* The smart positioners do not normally include built-in sensors for process parameters, so auxiliary sensors are used in conjunction with them if local process control is to be carried out.

Like the smart valve just discussed, they can also have an electronic signature in memory, allowing them to be identified remotely, so wires don't have to be traced. This can be a tremendous advantage at start-up. The inherent response can also be adjusted to characterize the device in just about any fashion to improve overall valve assembly linearity. The dynamic characteristics can be tuned, as well, to provide optimum response between the positioner and the valve. Most models can also be remotely calibrated from the control room.

Since they measure and transmit valve position, they can be set up to provide alarm signals much like limit switches to indicate when the valve has reached a particular point in travel. Some models even keep track of total stem movement to provide an early indication of when a valve and related hardware might need to be repaired. Of course, these built-in stem position sensors eliminate the need for a separate stem position transmitter.

17.2 Fieldbus

For those of us with mechanical backgrounds, the term *fieldbus* can be fairly confusing. This is only made worse by the tendency of those with electrical backgrounds (you know who you are) to respond to any inquiries on this subject with an endless stream of buzzwords that only those with state-of-the-art pocket protectors can understand. In an effort to change all this, this section will attempt to explain what fieldbus is and why the industry seems to be so excited over its development. More importantly, the explanation will be carried out in plain English.

Fieldbus, very simply, is a new way for *communication* to take place between the control system in a process plant and control equipment out in the field. It is expected that this new approach will become available sometime in the next 2 years. To truly get a feel for what fieldbus is and how it works, we need to understand how the current systems operate.

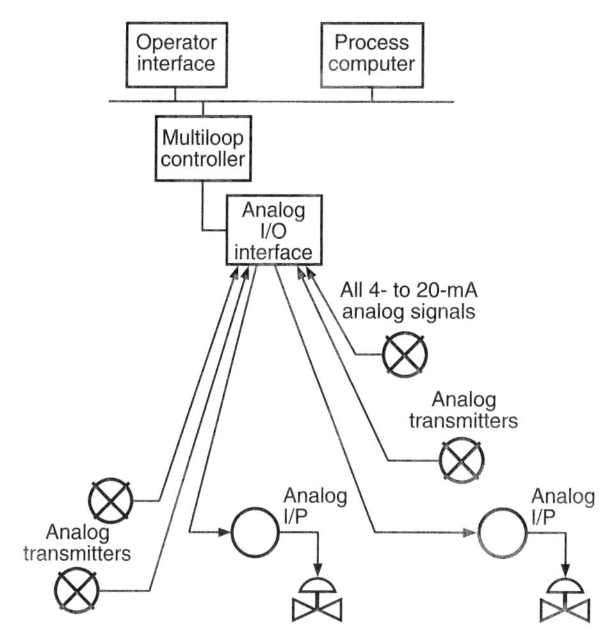

Figure 17.3 Conventional analog process communication network. (*Courtesy of Fisher Controls International, Inc., Marshalltown, Iowa.*)

In a conventional system, the operators interface with the control system through a process computer that, in turn, works with a controller network that sends and receives signals to and from the field. This is illustrated in **Fig. 17.3.** In this *analog system,* 4- to 20-mA signals are sent from the controller (through the interface) to the valves to control their position, and process variable signals are sent back to the controller from the transmitters. Each set of wires only carries one signal, and the signal only goes in one direction. This is the traditional approach, which has been used for a number of years. It works well enough and has a big advantage in that it is an *accepted standard.* Any vendor's equipment will work with anyone else's, so interchangeability is not a problem. Now let's look at some of the changes that are occurring in the industry with respect to this instrumentation and see where improvements are being made.

Over the last several years, field equipment has been developed that has integrated on-board microprocessors. This smart equipment (see previous section) can communicate digitally over the same wires that carry the analog signal. The process control signals are still of the 4- to 20-mA variety, but a *second layer of digital communication* is added on top of the analog. **Figure 17.4** shows how a typical system layout might look with smart devices. By adding this higher-level

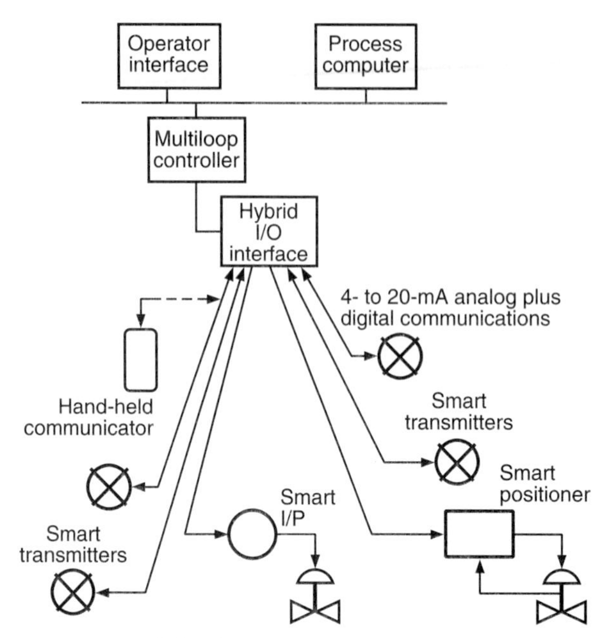

Figure 17.4 Hybrid process communication network. *(Courtesy of Fisher Controls International, Inc., Marshalltown, Iowa.)*

digital communication, a number of significant benefits are already available, including:

- *Remote diagnostics.* The devices can perform self-diagnostics and send the information back to the control room, or diagnostics can be triggered from the control room and the results reviewed online, without having to go out into the field. This makes maintenance easier and cheaper.
- *Remote calibration.* Many of the smart instruments can be calibrated and ranged from the control room by sending digital correction signals over the existing wires with the control system, a PC, or a hand-held device. This makes commissioning and routine maintenance much simpler and cheaper.
- *Single-ended checkout.* This applies to the commissioning process, as well. The smart devices have a built-in electronic ID that permits the control system and operator to know which device it is connected to, so the practice of tracing wires from the device back to the control room can be eliminated.
- *Improved reliability.* Because digital devices seem to be more reliable than their analog counterparts, the number of spare assemblies can be reduced.

One set of wires is still needed for each piece of this new equipment, but the operational advantages of smart devices are obvious. Unfortunately, the *language* that is used to communicate with them has not yet been standardized, so a control system from one manufacturer cannot necessarily communicate with the field device from another. This is a real hardship for the end user, but don't look for a solution coming about too soon. It's a little like the difference in VHS and BETA formats during the development of the VCR. The manufacturer who can make its approach to this communication link the industry standard will have a real competitive advantage, so there is a tendency for each one to defend its turf as the standards develop. This slows standards development but is something that the industry will have to continue to live with.

Even given these *interchangeability* problems, smart devices will make our life easier, but their real potential will only be fully realized when a fully digital communication link is established. The fieldbus is that link. Now let's take a look at how it will transform the way control systems will work, once it is fully implemented.

An example of what a future fieldbus arrangement with digital smart devices might look like is shown in **Fig. 17.5.** One of the major changes made possible with the fieldbus is that *multiple devices* may

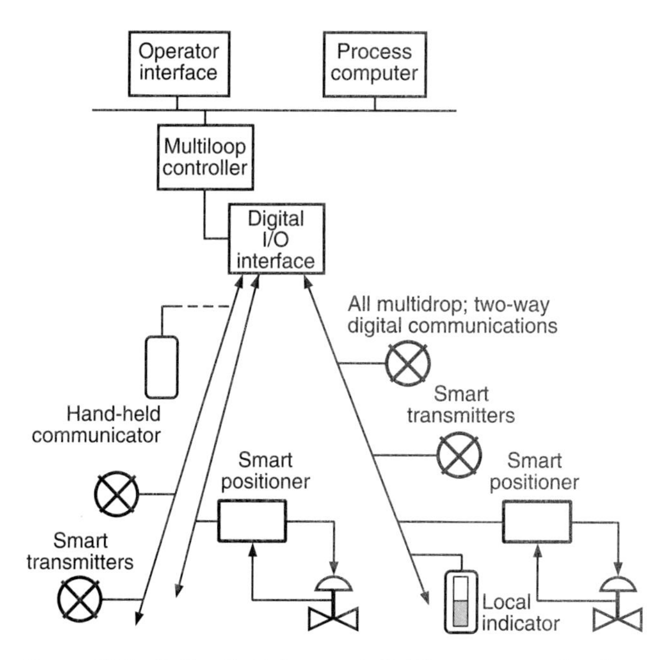

Figure 17.5 Fully digital, smart fieldbus communications. *(Courtesy of Fisher Controls International, Inc., Marshalltown, Iowa.)*

now be connected on a single pair of wires. This would not be possible with an analog line because the competing signals would overlap and interact. *Digital communications* can be separated so that the messages between the control system and the field devices are identified by device number; this means the messages can be properly sorted out and acted upon.

This change alone is a big advantage because it can significantly reduce the *number of wires* required for a given installation, but there are even bigger benefits possible if the fieldbus reaches its full potential. For instance, communications could occur between smart devices in the field without tying up the main control system. A smart valve coupled with a smart transmitter could provide local control, leaving the main processor for higher-level control activities. One smart device could take multiple measurements from one penetration in the pipeline and download the readings to whatever device in the network needs to have them. This would result in tremendous savings by eliminating a *number of penetrations* into the pipeline and by making one device do the work of four or five.

Remote diagnostics can be even more sophisticated. If a problem is detected at the loop level, each of the devices on the network that affect loop performance could be interrogated until the problem is found. One other benefit would be the elimination of the *digital-to-analog interface* since the digital field devices could communicate directly with the digital control system.

If all these advantages are added to the benefits already listed for the smart devices alone, it is easy to see why there is so much enthusiasm and excitement surrounding the introduction of fieldbus. Unfortunately, it is coming along very slowly. Once again, the battle over interchangeability in the communication language is being fought in the standards committees (ISA SP50 and IEC TC65CWG6), and this has delayed the introduction. The good news is that the logjam is slowly breaking down through the efforts of end users and a few progressive vendors who are willing to compromise, and it now looks like a full release should happen sometime in the next 2 years. When it happens, it will truly change the fundamental basis of process control, and those end users who are ready to adapt to it will have a significant competitive advantage for the long term.

17.3 Loop Tuning

Loop tuning has been around for a long time and refers to the adjustment of the proportional, integral, and derivative settings on a standard three-mode controller. As was mentioned in Chap. 5, these settings can be adjusted based on loop characteristics to provide optimum response in terms of both speed and the reduction in the overall variations between the setpoint and the process variable. Quarter-amplitude damping technique, or the *Ziegler-Nichols method* as it is sometimes called, has long been considered the industry standard for tuning loops. While its application is not particularly difficult, many loops are not tuned very well. This is due in part to the scarcity of process control expertise in most plants and the concern that aggressive tuning will result in instability, coupled with the fact that the traditional method of loop tuning can be long and tedious, particularly on slow loops. In many cases, the loops are tuned using *rules of thumb* that guarantee stability but do not provide very good performance. It follows that if the loop tuning is not optimized, process control is still not as good as it could be, no matter how much care is taken in maintaining the control valve.

It has been estimated that over one-third of all automatic control loops are not tuned properly and are hurting process plant performance as a result. In response to this problem, the industry has come up with two potential solutions. The first is to automatically calculate the tuning parameters based on observing and recording the process variable versus the setpoint as the setpoint is "bumped." An example of such a test is illustrated in **Fig. 17.6,** with the bottom line showing the step bump, and the upper line showing the system's response to the bump. Without worrying about the math, these systems all use some type of algorithm to determine the frequency response characteristics based on the response to the bump, and then a mathematical model of the loop is developed. The model can then be used to determine the best tuning parameters, depending on how aggressive you want to be with the control.

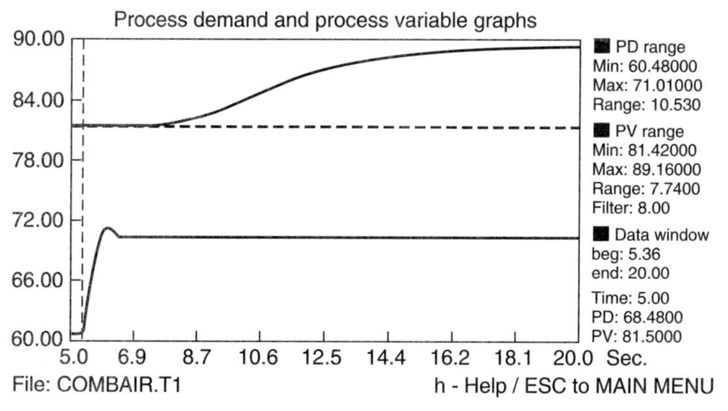

Figure 17.6 Tuning test showing "Bump" response. (*Courtesy of Techmation, Tempe, Arizona.*)

In practice, the systems are easy to use. They usually hook up to the controller I/O terminals to get their readings and then the data are downloaded into a PC for processing. It's all very fast and painless, and experience has shown them to work well. However, loop characteristics can change if process conditions change, so the settings found are really only optimized for the conditions present during the test. This may mean that *multiple readings* may have to be determined if a loop has to be operated at several sets of process conditions. In any case, the settings found using this equipment should be considerably better than those picked using whatever rule-of-thumb methods are in force at a particular site.

The other way to attack this problem is to use what are called *self-tuning controllers*. In this case, the software necessary to run a bump test is built into the controller, so no auxiliary equipment is required. If a tuning operation needs to be done, the signal is given to the controller, and it begins bumping the output very slightly over one or more cycles while recording the response. Like the device just described, it then determines optimum tuning parameters based on the response measured.

One of the side benefits to come out of this automated tuning activity is that *process modeling* techniques have improved to the point where process design can be tried out on paper before implementing it in hardware. This can help to improve process design at greatly reduced costs.

17.4 Software Packages

As with nearly every other area of business, a number of software programs have been developed over the last several years that can improve overall operations in a chemical plant. As just mentioned, *process simulation software* can help systems engineers to work out the bugs in a proposed process before investing in costly equipment.

As far as valve selection is concerned, nearly every vendor can now use, or supply to the end user, computerized *valve and actuator sizing software* that makes valve sizing and selection much simpler and faster. In addition to valve selection software, instruction manuals, spare parts lists, and cross-sectional drawings can all be supplied as *computer records* that can be stored, updated, and accessed much more easily than the traditional hard copies. In fact, some companies are putting together *interactive CD-ROM packages* for their instruction manuals that permit the user to very quickly find the procedures that are needed for a planned intervention on a valve.

Maintenance records are another area where software packages can play a big role. Keeping track of the type and frequency of maintenance required for a given valve can provide valuable insights into overall reliability and can show where additional training might be required. As mentioned in Chap. 15, it can also provide clues regarding *life-cycle costs* and can help to make more informed purchasing decisions. There is also the regulatory aspect. All four of the *new regulations* covered in Chap. 16 stress the importance of maintaining records on what kind of work is carried out on a valve and how it might affect the overall operation from a health and safety standpoint. In response to this, several vendors have come out with software packages aimed at tracking control valve documentation, maintenance activity, and performance records from the time the valve is purchased until it is taken out of service for good. Any plant with world-class aspirations needs to make this type of software package a part of its long-term plans.

17.5 New Stem Sealing Designs

New regulations have forced users and vendors to concentrate on reducing the amount of *external leakage* from control valves into the surrounding atmosphere. The result of this work has been the introduction of the new high-performance packing designs described in detail in Chap. 7. These systems, if installed and adjusted properly, will provide leak-free service for several years in a standard application, permitting the end user to avoid the heavy sanctions outlined in the EPA regulations. For a complete description of how the packing systems work, see Sec. 7.3.5. The *EPA regulations* are covered in Sec. 16.2.

17.6 Control Valve Diagnostics

As was stated in Chap. 13, control valve maintenance techniques have traditionally been *reactive,* either waiting for failure to occur and then correcting the problem or rebuilding valves on a sampled basis during major outages with the hope that this will cut down on expensive on-line failures. New technology has been developed over the last 3 years that permits the user to effectively *troubleshoot* a control valve without taking it out of the line and also to look for signs of degradation in performance so that *corrective action* can be taken on a selective basis.

This new technology is commonly referred to as *control valve diagnostics* and is addressed in Sec. 13.5, where it is compared to the traditional reactive approach. These techniques are now being used extensively in the chemical processing industry to not only determine which valves need to be maintained but to give the end user a *performance standard* that can be used to ensure that valve performance is optimized. This is in keeping with the overall philosophy of this book that says that valves need to be doing more than just getting by if the end user wants true world-class process performance.

17.7 References

1. *Fieldtalk,* April/May FLY 1045, no. 5, Interoperable Systems Project, Austin, Tex., 1993.
2. Ritz, George, "Advances in Control Valve Technology," *Control Magazine,* p. 32, March 1993.
3. Gerry, John, "Tune Loops for Load Upsets, not Setpoint Changes," *Control Magazine,* Sept. 1991.

GLOSSARY

The glossary is split into two parts. The first section includes a number of figures that show cross-sectional views of various valves, actuators, and accessories. The major parts are labeled to aid in their identification. The second section is an alphabetical listing and definition of common terms used in the control valve industry.

G.1 Control Valve Cross Sections (*Figs. G.1* through *G.18*)

G.2 Common Control Valve Terms

Note that some of these definitions are reprinted with permission from the following standards: ASME Standard 112, "Diaphragm-Acutated Control Valve Terminology" and ANSI/ISAS51.1, "Process Instrumentation Technology." Copies of the ISA Standard may be ordered from the Instrument Society of America, 67 Alexander Drive, P.O. Box 12277, Research Triangle Park, N.C. 27709. Copies of the ASME Standard may be ordered from the American Society of Mechanical Engineers, 345 E. 47th St., New York, N.Y. 10017.

Accuracy In process instrumentation, the degree with which an indicated value conforms to a recognized accepted standard value.

Actuator lever An arm attached to the rotary valve shaft to convert linear actuator stem motion to rotary force to position the disc or ball of the rotary-shaft valve. (The lever normally is positively connected to the rotary shaft by close tolerance splines or other means to minimize play and lost motion.)

Actuator spring A spring used in the actuator to move the actuator stem in one direction or the other.

Actuator stem A rod-like extension coming out of the actuator that permits

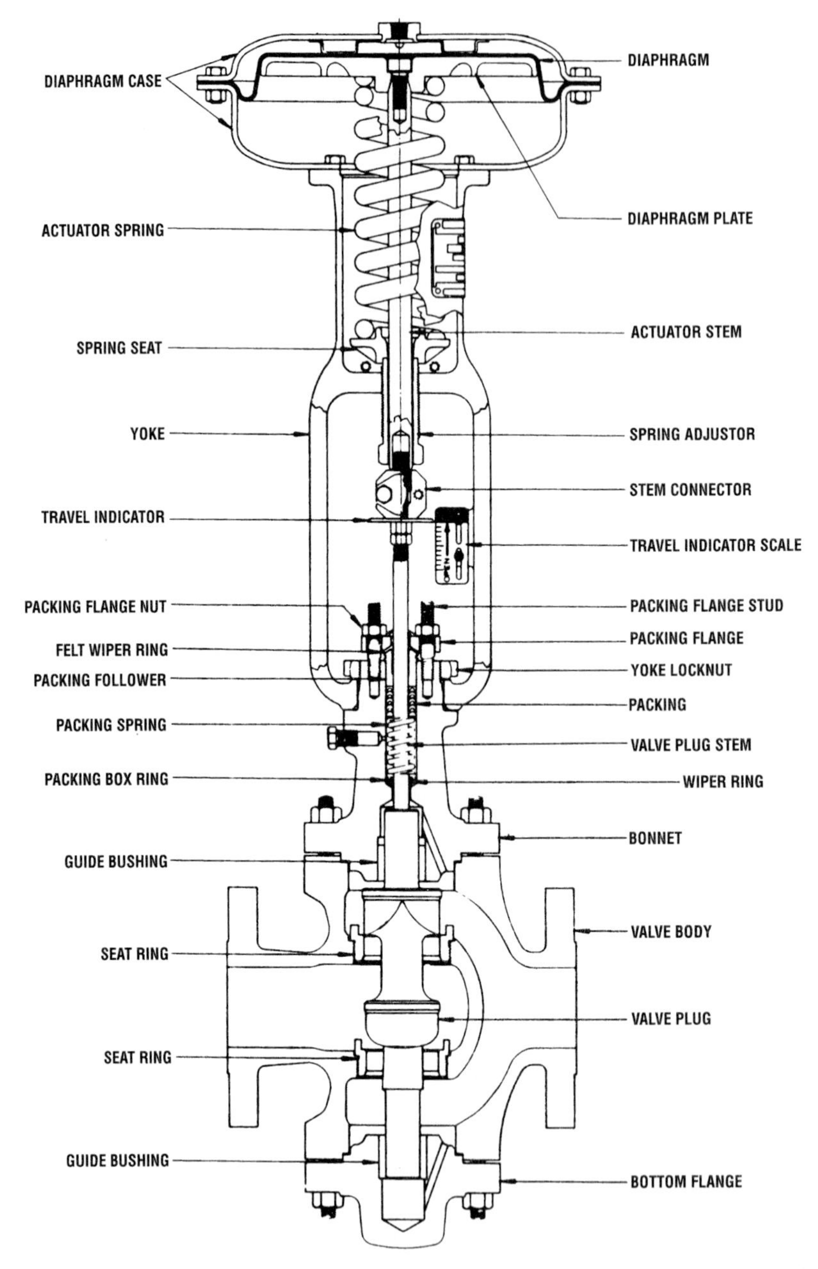

Figure G.1 Typical double-ported, direct-acting, spring and diaphragm control valve assembly. (*Courtesy of Fisher Controls International, Inc., Marshalltown, Iowa.*)

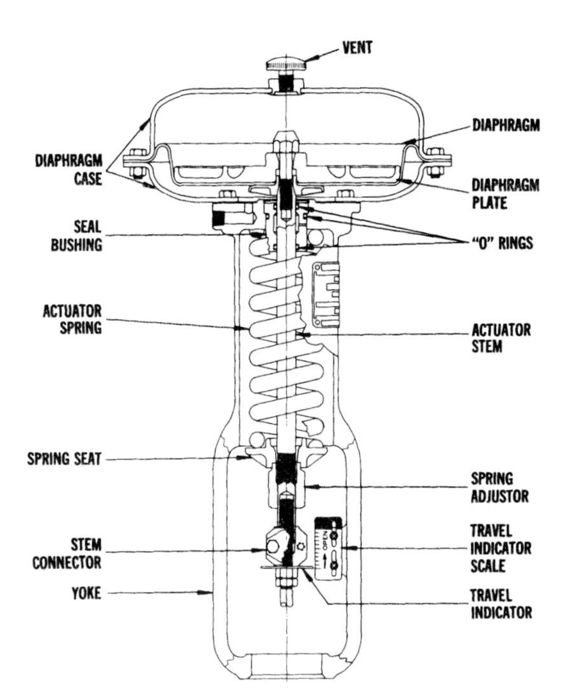

Figure G.2 Typical reverse-acting spring and diaphragm actuator. (*Courtesy of Fisher Controls International, Inc., Marshalltown, Iowa.*)

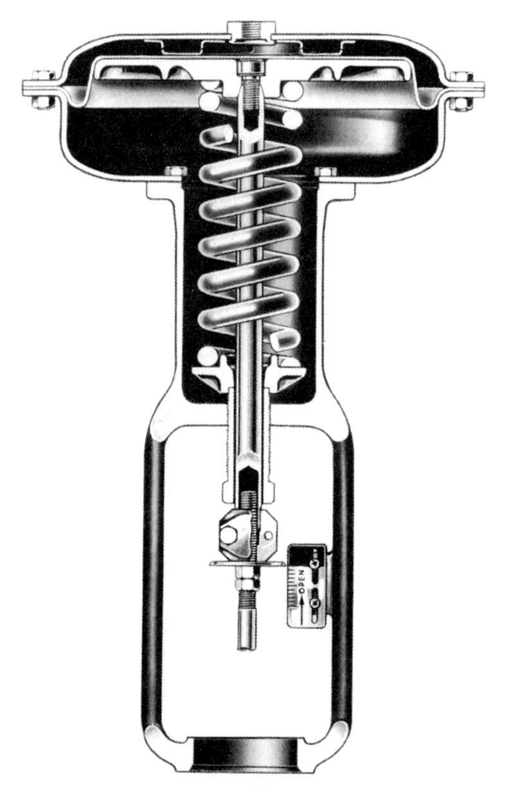

(a)

Figure G.3 Typical single-port, direct-acting, spring and diaphragm control valve assembly. (*Courtesy of Fisher Controls International, Inc., Marshalltown, Iowa.*)

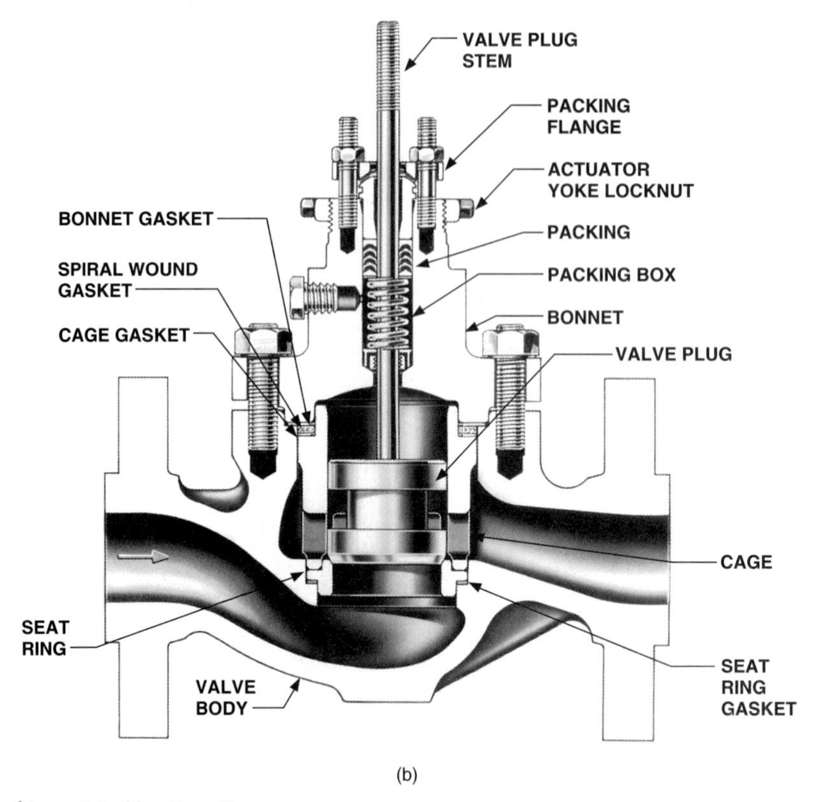

Figure G.3 *(Continued)*

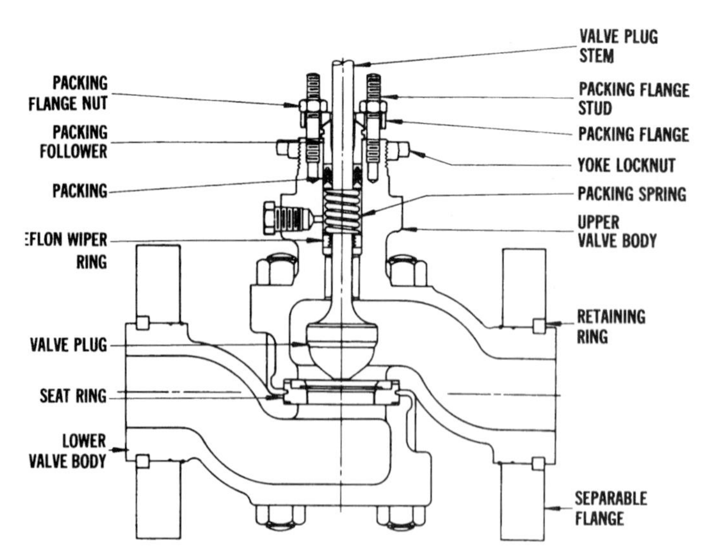

Figure G.4 Typical split-body valve. (*Courtesy of Fisher Controls International, Inc., Marshalltown, Iowa.*)

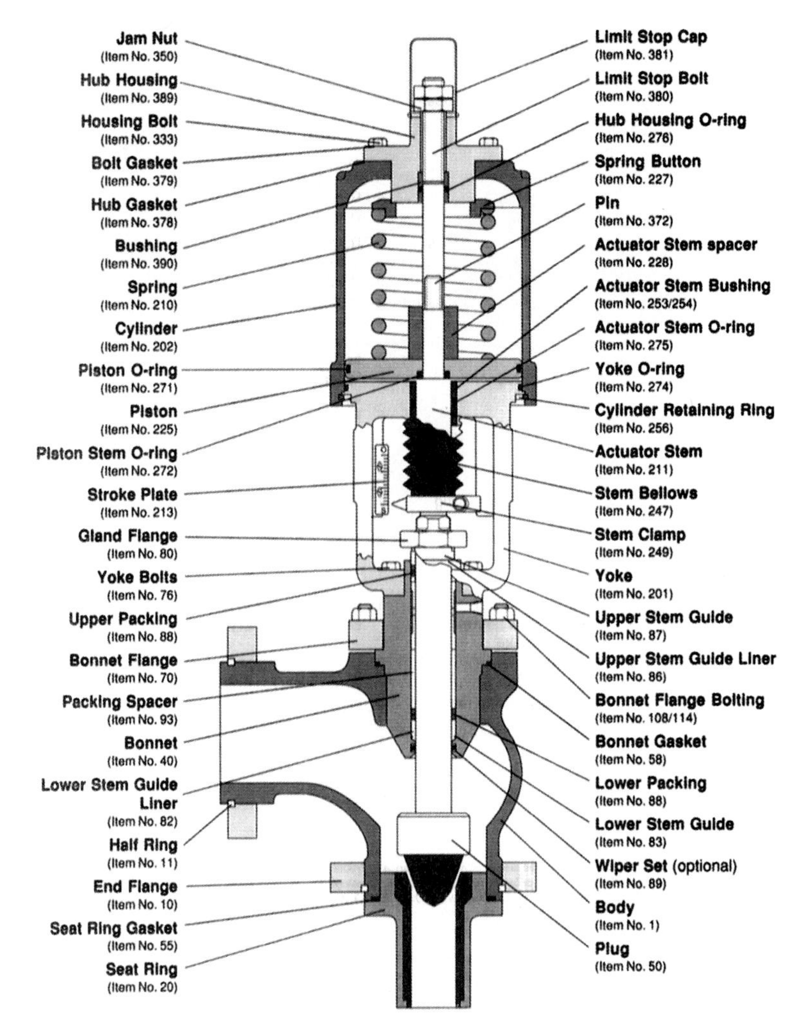

Figure G.5 Typical angle body with piston actuator. (*Courtesy of Valtek International, Inc., Springville, Utah.*)

a convenient external connection.

Actuator stem force The net force from an actuator that is available for actual positioning of the valve plug.

Air-to-diaphragm range The range of pressures that must be present and acting on the actuator for the valve to reach full open position and also close the properly shut off.

Automatic controller In process control, a device which operates automatically to regulate a controlled variable in response to a command and a feedback signal. The automatic controller is the device in a control system that

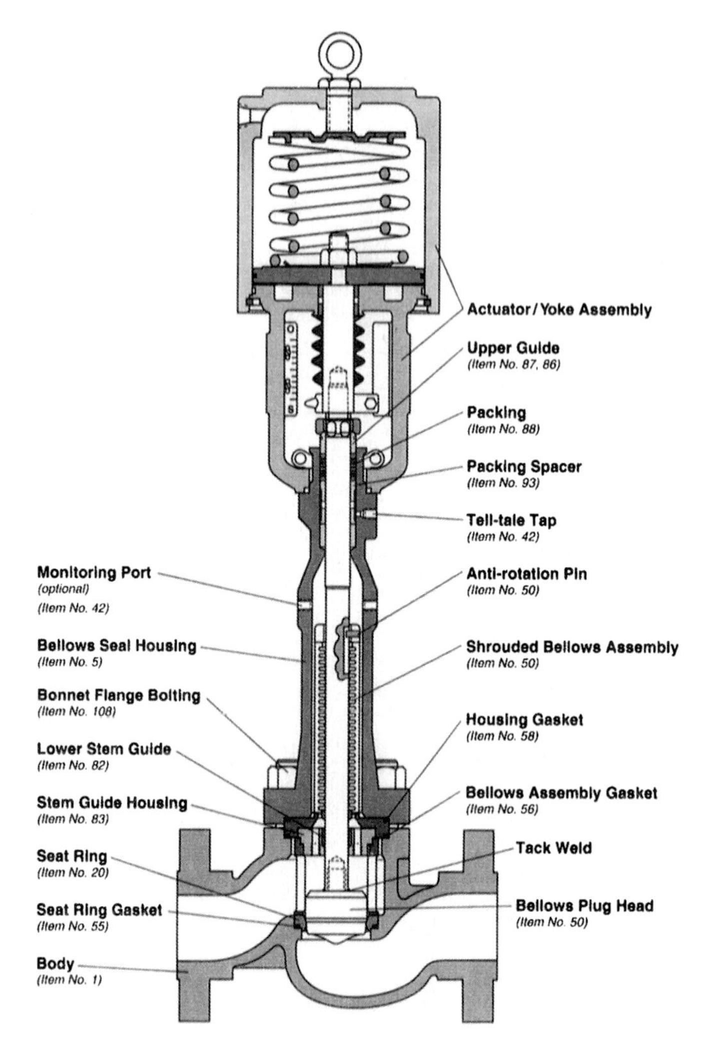

Figure G.6 Typical globe-style valve with bellows seal bonnet. (*Courtesy of Valtek International, Inc., Springville, Utah.*)

senses a change in the process variable (pressure, liquid level, flow, etc.) and transmits an appropriate signal to operate the final control element, which is usually a control valve.

Backlash In process instrumentation, a relative movement between interacting mechanical parts, resulting from looseness.

Ball, full The flow-controlling member of rotary-shaft control valves utilizing a complete sphere with a flow passage through it. (Many varieties use a trunnion-mounted, single-piece ball and shaft to reduce torque requirements and lost motion.)

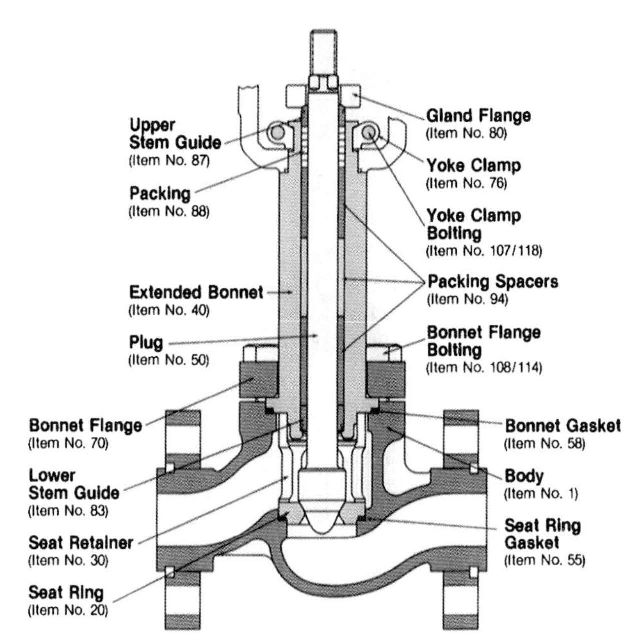

Figure G.7 Typical extension bonnet design. (*Courtesy of Valtek International, Inc., Springville, Utah.*)

Ball, V-notch The flow-controlling member for the most popular styles of throttling ball valves. The V-notch ball includes a polished or plated partial-sphere surface that rotates against the seal ring throughout the travel range. The V-shaped notch in the ball permits wide rangeability and produces an equal percentage flow characteristic.

Bellows seal A bonnet which uses a bellows for sealing against leakage around the valve plug stem.

Benchset The high and low values of pressure applied to an actuator with a spring that produce rated valve plug travel with no external loads present on the actuator stem. The benchset should normally be set and verified with no friction or flow loads present and should usually be checked from the upper actuator travel stop down to rated travel for the valve.

Bonnet The major part of the bonnet assembly, excluding the sealing means.

Bonnet assembly An assembly, including the part through which a valve stem or shaft moves, and a means for sealing against leakage along the stem. It usually provides a means for mounting the actuator.

Bottom flange A part which closes a valve body opening opposite the bonnet assembly. In a three-way valve, it may provide an additional flow connection.

Cage A hollow cylindrical trim element that is used as a guide to align the movement of a valve plug with a seat ring and/or retain the seat ring in the

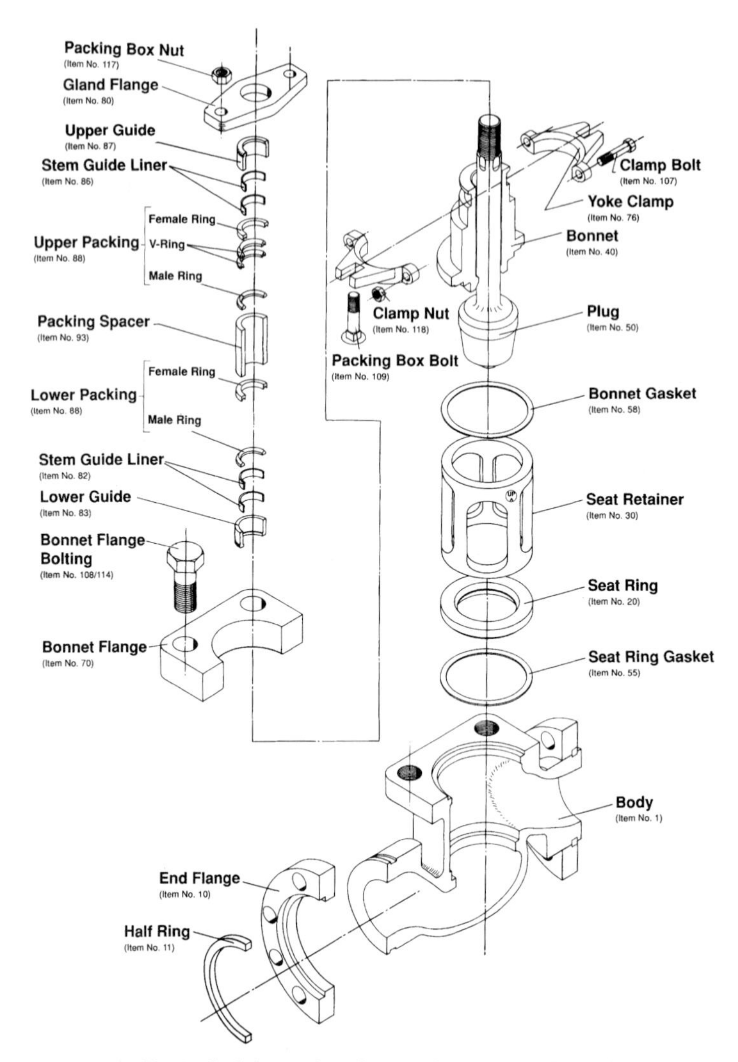

Figure G.8 Typical globe-style, chemical service valve. (*Courtesy of Valtek International, Inc., Springville, Utah.*)

valve body. (The walls of the cage contain openings which can sometimes determine the flow characteristic of the control valve.)

Cage windows Holes in the side of the cage that permit flow to pass from one side of the cage to the other. They may be specially shaped to provide a special relationship between the valve plug travel and flow.

Clearance flow That flow below the minimum controllable flow with the valve plug not closed.

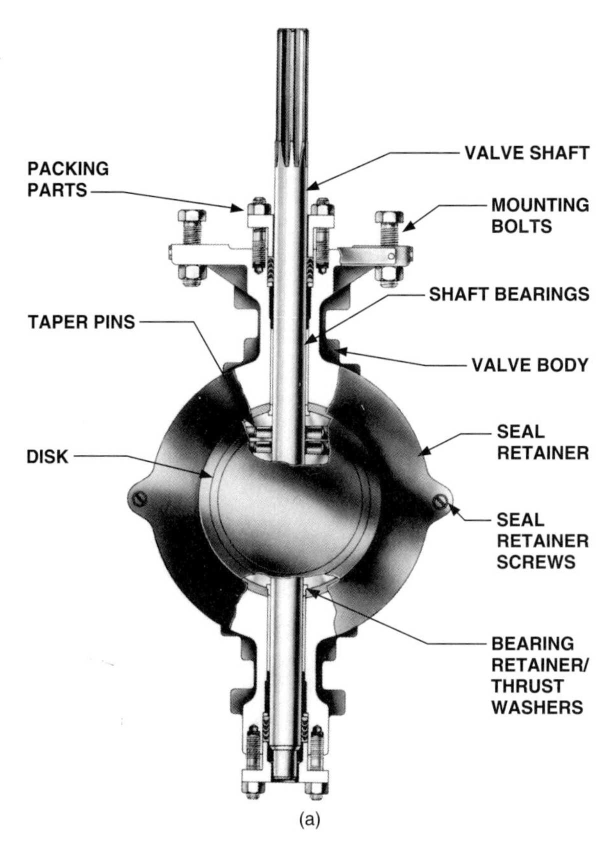

Figure G.9 Typical eccentric disc valve (high-performance butterfly valve). (*Courtesy of Fisher Controls International, Inc., Marshalltown Iowa.*)

Control, feedback Control in which a measured variable is compared to its desired value to produce an actuating error signal which is acted upon in such a way as to reduce the magnitude of the error.

Control mode A specific type of control action such as proportional, integral, or derivative.

Control valve A final controlling element, through which a fluid passes, which adjusts the size of flow passage as directed by a signal from a controller to modify the rate of flow of the fluid.

Controller, direct-acting A controller in which the value of the output signal increases as the value of the input (measured variable) increases. *See* Controller, reverse-acting.

Controller, reverse-acting A controller in which the value of the output variable decreases as the value of the input (measured value) increases.

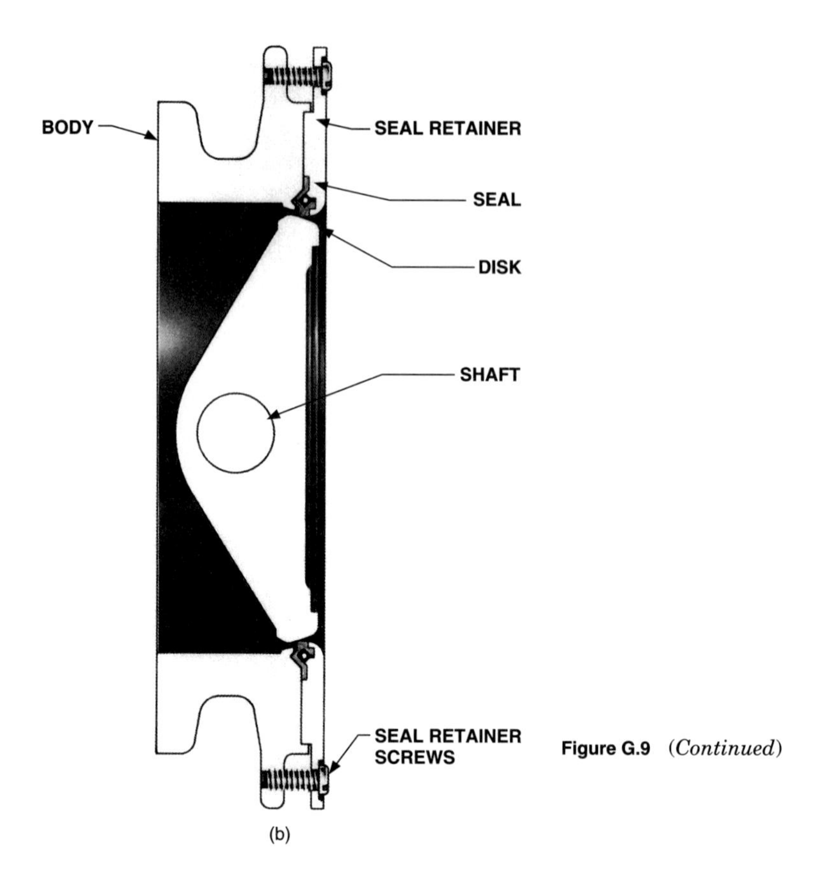

Figure G.9 *(Continued)*

Cross-over pressure For positioners on double-sided actuators, this is the pressure at which the opposing pressures equalize upon reversal of the stroke. This should normally occur at about 75 percent of the supply pressure.

C_v A sizing coefficient used to determine the liquid flow capacity of a valve under a given set of circumstances.

Cylinder The chamber in a piston actuator in which the piston moves.

Deadband The range through which an input can be varied, upon reversal of direction, without initiating a change in the output signal.

Diaphragm A flexible pressure-responsive element which transmits force to the diaphragm plate and actuator stem. The force is equal to the diaphragm effective area multiplied by the pressure.

Diaphragm actuator A fluid-pressure-operated spring opposed or fluid-pressure-opposed diaphragm assembly for positioning the actuator stem in relation to the operating fluid pressure or pressures.

Diaphragm case or casing A housing, consisting of top and bottom sections, used for supporting a diaphragm and establishing one or two pressure chambers.

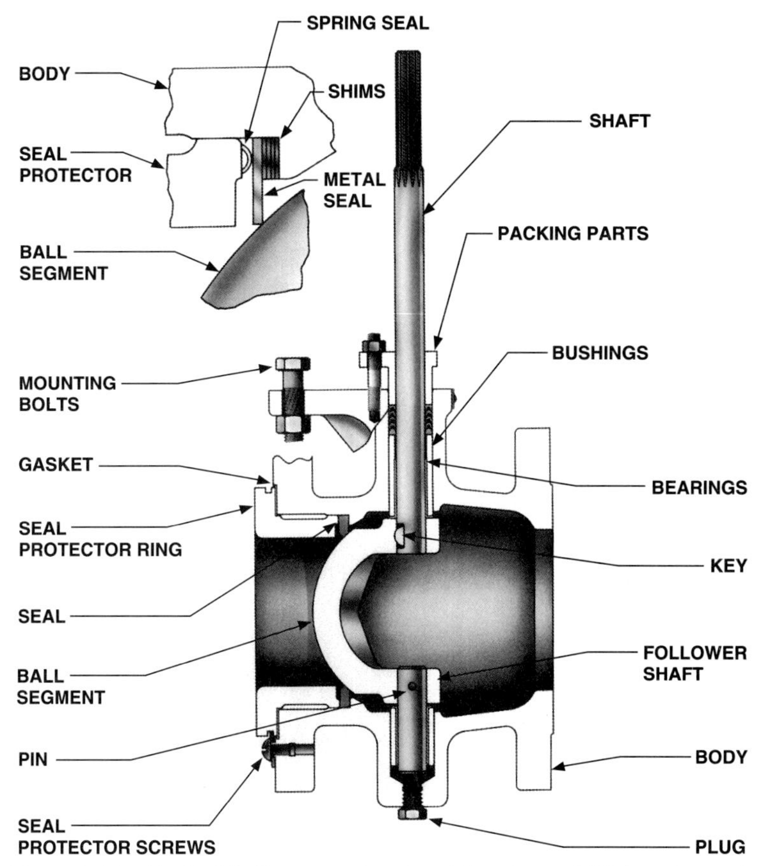

Figure G.10 Typical ball segment valve. (*Courtesy of Fisher Controls International, Inc., Marshalltown, Iowa.*)

Diaphragm plate A plate concentric with the diaphragm for transmitting force to the actuator stem.

Direct-acting actuator An actuator in which the actuator stem extends with increasing pressure.

Disc, conventional The flow-controlling member used in conventional butterfly rotary valves. High dynamic torques normally limit conventional discs to 60° maximum rotation in throttling service.

Disc, dynamically designed A butterfly valve disc contoured to reduce dynamic torque at large increments of rotation, thereby making it suitable for throttling service with up to 90° of disc rotation.

Disc, eccentric The common name for a valve design in which the positioning of the valve shaft and disc connections causes the disc to take a slightly eccentric path on opening. (This allows the disc to be swung out of contact with the seal as soon as it is opened, thereby reducing friction and wear.)

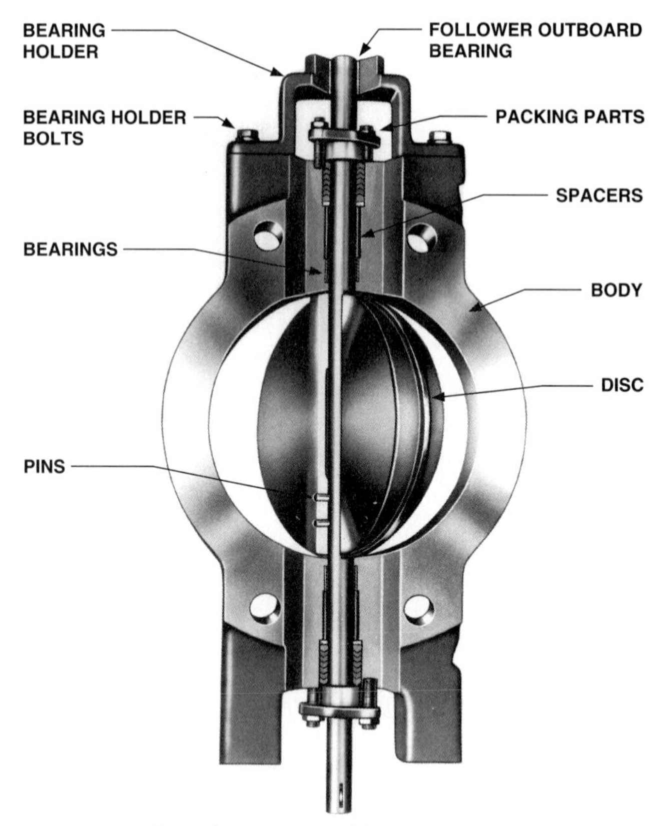

Figure G.11 Typical conventional butterfly valve. (*Courtesy of Fisher Controls International, Inc., Marshalltown, Iowa.*)

Double-acting actuator Sometimes called double sided. An actuator including a switching mechanism to permit powered operation in either direction, extending or retracting the actuator stem as dictated by the controller.

Double sided A term applied to positioners and actuators that implies dual input or outputs.

Drift An undesired change in output over a period of time, unrelated to the input, environment, or load.

Dynamic unbalance The net force produced on the valve plug in any stated open position by the fluid pressure acting upon it.

Effective area In a diaphragm actuator, that part of the diaphragm area which is effective in producing a stem force. The effective area of a diaphragm may change as it is stroked, usually being a maximum at the start and a minimum at the end. Most molded diaphragms have less change in effective area than flat-sheet diaphragms, and as a result, they are recommended over the flat-sheet designs.

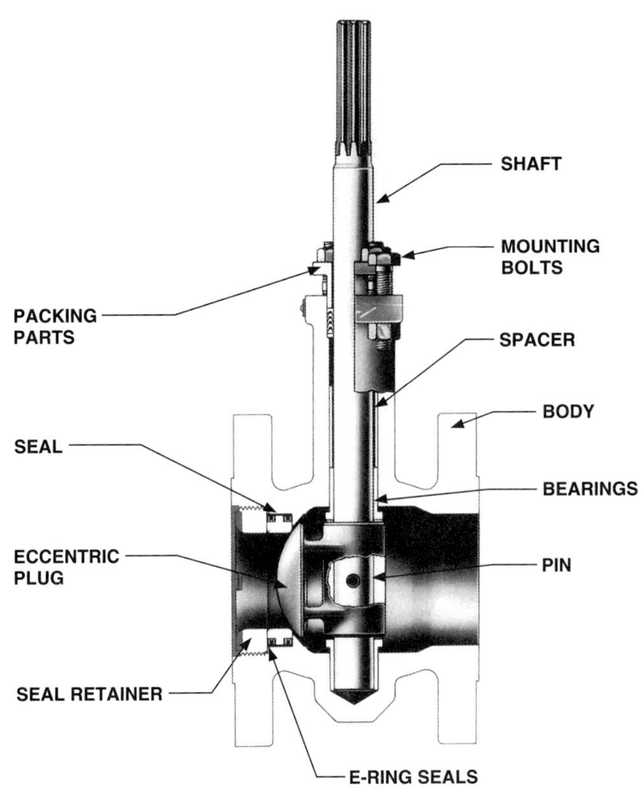

Figure G.12 Typical eccentric plug valve. (*Courtesy of Fisher Controls International, Inc., Marshalltown, Iowa.*)

Equal percentage flow characteristic A flow characteristic which, for equal increments of rated travel, will ideally give equal percentage changes of the existing flow (i.e., the slope of the flow characteristic curve at any point in travel is proportional to the flow at that point).

Extension bonnet A bonnet with an extension between the packing box assembly and bonnet flange for hot or cold service.

Fail-closed A condition wherein the valve port closes if the actuating power fails.

Fail-open A condition wherein the valve port opens if the actuating power fails.

Fail-safe A characteristic of a particular type of actuator, which upon loss of power supply, will cause the valve plug, ball, or disc to fully close, fully open, or remain in fixed position. (Fail-safe action may involve the use of auxiliary controls connected to the actuator.)

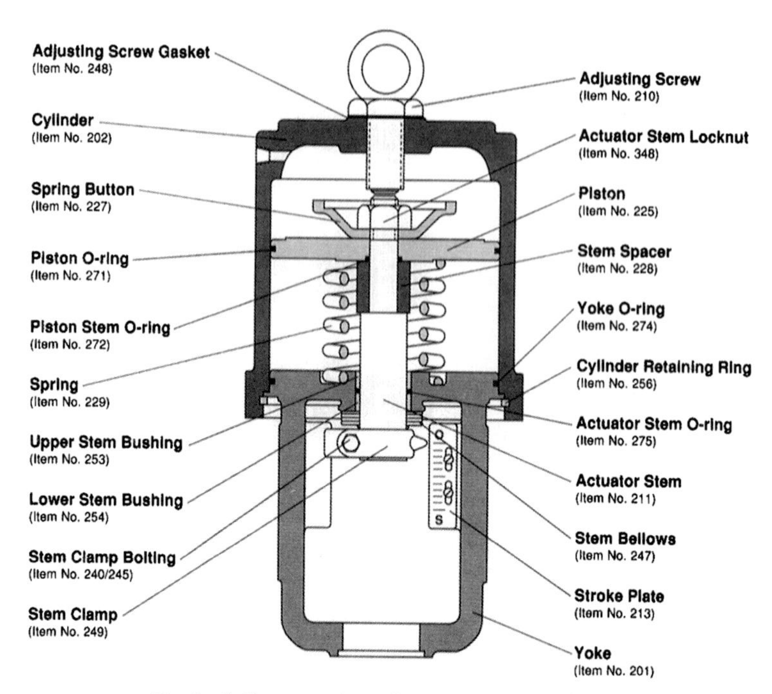

Figure G.13 Typical direct-acting piston actuator. (*Courtesy of Valtek International, Inc., Springville, Utah.*)

Feedback signal The return signal which results from a measurement of the directly controlled variable. (For a control valve with a positioner, the return signal is usually a mechanical indication of valve plug stem position which is fed back into the positioner.)

Flangeless body A body style common to rotary-shaft control valves. Flangeless bodies are held between ANSI-class flanges by long through-bolts. (Sometimes also called wafer-style valve bodies.)

Flow capacity The rate of flow through a valve under stated conditions.

Flow characteristic The relation between flow through the valve and percent-rated travel as the latter is varied from zero to 100 percent. This is a general term. It should be designated as either inherent flow characteristic or installed flow characteristic.

Flow control element The part of the valve trim that moves to regulate flow. Can be called a plug, ball, or disc depending on valve type.

Flow ring A heavy-duty ring used in place of a ball seal ring for rotary valves in severe service applications where some leakage can be tolerated.

Fouling The buildup of process fluid or debris on the trim parts in the valve.

Frequency response A test in which a device is subjected to a constant amplitude sinusoidal disturbance. The output amplitude and phase shift are

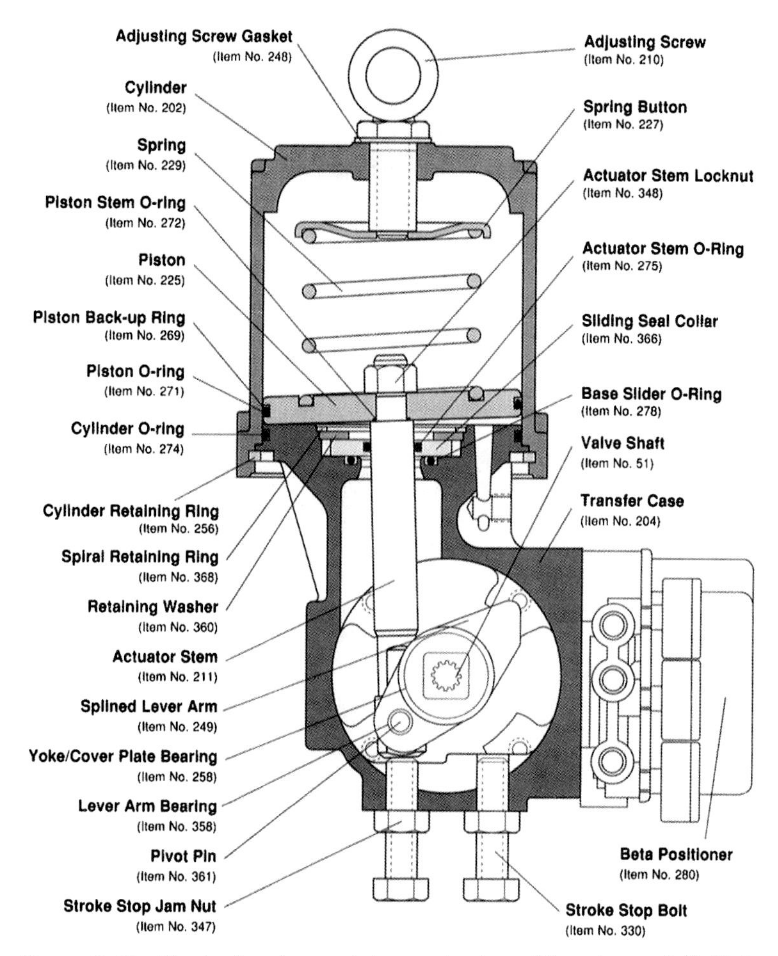

Figure G.14 Typical rotary piston actuator. (*Courtesy of Valtek International, Inc., Springville, Utah.*)

observed as functions of the input test frequency. In general, the higher the frequency response, the better the dynamic performance.

Gain, closed loop The gain of a closed loop system expressed as the ratio of the output change to the input change at a specified frequency.

Gain, dynamic The ratio of the steady-state amplitude of the output signal from an element or system to the amplitude of the input signal to that element or system, for a sinusoidal signal.

Gain, open loop The gain of the loop elements measured by opening the loop. (The product of all the individual gains in the forward and feedback paths.)

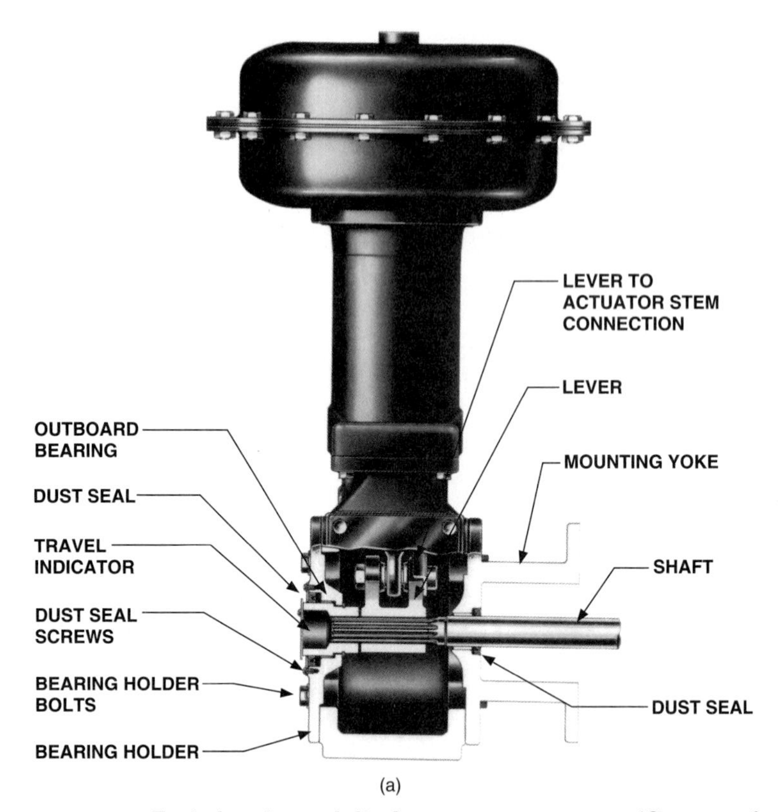

Figure G.15 Typical spring and diaphragm rotary actuator. (*Courtesy of Fisher Controls International, Inc., Marshalltown, Iowa.*)

Gain, static Of gain of an element, or loop gain of a system, the value approached as a limit as frequency approaches zero. (The ratio of a change in output to a change in input.)

Globe valve A valve construction style with a linear motion flow controlling member with one or more ports, normally distinguished by a globular-shaped cavity around the port region. (Globe valves can be further classified as: two-way single-ported, two-way double-ported, angle-style, three-way, split-style, unbalanced cage-guided, and balanced cage-guided.)

Guide bushing A bushing in a bonnet, bottom flange, or body to align the movement of a valve plug with a seat ring.

High-recovery valve A valve design that dissipates relatively little flow-stream energy due to streamlined internal contours and minimal flow turbulence. Therefore, pressure downstream of the valve vena contracta recovers to a high percentage of its inlet value. (Straight-through flow valves, such as rotary-shaft ball valves, are typically high-recovery valves.)

Hunting (limit cycling) In a linear system, an undesirable oscillation persisting after external stimuli disappear. This is sometimes called cycling. For

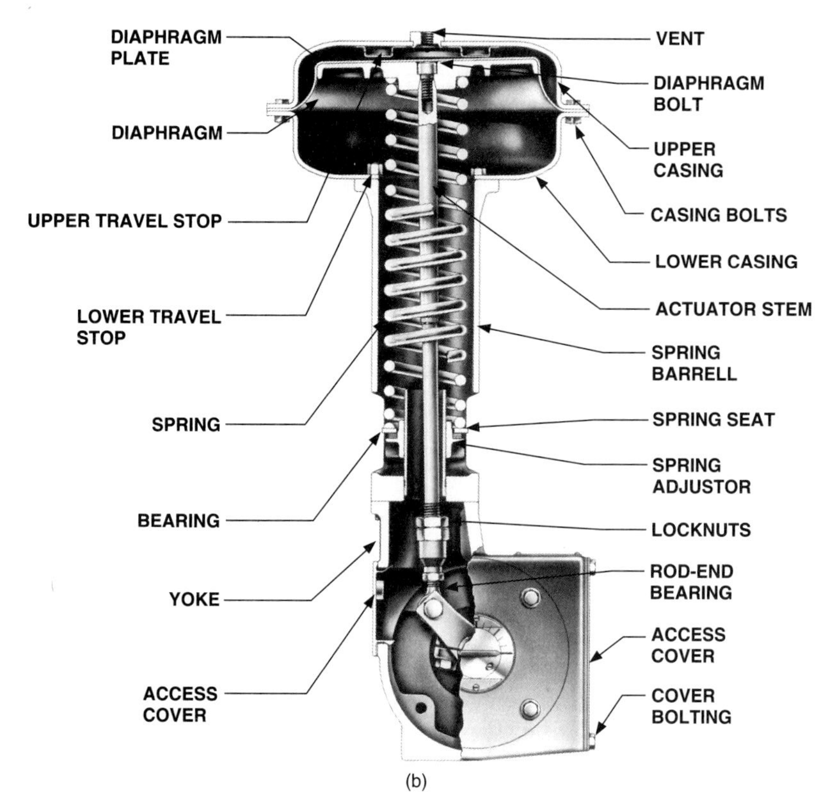

(b)

Figure G.15 (*Continued*)

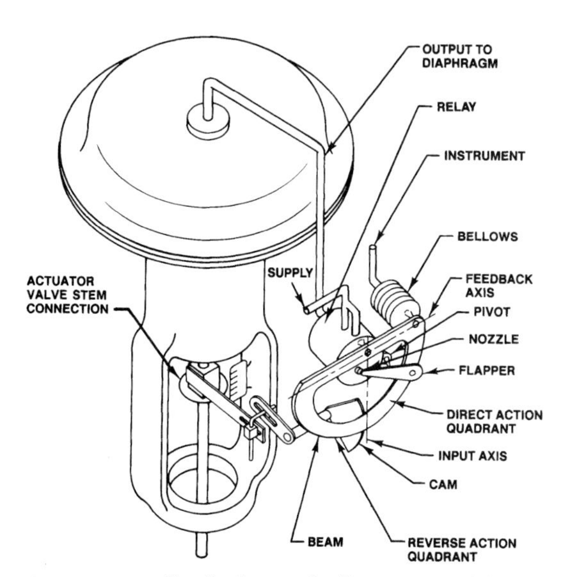

Figure G.16 Typical nozzle-flapper positioner schematic. (*Courtesy of Fisher Controls International, Inc., Marshalltown, Iowa.*)

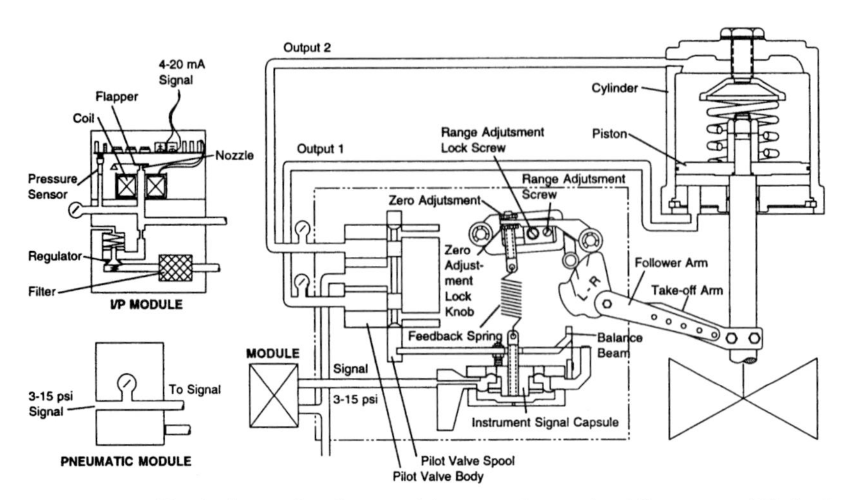

Figure G.17 Typical spool valve positioner schematic. (*Courtesy of Valtek International, Inc., Springville, Utah.*)

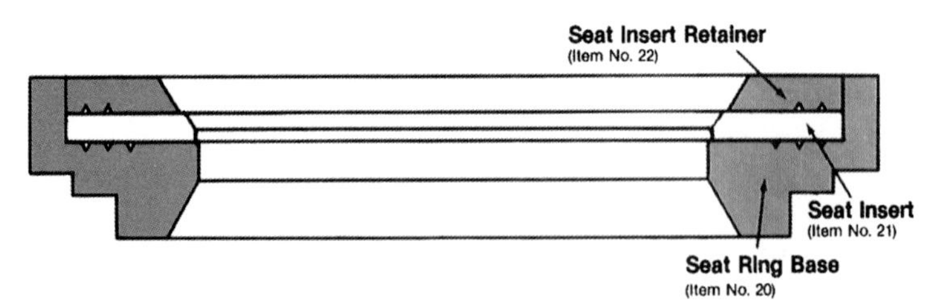

Figure G.18 Typical soft-seat construction. (*Courtesy of Valtek International, Inc., Springville, Utah.*)

control valve work, this would be considered an oscillation in the loading pressure to the actuator caused by instability in the control system or valve positioner.

Hysteresis The maximum difference that can occur in input values for any single output value during a calibration cycle, excluding errors due to deadband. Preferably expressed as a percentage of the calibration cycle amplitude. Hysteretic error is usually determined by subtracting the value of deadband from the maximum measured separation between upscale going and downscale going indications of the measured variable. Some reversal of output may be expected for any small reversal of input, which distinguishes hysteresis from deadband.

Inherent flow characteristic A flow characteristic in which a constant pressure drop is maintained across valve.

Inherent rangeability A ratio of maximum to minimum controllable flow with which the deviation from the specified inherent flow characteristic does not exceed some stated limit. A control valve that still does a good control job when the flow increases to 100 times the minimum controllable flow; has a rangeability of 100 to 1. Rangeability might also be expressed as the ratio of the maximum to minimum controllable flow coefficients.

Installed flow characteristic A flow characteristic when the pressure drop across the valve varies as dictated by the flow and related conditions in the system in which the valve is installed.

Lapping A procedure used to grind the mating surfaces of the valve plug and seat to improve shutoff (see Chap. 13).

Leakage A quantity of fluid passing through an assembled valve when the valve is in a fully closed position under stated closure forces, with the pressure differential and temperature as specified. Leakage is generally expressed as a percentage of maximum capacity.

Linear flow characteristic A flow characteristic which can be represented ideally by a straight line on a rectangular plot of percentage of flow versus percentage of rated travel.

Linearity The closeness to which a curve approximates a straight line. *Note 1*: It is usually measured as a nonlinearity and expressed as linearity (e.g., a maximum deviation between an average curve and a straight line). The average curve is determined after making two or more full-range traverses in each direction. The value of linearity is referred to the output unless otherwise stated. *Note 2:* As a performance specification, linearity should be expressed as independent linearity, terminal-based linearity, or zero-based linearity. When expressed simply as linearity, it is assumed to be independent linearity.

Linearity, independent The maximum deviation of the calibration curve (average of upscale and downscale readings) from a straight line so positioned as to minimize the maximum deviation (**Fig. G.19**).

Linearity, terminal-based The maximum deviation of the calibration curve (average of upscale and downscale readings) from a straight line coinciding with the calibration curve at upper and lower range values (**Fig. G.20**).

Linearity, zero-based The maximum deviation of the calibration curve (average of upscale and downscale readings) from a straight line so positioned as to coincide with the calibration curve at the lower range value and to minimize the maximum deviation (**Fig. G.21**).

Loading pressure That pressure employed to position a pneumatic actuator. This is sometimes referred to as the diaphragm pressure or piston pressure for diaphragm actuated control valves and piston actuated control valves, respectively. It is the pressure that actually acts on the diaphragm or piston, and it may be the *instrument pressure* if a valve positioner is not used.

Low-recovery valve A valve design that dissipates a considerable amount of flowstream energy due to turbulence created by the contours of the flowpath. Consequently, pressure downstream of the valve vena contracta recovers to a lesser percentage of its inlet value than is the case with a valve having a more streamlined flowpath. (Although individual designs vary, conventional globe-style valves generally have low pressure recovery capability.)

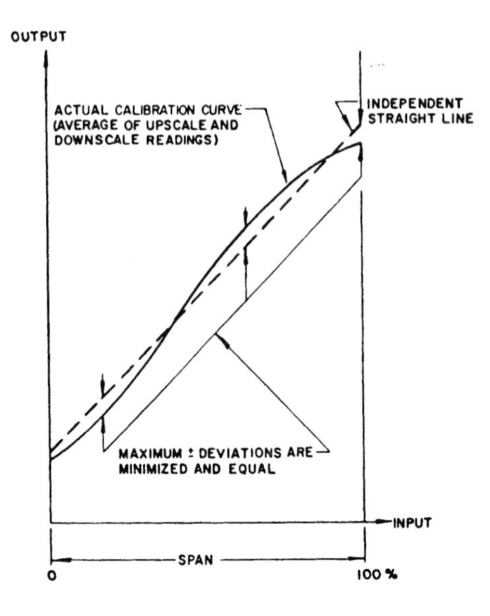

Figure G.19 Independent linearity. (*Courtesy of Instrument Society of America, Research Triangle Park, N.C.*)

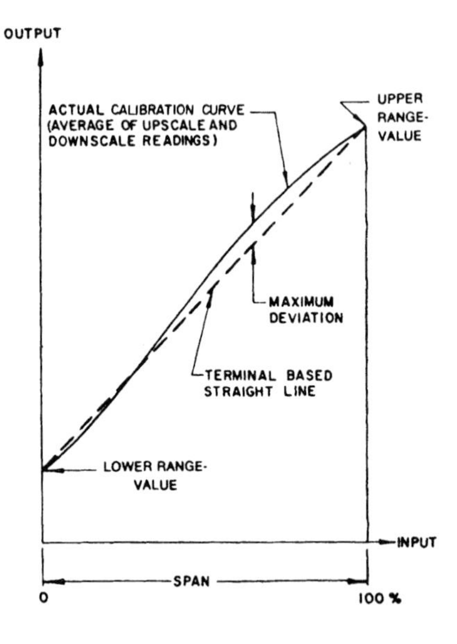

Figure G.20 Terminal-based linearity. (*Courtesy of Instrument Society of America, Research Triangle Park, N.C.*)

Normally open valve assembly One which opens when the actuator pressure is reduced to atmospheric.

Operating stroking range The high and low values of pressure applied to the actuator to produce rated valve travel with friction and service loads present. Has two sets of values, one for the opening stroke and one for the closing stroke.

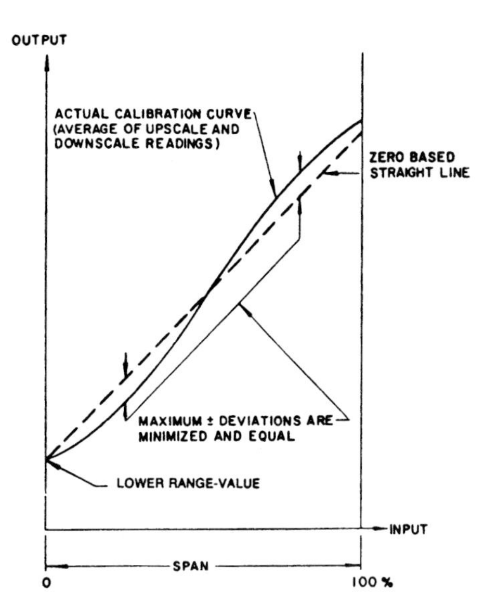

Figure G.21 Zero-based linearity. (*Courtesy of Instrument Society of America, Research Triangle Park, N.C.*)

Packing The material found inside the packing box that forms the seal against the valve stem. The packing can be any one of several materials but the most common are Teflon and graphite.

Packing box The cavity normally found in the bonnet that contains the dynamic seal that the valve stem passes through to enable it to control the internal parts that regulate flow. The name refers to the packing found in the cavity that is compressed against the stem to form the seal.

Packing box assembly The parts in the bonnet assembly used to seal against leakage around the valve plug stem.

Piston A movable, pressure-responsive element which transmits force to the actuator stem.

Piston actuator A fluid-pressure-operated piston and cylinder assembly for positioning the actuator stem in relation to the operating fluid pressure. The best piston actuators are double acting so that full power can be developed in either direction.

Port A fixed opening, normally the inside diameter of a seat ring, through which fluid passes.

Port guided A design in which the valve plug is aligned by the body port or ports.

Proportional band The change in input required to produce a full-range change in output due to proportional control action. The preferred term for this is proportional gain. *Note 1:* It is reciprocally related to proportional gain. *Note 2:* It may be stated in input units or as a percentage of the input span (usually the indicated or recorded input span).

Push-down-to-close construction A valve construction in which the valve plug (or disc or ball) moves toward the closed position when the actuator stem moves down (also called direct acting).

Push-down-to-open construction A valve construction in which the final control element moves toward the open position when the actuator stem moves down (also called reverse-active).

Quick-opening flow A flow characteristic that provides maximum flow for a given valve size and configuration.

Range The region covered by either the input or output signal of a device and expressed by stating the end-scale values.

Rated C_v The valve of C_v at the rated full open position.

Rated travel The linear movement of the valve plug from the closed position to the full-open position, which is determined by the valve design and the application.

Resolution The smallest interval between two adjacent discrete details which can be distinguished. Output resolution is the minimum possible output change which a device can produce. Input resolution is the smallest change in input that a valve will respond to.

Response, dynamic The behavior of the output of a device as a function of the input, both with respect to time.

Reverse-acting actuator An actuator in which the actuator stem retracts with increasing diaphragm pressure.

Reverse flow The flow of fluid in the opposite direction from what is normally considered the standard direction.

Rod end bearing The connection often used between actuator stem and actuator lever to facilitate conversion of linear actuator thrust to rotary force with minimum of lost motion.

Rotary-shaft control valve A valve style in which the flow closure member (full ball, partial ball, or disc) is rotated in the flowstream to modify the amount of fluid passing through the valve.

Seal ring The portion of a rotary-shaft control valve assembly corresponding to the seat ring of a globe valve. The positioning of the disc, ball, or plug relative to the seal ring determines the flow area and capacity of the unit at that particular increment of rotational travel.

Seat That portion of the seat ring or valve body which a valve plug contacts for closure and shut off.

Seat load The contact force between the seat and valve plug.

Seat ring A separate piece inserted in a valve body to form a valve body port.

Sensitivity The ratio of the change in output magnitude to the change of the input which causes it after the steady-state has been reached.

Separable flange A flange which fits over a valve body flow connection. It is generally held in place by means of a retaining ring.

Setpoint An input variable which sets the desired value of the controlled variable. (Setpoint should be expressed in the same terms as the process variable.)

Shaft The portion of a rotary-shaft control valve assembly corresponding to the valve stem of a globe valve. Rotation of the shaft positions the disc or ball in the flowstream and thereby controls the amount of fluid which can pass through the valve.

Signal A physical variable, one or more parameters of which carry information about another variable (which the signal represents).

Signal amplitude sequencing (split ranging) An action in which two or more signals are generated or two or more final controlling elements are actuated by a single input signal, each one responding consecutively, with or without overlap, to the changing magnitude of the input signal.

Single sided A term applied to positioners and actuators that implies single inputs or outputs.

Sliding seal The lower cylinder seal in a pneumatic piston-style actuator designed for rotary valve service. (This seal permits the actuator stem to move both vertically and laterally without leakage of lower cylinder pressure.)

Span The range covered by either the input or output signal of a device and expressed as the algebraic difference between end-scale values.

Speed of response In control valve operation, the rate of travel of the actuator in inches per second. This is also called stroking speed.

Spring adjustor A fitting, usually threaded on the actuator stem or into the yoke used to adjust the initial spring compression or benchset.

Spring seat A plate used to hold the spring in position and also to provide a flat surface for the spring adjustor to contact.

Standard flow direction For those rotary-shaft control valves having a separate seal ring or flow ring, the flow direction in which fluid enters the valve body through the pipeline adjacent to the seal ring and exits from the side opposite the seal ring. (Sometimes called forward flow.) *See also* reverse flow.

Static unbalance The net force produced on the valve plug in its closed position by the fluid pressure acting upon it.

Steady-state deviation Any departure of the calibration curve from the ideal value.

Stem guided A special case of a top guide in which the valve plug is aligned by a guide acting on the valve plug stem.

Stem connector A clamp, sometimes in two pieces, used to connect the actuator stem to the valve plug stem.

Stem unbalance The net force produced on the valve plug stem in any position by the fluid pressure acting upon it. Normally acts to push the stem out of the body.

Stroking range The high and low values of pressure that produce full-rated valve travel with friction but no pressure loads. The range differs depending on stroking direction.

Supply pressure For automatic controllers and valve positioners, the operating medium pressure supplied to the control device.

Time, constant For the output of a first-order system forced by a step or an impulse, time, constant, or T, is the time required to complete 63.2 percent of the total rise or decay. At any instant during the process, T is the quotient of the instantaneous rate of change divided into the change still to be completed. In higher-order systems, there is a time constant for each of the first-order components of the process.

Time, dead The interval of time between initiation of an input change or stimulus and the start of the resulting observable response.

Time, step response Of a system or an element, the time required for an output to change from an initial value to a large specified percentage of the final steady-state value either before or in the absence of overshoot as a result of a step change in the input.

Top and bottom guided A design in which the valve plug is aligned by guides in the body or in the bonnet and bottom flange.

Top and port guided A design in which the valve plug is aligned by a guide in the bonnet or body and the body port.

Top guided A design in which the valve plug is aligned by a single guide in the body adjacent to the bonnet or in the bonnet.

Transducer An element or device which receives information in the form of one quantity and converts it to information in the form of the same or another quantity.

Travel indicator A pointer, normally attached near the stem connector or on the end of the shaft, to indicate the position of the final control element.

Travel indicator scale A graduated scale attached to the yoke that gives an indication of valve travel.

Trim The internal parts of a valve which are in flowing contact with the controlled fluid. (In a globe valve body, trim would typically include a valve plug, seat ring, cage, stem, and stem pin.)

Trim, soft-seated A globe valve trim with an elastomer, plastic, or other readily deformable material used as an insert, either in the valve plug or seat ring, to provide very tight shutoff with minimal actuator force.

Trunnion mounting A style of mounting the disc or ball on the valve shaft or stub shaft with two bushings diametrically opposed.

Valve body A housing for internal parts having inlet and outlet flow connections.

Valve body assembly Commonly called a valve body or body, or more properly, valve body assembly. An assembly of a body, bonnet assembly, bottom flange (if used), and trim elements.

Valve flow coefficient (C_v) The number of U.S. gallons per minute of 60°F water that will flow through a valve with a 1-psi pressure drop.

Valve plug (flow control element) The name *plug* is used for sliding-stem valves. A more generic term for all valve types is *flow control element.* A movable part which provides a variable restriction in a port. Because of desired characteristics and for functional reasons, there are many forms of valve plugs.

Valve plug guide That portion of a valve plug which aligns its movement in either a seat ring, bonnet, bottom flange, or any two of these.

Valve plug stem A rod extending through the bonnet assembly to permit positioning the valve plug.

Valve positioner A control valve accessory which transmits a loading pressure to an actuator to position the flow control element such that its position is proportional to the input signal.

Variability A general term that provides a quantitative measure of the error or deviation over time between the setpoint and process variable. It may be measured in a number of different ways.

Vena contracta The location, for liquid flow, where cross-sectional area of the flowstream is at its minimum size, where fluid velocity is at its highest level, and fluid pressure is at its lowest level. (The vena contracta normally occurs just downstream of the actual physical restriction in a control valve.)

Wafer-style valve body A flangeless type of valve body. Also called a flangeless valve body; it is clamped between pipeline flanges.

Yoke A structure that connects the pneumatic element of the actuator to the bonnet assembly.

Appendix

1

Conversion Table

TABLE A2.1 U.S. Customary Units

Specific gravity of air (G) = 1 (reference for gases)	Specific gravity of water = 1 (reference for liquids)
U.S. gal of water = 8.33 lb @ std. cond.	1 ft^3 of water = 62.34 lb @ std. cond. (= density)
1 ft^3 of water = 7.48 gal	1 ft^3 of air = 0.076 lb @ std. cond. (= air density)
Air specific volume = 1/density = 13.1 ft^3/lb	Air molecular weight (M) = 29
G of any gas = density of gas/0.076	G of gas at flowing temp. = (G × 520)/(T + 460)
G of any gas = molecular wt. of gas/29	
Standard conditions (U.S. customary) are at 14.69 psia and 60°F	

Flow conversion of gas

$$\text{SCFH} = \frac{\text{lb/h}}{\text{density}} \qquad \text{SCFH} = \frac{\text{lb/h} \times 379}{\text{M}}$$

Flow conversion of liquid

$$\text{GPM} = \frac{\text{lb/h}}{500 \times \text{G}}$$

Temperature conversion

$$\text{F (Fahrenheit)} = \frac{\text{C } 9}{5} + 32$$

$$\text{C (Celsius)} = (\text{F} - 32)\,\frac{5}{9}$$

Multiply	By	To obtain
	Length	
Millimeters	0.039	Inches
Centimeters	0.394	Inches
Inches	2.54	Centimeters
Feet	30.48	Centimeters
Feet	0.304	Meters
	Area	
Square centimeters	0.155	Square inches
Square centimeters	0.001076	Square feet
Square inches	6.452	Square centimeters
Square inches	0.00694	Square feet
Square feet	929	Square centimeters
	Flow rates	
Gallons U.S./minute (gpm)	3.785	Liters/minute
Gallons U.S./minute	0.133	Cubic feet/minute
Gallons U.S./minute	0.227	Cubic meters/hour
Cubic feet/minute	7.481	Gallons per minute
Cubic feet/hour	0.1247	Gallons per minute
Cubic feet/hour	0.01667	Cubic feet/minute
Cubic meters/hour	4.403	Gallons per minute
Cubic meters/hour	35.31	Cubic feet/hour
	Velocity	
Feet per second	0.3048	Meters/second
Feet per second	1.097	km/hour
Feet per second	0.6818	Miles/hour
	Volume and capacity	
Cubic feet	28.32	Liters
Cubic feet	7.4805	Gallons
Liters	61.02	Cubic inches
Liters	0.03531	Cubic feet
Liters	0.264	Gallons
Gallons	3785.0	Cubic centimeters
Gallons	231.0	Cubic inches
Gallons	0.1337	Cubic feet
	Weight	
Pounds	0.453	Kilogram
Kilogram	2.205	Pounds
	Pressure and head	
Pounds/square inch	0.06895	Bar
Pounds/square inch	0.06804	Atmosphere
Pounds/square inch	0.0703	Kilogram/square centimeter
Pounds/square inch	2.307	Feet of H_2O (4°C)
Pounds/square inch	0.703	Meter of H_2O (4°C)
Pounds/square inch	5.171	Centimeter of Hg (0°C)
Pounds/square inch	2.036	Inch of Hg (0°C)
Atmosphere	14.69	Pounds per square inch
Atmosphere	1.013	Bar
Atmosphere	1.033	Kilograms/square centimeter
Atmosphere	101.3	kilopascal centimeter
Bar	14.5	Pounds per square inch
Kilogram/square centimeter	14.22	Pounds per square inch
kilopascal	0.145	Pounds per square inch

Index

ABOUT THE AUTHOR

Bill Fitzgerald is Director of Aftermarket Services for Fisher Service Company in McKinney, Texas. He previously worked in Fisher's design and marketing departments in both the United States and Europe. A registered professional engineer in the State of Iowa, Mr. Fitzgerald holds a B.S. and M.S. in engineering from Iowa State University and is a member of I.S.A. and A.S.M.E.